ARTIFICIAL INTELLIGENCE

Military Tactics, Bridges and Aspirations

ARTIFICIAL INTELLIGENCE
MILITARY TACTICS, BRIDGES AND ASPIRATIONS

Brigadier Pawan Bhardwaj

UNITED SERVICE INSTITUTION OF INDIA
NEW DELHI

PENTAGON PRESS LLP

Copyright © UNITED SERVICE INSTITUTION OF INDIA, New Delhi, 2025

All rights reserved. No part of this publication may be reproduced, stored in a retrieval system, or transmitted in any form or by any means, electronic, mechanical, photocopying, recording or otherwise, without the prior written permission of the Publisher.

First published in 2025 by
PENTAGON PRESS LLP
206, Peacock Lane, Shahpur Jat
New Delhi-110049, India
Contact: 011-26490600

Typeset in AGaramond, 11.5 Point
Printed by Aegean Offset Printers, Greater Noida, U.P.

ISBN 978-81-982857-1-3

Disclaimer: The views expressed in this book are those of the author and do not necessarily reflect those of the United Service Institution of India, New Delhi, or the Government of India.

www.pentagonpress.in

A Thankful and Proud Dedication

I offer my heartfelt gratitude to those gallant Indian soldiers who no longer walk with us but whose stories of supreme courage continue to motivate. To the Indian Army that stands as an omnipotent source of strength, not just for me, but for all Indians. I extend my special appreciation to my unit, 6 JAK RIF, for bringing out the best in me and setting me on the path of knowledge and exploration.

Contents

1. **AI's Ascendance: Transforming Societal Dynamics and Shaping the Future** — 1
   - What is AI? — 1
   - The AI Domains — 16
   - It is not always about Database or Big Data — 27
   - Ethical Use and Legal Frameworks — 28
   - Key Takeaways — 31

2. **AI on the Frontlines: Case Studies in Military Innovation** — 36
   - Military and AI — 36
   - Military AI Case Studies — 42
     - Russia — 43
     - China — 56
     - Türkiye — 65
     - Iran — 68
     - Australia — 71
     - United Kingdom — 76
     - France — 83
     - United States of America — 85
     - Taiwan — 90
     - Israel — 93
     - Non-State Actors — 98
   - Key Takeaways — 98

3. **India's AI Odyssey: Navigating Challenges, Fostering Innovation, and Shaping a Future** — 113
   - Indian AI Journey — 113
   - Ministry of Defence Initiatives — 119

- National AI Efforts — 127
- Summary of Military AI Development in India — 135
- National AI Mission and Indian Military — 146
- Key Takeaways — 152

4. Future Indian Military Conflict Environment — 158
   - Methodology — 159
   - Content Analysis — 162
   - Survey Analysis — 168
   - Subject Matter Experts (SME) — 170
   - Summarising Conflict Environments — 175
   - Key Takeaways: Indian Military in Future Conflicts — 187
   - Appendix A: Survey Form: Future Conflict Environments for India — 192

5. Tailored Tactical AI Solutions for Indian Army — 194
   - Tactical Battles — 194
   - Indian Tactical Battles — 197
   - Tactical Battle Priority Matrix — 206
   - Indian Army Rapid AI Design Architecture (IRADA) — 209
   - Designing AI Tools for Indian Army — 219
   - Key Takeaway: Tactical AI Tool Designing Framework — 227
   - Appendix B: Peas Analysis Framework — 230
   - Appendix C: Language Translator — 231
   - Appendix D: Information Retrieval (IR) System — 232
   - Appendix E: Urban Combat Buddy — 233
   - Appendix F: Autonomous Tracking of Person of Interest (POI) — 234
   - Appendix G: Tactical Logistic Bot — 235
   - Appendix H: Tactical Support Drone — 236
   - Appendix I: Communication Relay Bot — 237
   - Appendix J: Autonomous Weapon System — 238
   - Appendix K: Autonomous Prediction of Landslides and Rock Falls — 240

6. AI Revolution: Transforming Indian Military Operations — 241
   - Indian Army Realities — 241
   - Doctrinal Aspects — 242
   - Organisational and Personnel Aspects — 243
   - Training Aspects — 246

- Material and Facility Aspects 249
- Leadership and Educational Aspects 256
- Policy Aspects 257
- Seminar Suggestions 260
- Key Takeaways 262
- Action Plan 266

Epilogue: Harmony of Military AI Integration 272
- Unfathomable and Unchartered AI Battlespace 272
- Traditional Battlefield 273
- A Functional Research on AI at Tactical Level 274
- Peer Relationships 276
- Future Soldiering 277
- Mass or Effect 277
- Man-Machine Teaming/Man Unmanned Machine Teaming .. 278
- AI Artefacts 279
- Generative AI 280
- Securing AI through Legal Acts and Advisories 281
- Ethics: User or Developer Driven 282

Bibliography 285

Index 308

Lt Gen V G Khandare
PVSM,AVSM,SM (Retd)

Principal Adviser
Ph No: 23010168-69

भारत सरकार
रक्षा मंत्रालय
नई दिल्ली - 110 011
Government of India
Ministry of Defence
New Delhi - 110 011

No 01/Gen/Pr Adv/2024

14th August, 2024

FOREWORD

Brigadier Pawan Bhardwaj has undertaken an insightful exploration of Artificial Intelligence (AI) in the context of Indian military operations. As a Principal Adviser to the Ministry of Defence on national policies, I have long advocated for leveraging emerging disruptive or empowering technologies like AI to enhance our nation's defence capabilities.

With the intersection of future battlespaces and technological advancements, India should focus on developing tailored AI solutions. There is a need to prepare a generation of Techno Commanders and military leaders who can bridge the critical gap by providing a theoretical construct with applications for Indian Army AI Systems. Well-read and pragmatic military leaders will be essential to align tactical requirements with technological capabilities and vice versa. Furthermore, managing the technical workforce is crucial to balance personal aspirations with organizational goals. Integrating AI into the National AI Mission will benefit from the expertise of various national organisations and peer ministries. Creating, cleaning and sharing non-fungible data will remain a challenging task, which can only be accomplished through a concerted and coordinated effort, rather than a competitive one. A whole of nation, whole of Govt, whole of tri services and a whole of service approach is needed.

This work is well timed to support a unique military discourse: the integration of AI at the tactical level. It presents a compelling hypothesis: that India must conduct a pragmatic analysis of future battlespaces and leverage its indigenous technology development capability to identify, define and demand its AI tools. This perspective challenges conventional wisdom and underscores the necessity for India to chart its own course in harnessing AI's transformative capabilities.

This book offers a comprehensive exploration of AI's implications for the

Indian military, emerging as a valuable resource for military planners, technology partners and decision-makers alike. From simplifying the technical intricacies of AI to extrapolating military necessities and providing actionable insights, each chapter contributes to a deeper understanding of AI's role in shaping the future of warfare. Importantly, it offers an Action Plan to be debated and critiqued and hopefully implemented.

However, it is crucial to acknowledge the ethical considerations that accompany such advancements. The development and deployment of AI in warfare must be guided by principles that ensure accountability, minimize civilian harm and adhere to international law. It's important to remember that AI is a tool to augment human capabilities, not a replacement for soldiers themselves. Effective integration of AI will require fostering a strong human-machine partnership, where soldiers leverage AI's strengths while maintaining critical decision-making authority.

This seminal work represents a significant contribution to the discourse on AI in the Indian military context. It serves as a call to action for policymakers and military leaders to embrace AI as a strategic enabler while highlighting the importance of developing indigenous solutions to meet India's unique security challenges.

(Lt. General Vinod G. Khandare) PVSM, AVSM, SM

Preface

As part of Progressive Military Education, Indian Army officers are provided opportunities to undertake dedicated research on military aspects. In the exploration of AI, discussion on employment of AI at the tactical level is rarely ventured. Contextual research on Indian solutions is yet to be seen, and hence it became the main motivation for this book.

The developed economies and their supporting militaries benefit from a close relationship between technology development and military acquiesce to utilise it. Such collaborative technology experiments including AI, stands out as a major game changer in their global military forays. 'Out of continent' military explorations, wide ranging, all-out national effort, forces creation of huge AI systems capable of inter-country, inter-agency, inter-military cooperation; with omnipotence as a benchmark. This 'rule' fails in the Indian context – leading on to the hypothesis that India has to undertake a pragmatic analysis of future battlefields, combined with Indian technology development speeds and only then identify, define and demand its IA tools. Mimicking the Western concept of AI may not prove useful.

Insufficiency of dedicated Indian literature forces researchers and planners to rely on inorganic theories and constructs. AI functions in a controlled environment with a decidedly well-articulated result goal and thus a non-Indian construct will not be successful. This research provides a theoretical construct for defining and designing Indian Army AI systems and this book is born out of a deep-seated motivation to connect the intersections of future battlefields and technological advancements, in developing AI tools and fostering the development of indigenous solutions tailored to India's unique challenges.

The primary objective is to explore future battlefields and the pace of technology development to empower the user and developer. The actionable insights and recommendations will automatically align with India-specific national security imperatives. It is a logical progression of the world military AI experiences and its lessons in the Indian context to understand the pitfalls of military technology imitation. The book is divided into six chapters, not linked, but sequenced so that the readers may revisit chapters and understand the key takeaways for better Indian solutions.

The first chapter 'simplifies' the technical intricacies of artificial intelligence (AI), providing an understanding of its various pillars and connections within the technology landscape. Readers will explore the fundamental aspects of AI and readiness for deployment. By identifying the typical characteristics of different AI applications and a comparative analysis of competing technologies, this chapter sets the technical benchmark for the book. It serves as the starting point to delve into the nature of AI, not necessarily technical.

The second chapter examines the military applications of AI, focusing on technologically savvy countries with military adventurist tendencies. Through an analysis of selected countries with varying levels of military experience and experimentation, readers will gain insights into the strategic alignment of AI initiatives with national aspirations and goals. Researching widely between nations seeking technical advantages and those with extensive combat experience, this chapter sheds light on the diverse approaches to integrating AI into military operations, offering valuable lessons and perspectives for India.

The third chapter explores India's emergence as a rising giant in technology, with impressive growth driven by national policies and multi-agency coordination. By examining the unique characteristics of Indian AI development and the surrounding ecosystem, this chapter provides a comprehensive understanding of India's trajectory, highlighting both successes and challenges along the way. Analysis of the National AI Mission in the light of military AI provides a unique insight into the niche field.

Against the backdrop of India's economic growth and evolving geopolitical landscape, the fourth chapter reads into potential conflict scenarios. From contested borders to maritime disputes and internal security challenges, readers will gain a nuanced understanding of the complex conflict environments with inescapable military involvements. By predicting and analysing future conflicts, this chapter clarifies the operational environment where military AI role can be postulated.

Drawing on the technical development capabilities and India's unique conflict environments, the fifth chapter extrapolates military necessities to evolve typical Indian solutions to indigenous problems. Inferences from global military attempts, stitched with usable Indian technology capabilities, provides readers with insights into the AI tool design framework. Specifically, it provides a framework for unambiguous Indian military AI tools, offering valuable insights for military planners, technology partners, and decision-makers. This chapter sets a common standard for practitioners and technical experts, facilitating the development and deployment of AI tools tailored to Indian military requirements.

The final chapter focuses on providing the necessary support structures, policies, doctrinal discussions, and documentation vital for the effective utilisation of AI tools in tactical military operations. It concludes with a seminal development framework, a broad action plan and recommendations for further research, paving the way for future military AI development in India.

Ethics, Compliance and Regulations

The study of AI cannot be divorced from ethics, compliances, and regulations. This book however deliberately intends to steer clear and the following content will attempt to clarify why. As a responsible democracy with no revisionist intent, India stands out as a mature nation with an efficient military. This *sutra* (a literary rule or aphorism, or a collection of them) is the foundation of this intent. The author was posed with a question "Where will India fight her next war?" at an international course in Taiwan in June 2023. In an era where global dynamics often demand dramatic, highly visible actions, the international community appears to be awaiting India's strategic move. Historically, the Western world's international relations calculus has often been punctuated by events of profound visibility and destructiveness, from Pearl Harbour to the bombing of Japan, and the invasions of Iraq and Afghanistan. Such events are etched in the collective memory of humanity, often epitomising a nation's 'arrival on stage.' How will India, a nation now notably present on the global stage, define its role? It is a question of assessing whether India's rise embodies a hegemonic ascent or a transformational growth. India's geopolitical evolution and its global impact is connected with its historical journey, its current strategic posture, and its aspirations for the future; all non-reservationist in character.

India's history is marked with centuries of cultural richness, diversity, and resilience. Its ancient civilisations, the Saraswati, followed by the Indus Valley, laid the foundation for a multifaceted heritage. While the Mauryan Empire spread across 50 lakh square km during 250 BCE, the Cholas a thalassocracy governed 36 lakhs square kilometres at their peak during 1030 CE, the Gupta Empire extended to 3.5 lakh square km during 400 CE. Relevantly, none of the ancient Indian kings annihilated any society or decimated a group of people. There were no crusades or dark ages and India never saw any religious war.

Since Independence in 1947, the military was never deployed to defeat a country; instead done to save a country and its people. In 1988, Indian paratroopers thwarted a coup attempt against the government of President Maumoon Abdul Gayoom of Maldives, flying over 2,000 km and restoring control of the capital to the government within hours. The coup leaders and rebels were handed over to the Maldivian authorities. The Indian paratroopers returned on 17 November – two weeks after landing. Bangladesh, known as East Pakistan since 1947, was a province in the Dominion of Pakistan. The Pakistani military junta refused self-rule aspirations to Sheikh Mujibur Rahman's Awami League in the 1970s. This triggered the Bangladesh liberation war by the indigenous Mukti Bahini in 1971 leading to Pakistani genocidal retribution and the massacre. 150 lakh refugees crossed into Indian territory by mid-1971 forcing Indian military intervention for humanitarian reasons on the request of the liberation fighters. The war lasted for only two weeks resulting in a Pakistani surrender and the birth of a new country – the People's Republic of Bangladesh. Indian forces returned after that.

Nations fight for economic or territorial gains, religion, nationalism and revenge leading to civil war, revolutionary war, or a defensive war. No Indian war has been triggered by economy or religion. India faces territorial disputes, cross-border terrorism, and regional power dynamics. Its border disputes with Pakistan and China sometimes escalate into military standoffs. Demonstrating reliance on self-worth, India avoids international mediation and pressure points. Confident in its own capabilities, India has no intention of joining NATO, citing that the military alliance is not suitable for India. Defensive wars to protect territory will always intensify nationalism and call for revenge. For Indian watchers, it is obvious: the 1947 Indo-Pak war was surreptitiously planned and led by British officers and India lost 78,000 square km of land; the 1962 war with China and in 1965 with Pakistan was not initiated by India, and finally, the 1999 Kargil war was a deceitful venture by Pakistan.

There is also a discourse on civil or revolutionary war. Consider this; out of the ten largest world economies, India is the only country that has never had a civil war. Though, strictly speaking Japan also never had one, if the Boshin War (1868-69) is disregarded – fought between forces of the ruling Tokugawa shogunate and a coalition seeking to seize political power in the name of the Imperial Court. India did have one civil war – the Mahabharat (Great War) in 3139 BCE!

India, as a nation with a rich and diverse history, remains committed to peace and stability on the global stage. The country's foreign policy has consistently emphasised diplomacy, cooperation, and non-aggression. War is never a desired course of action, and India's historical experiences, have instilled a deep commitment to resolving conflicts through peaceful means.

And that is the reason this book steers away from the discussion on AI Ethics, Compliance and Regulations.

However, it is essential to note that India, like any other sovereign nation, places the highest priority on safeguarding its national integrity and security. In the face of genuine threats to its sovereignty, India has demonstrated its readiness to defend itself. In the foreseeable future, India's path is one of responsible global citizenship, promoting regional stability, and contributing to international development and peace, while reserving the right to protect its sovereignty and national integrity if ever challenged. This book on Military AI bridges the gap between tactician's desires and technology manager's capabilities. With an attempt to delineate the actuality from fiction, it will intend to become a guidebook for the entire ecosystem.

Acknowledgements

I am thankful to the United Service Institution of India for providing a haven for study and reflection, and a special thanks to my mentor, Major General P K Mallick (Retired), for his grounding guidance. Gratitude is also extended to all specialists; military, engineers, scientists, and developers whose insights have broadened my perspective and fuelled my curiosity.

To my esteemed colleagues, thank you for sharing in my enthusiasm and turning ideas into reality. Our collaboration has enriched this endeavour beyond measure, and I am deeply grateful for our shared journey.

To my beloved wife, Pooja, whose unwavering love and understanding has been my anchor throughout this journey. To my sons, Shouryha and Avvi; your encouragement and motivation fuelled my passion.

And to Bingo, our beloved Labrador; your loyal companionship would have brought solace during the long nights of writing. Though you are gone, your memory remains a beacon of joy, reminding me of the simple pleasures found in unconditional love.

I also acknowledge Generative Artificial Intelligence as a teacher, elucidating difficult concepts in simple terms, streamlining a longer endeavour. Of course, it was my librarian too, recommending supplementary reading along the research path.

Finally, to you, the readers, this work is for your benefit. It is my hope that it may inspire, provoke thoughts, and ignite dialogue.

Pawan Bhardwaj

Chapter One

AI's Ascendance: Transforming Societal Dynamics and Shaping the Future

WHAT IS AI?

When computer systems simulate human intelligence processes of learning, reasoning, and self-correction, they create Artificial Intelligence (AI). It simulates how humans sense, learn, process, react, and interact to information in the environment.[1] This human-like simulation, matched with machine-like meticulousness, gives AI deep insight than humans. AI performs well in repetitive, detail-oriented tasks like analysing data. It commits fewer errors, and completes job quicker.

Researchers have discussed – 'How does one classify a machine as Intelligent?'[2] They will continue to do so routinely as the understanding increases or new questions arise. It is an ongoing debate because this interfacing of cognitive to computational and logic, is a common problem for science, mathematics, psychology and biology. For the sake of this research, it can be summarised into four of which only the first (the Turin Machine) can be seen in existence.

- **Machines are intelligent if they act humanly.** In his thesis Allan Turin stated that the human cortex at birth is an 'unorganised machine'[3] and through some form of training by the parents, teachers and the environment it becomes organised into a better machine which helps humans take decisions. He further propounded that this training, if given to machines, can enable machines to act as humans. His Turing Test is used to test the intelligence of a machine and is a gold standard for AI testing now. This is the most followed premise and researchers continue to find ways to 'train' a machine. Machine Learning (ML) uses stored information to train and retrain the machines. The aim is to make the machines draw conclusions to adapt, to detect, and to deduce patterns aping human decision making.

- **Machines are intelligent if they act rationally.** This approach presumes that machines can be made to operate autonomously once they have perceived (thus, perception of the environment is the primary focus of this approach) their environment long enough and are capable of adapting to the changes in the environment. This approach also presupposes that such type of an intelligent machine will either do something or inform its inability to complete a job.
- **Machines are intelligent if they are thinking humanly.** This approach is still a research subject; much literature, dealing with superintelligence[4] is available but tangible machines don't. It expects machines to work in a cognitive mode. The research is focussed on observing and modelling a brain in a hope of creating a computation model of firing neurons.
- **Machines are intelligent if they are thinking rationally and think right(?) like Aristotle.** This approach proposes congruence of Logic and Knowledge to make argument structures and give conclusions with a correct premise. Obviously, this approach is also highly theoretical in nature as yet.

AI can be summarised as a series of actions, always learning from previous steps. It is a complex network which consists of various pathways and applications, requiring specialised hardware to run software which trains the computer through self-correction.

- During training, a simplest AI system is fed large amounts of labelled training data,[5] which is analysed, correlated, and patterned into actionable information.
- AI focuses on creating rules (also called algorithms) to extract usable information from this data.
- Reasoning in such an AI system concentrates and instructs machines to choose the right algorithm to reach the desired outcome.
- The system continuously self-corrects and fine-tunes these algorithms to provide the most accurate results possible. This learning-reasoning–self correction makes the predictions about future results. It re-learns

for better predictions, if the output is indicated wrong (by humans).
- These processes are reiterated numerous times (duplicating human learning by observation and self-correction) to achieve near parity with human intelligence.

Categories of AI

There are three ways to classify AI, based on capabilities. Some theorists do not classify them into categories; rather as evolution stages leading from one to another.

- **Artificial Narrow Intelligence (ANI).** Also known as Weak AI, it operates in a limited context and merely simulates or tries to simulate human intelligence. Weak AI[6] is designed and trained to complete a specific task only; it just seems to behave intelligently. They actually operate under far more constraints and limitations than the most basic human intelligence. Industrial robots and virtual personal assistants are examples of Weak AI. Almost all AI in existence today is a Weak AI. Reactive machines in this category have no memory and cannot use past experiences for future ones. Deep Blue, the IBM chess program that beat Garry Kasparov in the 1990s, in an example. Limited memory machines of this category are used in some decision-making functions of a self-driving car.
- **Artificial General Intelligence (AGI).** Referred to as Strong AI, it is a machine that can apply the intelligence to solve any problem like humans, replicating the cognitive human abilities. During an unfamiliar task it can apply previously acquired knowledge to another domain and find a solution autonomously. Arendt Hints, assistant professor at Michigan State University calls it 'Theory of mind machine'.[7] When applied to AI, it means that the system would have the social intelligence to understand emotions, infer human intentions, and predict behaviour. Though still nascent, there are some systems like Maseru and GPT,[8] reaching the AGI benchmark. Maseru can master games even when it has not been taught to play, like chess and an entire collection of Atari games, playing games millions of times, learning on the job.
- **Artificial Super Intelligence (ASI).**[9] In this category, AI systems will have a sense of self, understanding their own current state, which gives them 'consciousness'. In addition to performing tasks, these machines

could also have emotions and relationships. This machine will be capable of surpassing the complex emotion and intelligence of human beings in every manner. Scientists believe that the invention will require not only sufficiently powerful hardware, but also 'adequate initial architecture' and a 'rich flux of sensory input.'[10] It is predicted that Human Level Machine Intelligence (HLMI) will exist in 10 per cent of AI applications by 2022, 50 per cent by 2040, and 90 per cent by 2075. Superintelligence or Intelligence implosion is most likely to present itself within two years from HLMI in 10 per cent of the applications and will touch 75 per cent within 30 years of HLMI. It seems that AI needs time to evolve, which it will surely, unless continually impeded by human induced winter – COVID 19 or the Ukraine-Russia War.

Birth and Ascendency

In the fifties, a John McCarthy study indicated that machines (including computers) were incapable of intelligence beyond Automata.[11] He was not satisfied with the results and proposed a new project, a conference with the brightest minds. His hypothesis was ground breaking.[12]

> "...that every aspect of learning or any other feature of intelligence can be so precisely described that a machine can be made to simulate it.
>
> An attempt will be made to find how to make machines use language, create an abstraction and a concept, solve kinds of problems now reserved for humans, and improve themselves.
>
> We think that a significant advance can be made if a carefully selected group of scientists work on it together for a summer."

This was the Dartmouth Summer Research Project on AI held for eight weeks in Hanover, New Hampshire, UK in 1956. The workshop had mathematicians, psychologists, electrical engineering and biological experts. They all believed that thinking is not something unique either to humans or biological beings. They believed that 'thinking' could be understood computationally and a computer is the best nonhuman instrument to do so. AI was also coined as a term during this conference.

- ELIZA[13] was developed in the 1960s as an early natural language processing program which laid the foundation for today's catbots. But

due to limitations in computer processing and memory shadowing, AI research and development soon slowed to a crawl.
- The 1990s saw an increase in computational power[14] and eruption of data, rebirthing AI brilliance.

Data

AI lives off data which is more important than the quality of learning algorithms. More options mean better choice, which leads to better finish. Larger volume of data creates a better dataset and the ML algorithms learn better. ML with Data Analytics collaborate to constantly push data usage beyond limits to develop smarter AI.[15] To express all the data into knowledge, developers have to either teach the machine by hand-coded knowledge engineering or huge datasets and learning algorithms to train AI. Either way, data availability precedes training.

Enterprises possess their own databases, which is generally organised as per organisational needs. This data may not be sufficient for creating goal-oriented AI; thus, the enterprises have to obtain additional data, which can be a formal customer feedback, a market survey report, or a specialist report from a consultancy firm. There are numerous sources to obtain training data, and the choice is mostly determined by the purpose of the project. Reaching out to the external vendor[16] is the most common and effective way for obtaining a training data. These specialised vendors provide accurate labelled data, balancing the image classification and categories, following regulatory guidelines.

Data when collected from multiple sources is usually in an unorganised format, which is not useful for machines to ingest; it can result in loss of useful information. But when such data is labelled or tagged with annotation it becomes well-organised data that can be used to train. A lot of unstructured data is generated daily by users around the world, which may not relate to the enterprise directly but AI systems are capable of generating insights where humans cannot. Text, videos, audio, and data coming in from social media are all unstructured. This valuable data cannot be ignored. It has to be tagged, labelled, and annotated.

Some developers also augment their data to increase the training examples and artificially generate additional training data from existing training data. This data augmentation is done by applying random transformations to the training data, such as cropping, flipping, and rotation. It is especially useful when training data is scarce.

The training data for AI is directly related to the performance of data models. It is a key input to algorithms that learn from such data and memorise the information for future prediction. AI or ML engineers have to make sure they have accurate datasets to get their result right. Annotated or labelled data helps machines through Computer Vision (CV) and Natural Language Processing (NLP) to detect various objects from a group and store the information for the future.

Algorithms

An algorithm is a set of instructions for solving a problem or accomplishing a task. The recipe for making a cocktail is an algorithm – follow the procedure. In computer programming terms, an algorithm is a well-defined instruction to solve a particular problem. A computer algorithm takes a set of input(s) and produces the desired output. Much like a cocktail recipe; cover the bottom of the shaker with four dashes of salt, two of black pepper, two of cayenne pepper, and a layer of worcestershire sauce; add a dash of lemon juice and some cracked ice, put in two ounces of vodka and two ounces of thick tomato juice, shake, strain, and pour in a glass lined with salt; there is no deviation.

Machine Learning

ML is different; it is the science of getting a computer to act without programming. Simply put, an ML algorithm is fed data by a computer, and uses statistical techniques by collecting, organising, and analysing data based on rules to identify patterns and trends to help it 'learn'. Basically, ML algorithms use historical data as input extracting relevant features to predict new output values. It is a set of algorithms which try to mimic human intelligence and outputs. But ML methods plateau at a certain level of performance, showing no improvement even when more examples and training data is fed to it. At this stage, Deep Learning (DL) comes into play. Before understanding DL, the following example explains ML.[17] For a simple algorithm of addition, the following stages are followed:

- Stage 1. Create a function: Add ()
- Stage 2. Accept Data or Input: 1 and 2
- Stage 3. Provide a result: Add (1, 2) = 3

What if you didn't know what 'function' ('Add' in Stage 1 in the example above) to use or create (it is a possible doubt when the programmer is dealing with an NP Complete class[18] of problems). In such a scenario, a human programmer creates a 'Learner Algorithm' with lots of examples for the computer to teach itself. The learner algorithm uses this training to create a suitable 'Function'. The ML thus continues to build its own 'cognitive capabilities' by creating mathematical functions that can distinguish one class from another and is suitable for the problem at hand. ML 'learns' in three major ways.

- **Supervised.** Datasets are labelled so that patterns can be detected. These patterns are used to label new datasets in the future. Here, the expected output for the input is thus known. Supervised ML requires classification[19] of objects in the training data. For a model to start recognising patterns, it has to evolve through multiple stages of dataset iterations.[20] Training a ML model also involves iteratively adjusting[21] the model parameters so that it makes accurate predictions on the training data. An algorithm is made to learn important distinguishing features between two classes to recognise them in future.
- **Unsupervised.** Data sets aren't labelled but are sorted according to similarities or differences. Here the ML learns by reading the differences and similarities, correlating the features which define these and then attempts to identify the same features in the new data. Though the expected outputs are unknown, the ML learns better and is capable of correlating the knowledge to previously unknown tasks.
- **Reinforcement.** Data sets aren't labelled but, after performing an action or several actions, the AI system is given feedback, much like a reward. The ML here attempts to collect most rewards and learns to train itself over a wide number of iterations.

Deep Learning (DL)

DL or Deep Neural Network is a ML technique that teaches computers to do what comes naturally to humans: learn by example. In this, a computer model learns to perform classification tasks from images, text, or sound features directly from the data without the need for manual feature extraction. Models are trained by using a large set of labelled data and many layered neural network architecture (traditional neural networks contain 2-3 hidden layers, while deep networks can have as many as 150). It is biologically-inspired neural network architecture.[22]

The neural networks contain number of hidden layers through which the data is processed, allowing the machine to go 'deep' in its learning, making connections and weighting input at all layers for the best results. The ML taxonomy chart gives out their interdependence.

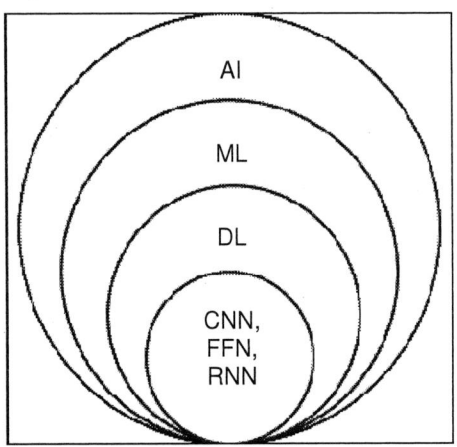

- Convolutional Neural Network (CNN) is a popular architecture for image processing. The network of layers trains on a dataset of images and every hidden layer increases the complexity, and clarity of the learned image simultaneously, where feature extraction and modelling steps are automatic. This automated feature extraction makes this architecture suitable and accurate for CV tasks (object classification).
- Feed Forward Neural Networks (FFN) is a variant that passes information in one direction, through various input nodes. In this model, a series of inputs enter the layer and are multiplied by the weights. The outputs of its nodes are compared with the intended values, thus allowing the network to adjust its weights during training to produce more accurate output values. A series of feed forward neural networks, intentionally running independently from each other, with an intermediary for moderation can be utilised to undertake complex actions. It is useful in facial recognition and CV technologies.
- Recurrent Neural Networks (RNN) save the output of processing nodes and feed the result back into the model; certain pathways are cycled. The model learns to predict the outcome of a layer. Each node in the RNN model acts as a memory cell, and remembers all processed information to reuse it in the future. If the prediction comes out

incorrect, then it self-learns and continues working towards the correct prediction during back propagation. It is frequently used in text-to-speech conversions.

Within the Neural Networks, the developers are use various techniques[23] to hasten the DL model training.

- Training from the Scratch technique is created for a new application or an application that requires a large number of output categories. Imagine a school project to create an AI system to recognise a cat. The students create a very large labelled dataset of numerous cat images (cat dataset) from all angles and sitting, walking, standing, sleeping positions. It is followed by an architecture that will learn the features (pointy ears, small muzzle, cat eyes, languid paws, legs, etc) and model. This is a less common approach and obviously has a slow rate of learning.
- To improve the gestation, the developers sometimes utilise transfer learning by fine-tuning a pre-trained model. A new data containing previously unknown classes are fed into the training dataset and an interface is created to the pre-existing network, teaching the machine the difference between the new classes from the existing one. Let us go back to the same group of students who want to differentiate dogs from cats. The new data (dog dataset) does not have to be as large as the cat dataset. The students tell the AI the difference in the features of cats and dogs (the ears can be droopy, they wag their tails often, have bushy tails, are not as languid as cats, have longer muzzle, etc). The transfer learning AI will generate new inputs but it has utilised lesser newer data aka Dog dataset and the smart students have tweaked the Cat AI system to identify dogs comparing the difference with the Cat dataset. The AI computation time has reduced remarkably.
- Feature Extraction is a more specialised approach, in which the network is used as a feature extractor. All the layers are tasked with learning certain specific features from images, so the required features can be pulled out of the network at any time during training. Imagine the same group of students nominating a layer each to extract all cat ears (erect ears) and another to extract all dog ears (droopy ear layer and pointy ear layer), cat muzzle layer (short) and dog muzzle layer (long). These features are used as inputs to a ML model which clearly and

widely separates the data between two classes, thus increasing the output accuracy of the model.

Intelligent Agent

In the field of AI, an Intelligent Agent[24] is a program that can make decisions or perform a service based on its environment, user input, and experiences. These can autonomously gather information on user defined parameters. An intelligent agent can also be a tangible product or a tool that perceives its environment through sensors and acts upon it through actuators.

An autonomous vehicle could be considered an intelligent agent as it uses sensors, GPS, and cameras to make reactive decisions based on the environment to manoeuvre through traffic. It is expected to perform the best action in a particular situation and thus there is a critical need to define the properties (of the environment – physical and virtual) and goals precisely. An agent's action choice is dictated and affected by quality of perception,[25] – better sensors improve perception. AI developers utilise Performance – Environment – Actuators – Sensor (PEAS) relationship with agents to create a 'logic map'. This logic map creates a value AI machine, since increased variables in the PEAS analysis will create complicated machines, requiring increased computational power and greater learning burden. Intelligent agents[26] are defined by their range of capabilities and degree of intelligence.

- Simplex Agents function in a current state only, totally ignoring past history. The agent is programmed to respond to a set of conditions and responses are based on the event-condition-action (ECA) rule, a list of pre-set rules and pre-programmed outcomes. Not so intelligent – 'if a cat sees a dog; then freeze the cat'.
- Model Based Reflex Agent chooses an action in almost the same way as a simplex agent, but has a more comprehensive view of the environment. A model of the environment is programmed into the internal system that incorporates the agent's history too. It is now enabled to keep track of visible or present environment, while maintaining a certain degree of internal state and can analyse unobserved environment (its history) too. If a cat sees a dog – then freeze and jump on the dumpster – because last time the dog turned and saw the frozen cat – not good for the cat (agent's history). The cat analyses its history and make a more intelligent decision.

- Goal-Based Agent is provided goal information, or information about desirable situations. It is a model based reflex agent; on a multivitamin. The agent is capable of reading the unobserved environment, correlates it with its goal definition, and generates better decisions or actions. The cat has a goal to cross the street – it jumps over the dumpsters lining the street to cross, without fear of the dog.
- Utility-Based Agent rates numerous possible scenarios to its desired goal and chooses the action that maximises the outcome. It can compare different states, identify, and resolve conflict goals with speed and safety. It can weigh decision actions against the importance of goals too. The cat decides to jump from the dumpster to the rain drain (additional effort and time) because one of the dumpsters is unstable and the cat may slip and fall into the dog's jaws.
- Learning Agents have the ability to gradually improve and become more knowledgeable about an environment over time through an additional learning element (multi-iterative process). The agent will use this feedback to determine how performance elements should be changed to improve gradually.

Cloud vs Fog vs Edge Computing

The data has to be stored somewhere; there is an option to store it in-house but it is limited to hardware and software costs and their security. Another option is to store it elsewhere. Cloud storage provides tremendous cost saving, running applications and processes online, without the hassle of day-to-day maintenance or licence renewals, protecting and recovering all crashed or loss data, if it occurs. Cloud computing data is stored on multiple servers and can be accessed online from any device. It caters high data computing, giving priority to high volume data in which security is managed by a third party.[27]

Edge computing processes data directly at the edge without communicating with the cloud, allows immediate response and provides unprecedented speed. In theory each edge device acts as a server and processes data independently, and as the name suggests, processes data on edge devices and gateways, immediately close to the organisation and away from centralised storage. To break into, hackers would need synchronized access to thousands of distributed devices, which is practically impossible, providing better data sovereignty.[28] It does not eliminate the need for cloud but is viewed as ideal for tasks with extraordinary inactivity or information security concerns.

Fog obviously hovers in between, in which the node decides whether to process it locally or send it to the cloud, a suitable feature when internet connection isn't always stable. It consists of multiple nodes (fog nodes) and creates a local network which makes it a decentralised ecosystem.

Biases in AI

> *"If we are to develop trustworthy AI systems, we need to consider all the factors that can chip away at the public's trust in AI. Many of these factors go beyond the technology itself to the impacts of the technology."*
>
> *– Riva Schwartz*[29]

An AI system is fed by the data sets it is provided to train and validate. It will reinforce whatever it has already learned; good or bad. Most AI tools are powered by ML which learn from a training data set, which defines the degree of smartness – poor training data set produces poor ML. A human being selects this training data and bias is inherent in it, which will creep into the ML if it is not monitored constantly.

A bias crept in inadvertently[30] when the Americans reported denial of bank loans or same-day purchase delivery to certain neighbourhoods in American cities. Investigations found that the areas from where training data was collected was not sufficiently inclusive. It was biased during collection of data about individuals in a specific community and neighbourhood. The AI was trained on this fractured dataset creating output bias since selection or exclusion datasets upsets a balanced informational analysis. Variety of biases is staggering; while a reporting bias tends to leave out 'uninteresting information', the detection bias overlooks certain data altogether. A CV feature extraction model of DL for example may induce collections compromise for samples from the real world, if the data has been augmented and disregarded for scale, occlusion, deformation, or rotation.

The developers have to make sure that the training data is always representative of the real-world data that the model will be used on. Imagine learning Cyrillic script to read English. They have to choose high quality and error free training data. They also have to make sure that the training data is sufficiently large. A good rule of thumb is to use at least 100,000 training examples[31] for each task of ML model to learn. If an AI system doesn't have access to enough categorised images, then it will not be able to provide accurate results and the overall system will fail. An ML algorithm needs certain specific inputs about the object to understand the things in its own way. The data, like images have to be annotated with precise metadata making the object recognisable to machines.

Explainability in AI gives rise to Contextual AI

An AI-recommended decision may not be fully explained to humans and may affect the final human relations, which is a potent stumbling block. This unexplained feature of AI termed as AI Black Box is making researchers study a newer field of AI in which the decision-making process has to be explained by the AI. Continued advances in AI and ML will produce autonomous systems that will perceive, learn, decide, and act autonomously. But ML programs are opaque and non-intuitive, difficult for people to understand and the machines are unable to explain their decisions and actions to humans. Research for a symbiotic system which can explain this 'unknown' is called 'Explainable AI (XAI).[32] This system is expected to explain its rationale and generate more trust in humans, allowing effective management of AI systems.

Contextual AI takes a human approach while processing content and explaining itself. It allows AI systems, to possess a near real-world interpretation of audio, video, and languages. It can analyse sentences and sentiments in a historical, cultural, and environmental situational aspect and behave less like traditional computers and more like humans. Contextual AI can pick up patterns and features in the data, extrapolate context clues from internal supervised learning cases and gain much deeper understanding of any situation. It improves unsupervised learning on a case-by-case basis. It is able to take what it learned from one context and apply it to another to perform better on a similar task. Basically, Contextual AI adds explainability[33] throughout the ML pipeline, from data absorption to extrapolation.

Accuracy in AI

The choice of an evaluation method depends on the task and the desired outcome, and it is often a good idea to use multiple evaluation methods to get a comprehensive understanding of a model's performance. Merely developing an AI model is not enough. The output by AI has to be validated for its correctness and accuracy. Giving priority to the quality of training data will definitely achieve best results, but the user has to be satisfied that the developer has created an accurate AI.

The simplest way to check the accuracy of an AI system is by having human experts evaluate the performance of the model and compare it to a baseline. Validation[34] of the model to check its accuracy is done by yet another dataset called validation data, which is used to check the accuracy level of a model in different scenarios. Common statistical metrics like F1-score, AUC-ROC are utilised for evaluating AI models for accuracy, precision, and recall, but they measure different aspects of AI performance. The aspect to check for depends on the role[35] of AI.

- Accuracy measures the overall accuracy of the model, but can be misleading in cases where the class distribution is imbalanced, so statistically it is supported by identification of Precision and Recall.
- Precision is the fraction of positive predictions that are actually correct. It measures the ability of the model to avoid false positive identifications. In a decision support system, the precision may be given higher value by the user.
- Recall (or Sensitivity) is the fraction of actual positive instances that are correctly identified by the model. It measures the ability of the model to identify all positive instances. In an information mining and visualisation system, recall or sensitivity may be considered of vital importance

Hardware in AI

A high-performance processing unit is a must for all types of AI which can speed up the method significantly. As systems become more sophisticated, they need more computation power to accelerate training and performance of neural networks and simultaneously reduce power consumption. The traditional solution is to reduce the size of logic gates to fit more transistors, but shrinking logic gates below about 5nm[36] can cause the chip to malfunction, so the challenge now is to find another way and develop specifically designed hardware. Developers are trying to target all the aspects of processing units – be it architecture, power consumption or application area – just to make it faster and efficient.

- Traditional Central Processing Units (CPU) runs logic operations in a sequence, the Graphic Processing Units (GPU), designed for gaming, on the other hand, has parallel processors that perform multiple tasks at the same time. A new chip, called an Intelligence Processing Unit (IPU),[37] or a Dataflow Processing Unit (DPU), can conduct massive parallel computing through its 1,000 processors. These are capable of holding complete ML models within; they are fast and energy efficient.
- A multi-core microprocessor[38] is a set of independent processors on a single silicon chip, all interconnected. The cores work in parallel, communicating with each other to speed up the processing. These can provide 2.3 Tera operations per second (TOPS) of dedicated deep neural network (DNN) computing.[39] Some of these 1,024-core chips can therefore perform real-time image processing, autonomous driving, and machine learning. Some developers provide them as plug and play version too – as a USB pen drive!
- Application Specific Integrated Circuits (ASICs) are manufactured to custom design specifications, with lower productions cost and smaller form factor. Combined with Fully Programmable Gate Array (FPGA) it permits quick testing, where, at the prototyping stage, FPGAs are used to verify the logic before ASIC is actually manufactured. System on Chip (SoC) like TDA4VM[40] has achieved 8 TOPS. It is embedded in self-driving cars which can handle a video stream of 8 MP and concurrent inputs from RADAR, LiDAR, and ultrasonic sensors. Tensor Processing Unit of Google[41] is benefiting AI at the Edge.
- In-Memory Computing (IMC),[42] RAM based data storage and index processing is being considered a disrupter in the field of SoC. It exploits register files and memories within processors, drastically reducing computing time and power usage. Both hardware and software solutions form part of IMC.
- Still a product of the future, Photonics (study of properties and transmission of photons) is suggesting radical structural designs – work on photons in visible light or infrared beams, rather than electric current.

Optical Computing Chip[43] being one, is expected to be faster than traditional processors and smaller in form – a supercomputer on a chip?
- France-based PropheSee[44] is analysing the way an eye works – reporting only 'an event' (a movement) and overlooking the irrelevant and huge background. The company believes that this process of capturing only the 'change' will reduce the load on the CV and make it efficient – why waste computing power on information that has not changed?
- Revolutionary researchers have found that 'Very Low Precision Computation' is sufficient for running trained networks and for training them. They have obtained good results; 10 bits for computing activations and gradients, and 12 bits for storing updated parameters.[45].
- Ultra-Low power processors can compute an 'always on application' for as low as 100 µW (Eta Compute ECM3532[46]). They are developed for IoT use, image processing and sensor fusion, vibration sensing, voice commands, enabling hands-free operation of personal devices.

THE AI DOMAINS

Computer Vision (CV)

The CV performs a deep dive into data analysis. It has a much greater processing capability of acquired visual data (which may not be through a camera always) with a greater emphasis on understanding and predicting. CV is a field of AI that trains computers to capture and interpret information from image and video data. Though CV has a neurology driven history, other concurrent invention increased the speed.

- In 1960, neurophysiologists discovered that cats' brains respond hierarchically to process an image – first hard edges or lines, then curves and finally the whole image. In 1980s, neuroscientist David Marr introduced algorithms to detect these basic shapes – creating the rule on which CV works today.
- Computer image scanning ability to transform two-dimensional images into three-dimensional forms also developed concurrently.
- Optical Character Recognition (OCR) technology (1974) led to development of Intelligent Character Recognition (ICR) capable of deciphering hand-written text.

- Standardisation of tagging visual data sets and annotations emerged in the 2000s, giving developers access to millions of tagged images across a thousand object classes.

> - CV utilises ML to improve itself
> - ML utilises CV to capture data and improve its predictions
> - It is both a development tool and an output product of AI

An earlier CV programmer would have to undertake a large amount of manual work (very little automation), using feature descriptors (SIFT, SURF, BRIEF, etc.) for object detection.[47] They would create a database to capture individual images, annotate them with several key data points (distance between eyes, distance between ears and eyes etc), all in an effort to define unique characteristics of an image. Then capture a new image and make the same measurements. After all this manual work the application would compare the two and certify it as similar or not.

ML in CV on the other hand now uses smaller applications to detect patterns in the picture and has replaced manual tagging and measurements. It uses statistical learning algorithms to detect patterns and classify images. Convolutional Neural Network (CNN) version of DL when provided with many labelled examples of a specific kind of data, can extract common patterns, transform them into a mathematical equation, and then classify them for future pieces of information. CNN breaks down the image to pixels and tags them (giving a context to each pixel). It then performs convolutions (a mathematical operation on two functions to produce a third function) and makes predictions about what it is 'seeing.' This accuracy of its predictions is done over a series of iterations until the predictions start to come true.[48] The AI system then recognises the image in a way similar to humans, remembering it for the future. Similarly, a face recognition application chooses a pre-constructed algorithm,[49] trains it with examples and after lots of examples, will detect faces without further instructions. IBM categorises the following as CV tasks:

- Image Classification to classify an object (grouping it to a certain class). Typical example is when a service provider segregates pornographic images uploaded by users.
- Object Detection can detect damages on an assembly line of a factory or used by a self-driving car to detect traffic signals or cross roads.

- Object Tracking follows or tracks images captured in sequence or real-time video feeds. Self-driving cars have to track objects in motion to avoid collisions.
- Video Motion Analysis uses CV to estimate the velocity of objects in a video.
- Image Retrieval browses, searches, retrieves and tags images from large data stores replacing manual tagging, increasing the accuracy.
- Scene Reconstruction creates a 3D model of a scene inputted through images or video.
- Image Restoration removed noise such as blurring from photos using ML filters.

Mobiles with built-in cameras have swamped the world with photos and videos, thereby improving analysis. Developers are training DL models using these freely available thousands of labelled or pre-identified images. CV is also playing an important role in Augmented and Mixed Reality. It allows the computing devices to overlay or embed virtual objects on real world images.

A sharp rise in CV usage is also attributed to the willingness of big companies to share their work.[50] Facebook, Google, IBM, and Microsoft, offer open-source use of some of their ML algorithms. For many real-world applications, this process happens continuously in microseconds, without human interference.

Natural Language Processing (NLP)

NLP is the processing of human language by a computer. It gives computers the ability to understand text and spoken words in the same way as human beings. It is a rule-based[51] modelling of the human language – with statistical,[52] machine learning, and deep learning models to 'understand' the full meaning, complete with the speaker or writer's intent and sentiment. A software writer cannot conceivably include all homonyms, homophones, sarcasm, idioms, metaphors, grammar and usage exceptions or variations in sentence structure to accurately determine the intended meaning of text or voice data. AI steps in here to understand the nuances which humans have developed over so many years, to fill the gap between human communication and computer understanding.

NLP breaks down language into shorter, elemental pieces, like children do when learning a new language. It includes breaking (Tokenisation) down text into small units, then removing common words, retaining unique words offering most information of the text (Stop Word Removal), followed by reducing the words to their root form (Lemmatization and Stemming), finally marking the words as part of speech – nouns, verbs and adjectives (Part of Speech Tagging). After this reprocessing, the algorithms are built to understand the language by studying a sentence to make grammatical sense, or meaning behind words.

- *NLP utilises ML to improve its ingestion of data created by humans*
- *ML utilises NLP to capture more data and improve its predictions*

The ML based system uses statistical methods, learning from training data, and adjusts and trains its methods as more data is processed. This is faster and more accurate, pulling in inputs from unstructured, unlabelled text and voice data sets too. NLP can analyse more language-based data than humans, without fatigue and in a consistent, unbiased way, scanning the large volumes of such data which is continually growing.

- NLP has numerous applications like Spam Detection, which detects and isolates undesirable emails.
- Machine Translation replaces words in one language with words of another, improving and creating distinctive the text – Google lens is a real example.
- Virtual agents and Chatbots, answer queries without delays (almost).
- Text summarisation consumes huge volumes of digital text and creates summaries and synopses.
- NLP can visualise data from complex documents (tables, PDFs, big data, etc.), including from unstructured data formats including HTML, web pages, reels, pictures and much more.
- Text Analytics[53] turns text into data for analysis and deep contextualised insights (see the Word Cloud below).[54] It can derive new variables from raw text that can be visualised, or used as inputs for predictive modelling.
- Social Media Sentiment Analysis is a specialist form of text analytics which discovers hidden data insights from social media channels. It can analyse informal data to extract attitudes, emotions, fears, and aspirations of the users and identify key influencers.

Progressing real time language around a noisy environment is still a challenge. The natural language is ever-evolving and always somewhat ambiguous. Human speech is not always precise and depends on many complex variables. The language rules are never constant and change with time. Hard computational rules of NLP will become obsolete[55] unless the developers utilise AI concurrently to catch with to the real-world language changes over time.

Robotics

A field of engineering which focuses on the designing and manufacturing robots to often perform tasks that are dangerous or difficult or extremely repetitive for humans to perform. Robotics is known as a self-operating machine which, with precisely trained and learned inputs can continue to perform consistently and accurately. In itself robotics is an old discipline but AI provides it with special abilities spatial relations, grasping objects, computer vision and motion control.

Robotics involves building physical robots[56] only, while AI involves programming intelligence as well. There are robots which can flourish without AI (spot welding, positioning the torch in the exact same orientation on every cycle, assembling smaller parts such as motors and pumps at high speed, screw driving, wheel mounting, and windshield installation are just a few instances). While robots mimic human functioning, AI robots try to imitate the human brain as well. A Robot interact with the physical world via sensors and actuators, they are programmable and can be autonomous or semi-autonomous. The table below compares the human and robot representations

Human	Robot
A biological body	Artificial body (metallic, mechanical, silicone, etc)
A brain that perceives information and controls the actions	AI component which gauges the environment and self-state and instructs the appendages or tools to operate.
A sensory system to receive information and shares with brain	Sensors (Camera, LIDAR, LASER, SONAR, GPS) For the work environment and internal state of the robot
A muscle system to move the body	Electrical components[57] which power and control the machine. Actuators, motors, and pneumatic systems to move the appendages or tools. Mobility devices (legs, wheels, tracks or skids)Robots could be static too.
A power source to activate the brain, muscles and sensors	Powerful energy source (electric, batteries, solar)

- *AI Robots can manipulate, themselves and objects flawlessly and safely in close proximity of humans, NLP enhanced human-robot interaction and CV perception of environment.*
- *Autonomy and efficiency of AI Robots is enhanced by ML, NLP and CV*

AI, ML, and CV provides motion control and perception to understand environment and make the robot interact with it. The sensors help the robots to learn, thus prevent constant human intervention[58] and parallel effort. All in all, AI improve the efficacy of robotics.

- Home vacuum robots are useful cleaning tools and now the new robots are machine learning and AI enabled to spot and avoid pet-excrement![59]
- Robots assisting medical practitioners will not get tired, which makes them a perfect substitute or interns, performing continuous medical tests or procedures.

The current tasks assigned to robots (manufacturing sector) are predictable and repetitive, so adding any form of AI would simply be overkill – not a necessity. So why create AI into it – a group of researchers believe. The discussion is endless, but robots will remain a good option to take on repetitive dangerous or undesirable jobs.

GAN

The Generative Adversarial Networks (GAN) is a most powerful algorithm in AI and interesting too. It is an unsupervised ML model in which two neural networks compete with each other to become more accurate in their predictions.[60] Both networks are trained on the same datasets and, this technique learns to generate new data with the same statistics as the training set. One model called a generator is a convolutional neural network (CNN), and the other – a discriminator, is a deconvolutional neural network (DNN), performing an inverse convolution. Since its inception in 2014, GAN is now used for semi-supervised, fully supervised, and reinforcement learning too.

- *The objective of the generator is to artificially produce outputs that could be easily mistaken for real data.*
- *The goal of the discriminator is to identify the artificially developed ones from the output.*
- *This feedback loop between the adversarial networks continues, and the generator will begin to produce higher-quality output. The discriminator will become better at flagging data that has been artificially created.*
- *Based on these results, the generator adjusts its parameters to create new images. And so it goes on until the discriminator can no longer tell what's genuine and what's fake.*

GAN is capable of combining its known information and generating almost novel output, which are difficult to identify as fake or not necessarily original. The outputs are theoretically endless – texts, images, software programs, music, architectural inspirations and de novo works of art. The data generated by a GAN is referred to as 'realistic' or 'synthetic' data, not 'real' data. The goal of a GAN is to generate data that is 'similar to a target data distribution', but it is not guaranteed to be an exact match. The quality of the generated data will depend on several factors, including the quality of the training data, the architecture of the generator and discriminator, and the training procedure. GANs exist for image, video, and audio syntheses too.

- StyleGAN (Generative Adversarial Network) well-suited for image synthesis, as it can generate high-resolution images with a high degree of control over fine-grained details.
- CycleGAN can perform image-to-image translation, such as converting

photographs of one person to another or summer landscapes into winter.
- Flow-GAN can generate high-quality videos with smooth, continuous motions.
- WaveGAN can generate raw audio signals as sound effects.
- CLGAN (Cycle-Consistent Adversarial Networks) can perform audio-to-audio translations, such as converting speech to singing or converting the style of talking from one to another.

In theory, GAN can work without a cloud or internet connection. GANs can be trained and run on local hardware, such as a desktop or laptop computer. The training process may take longer on a local machine compared to training on a cloud-based system with more resources, but it could be possible to train and run a GAN without a cloud or internet connection – it is still a theory. GANs have several applications in the military, limited by the training data sets and human imagination. A few could be:

- Image generation and manipulation to generate synthetic images and manipulate existing images to conceal sensitive information, or generate deceptive images or information.
- Object detection and tracking to improve object detection and tracking algorithms for surveillance and reconnaissance tasks.
- Fraud detection to detect fraudulent activity, such as fake identities or false documents.
- GAN can be used in cyber security to defend against cyber-attacks by generating realistic-looking decoys that can distract or mislead an attacker.
- GANs can be used in simulation and training to generate realistic simulations for training military personnel in a wide range of scenarios – from battlefield simulations to disaster response.

GPT

Though there are numerous Generative Pre-trained Transformer (GPT), one especially is considered the leader of the pack. ChatGPT (Chat Generative Pre-Trained Transformer) is a GAN chatbot launched by OpenAI in November 2022. It is known for detailed responses and articulate answers across many domains of knowledge. ChatGPT has learned through supervised and reinforcement learning by human trainers to improve performance. It also gathers inputs from ChatGPT users that are trained further and fine-tune it. It can only process and generate text based on the input provided via a chat box text inputs by the users. It is a cloud-based AI language model that is not designed to be run on a standalone system. When asked 'How do you work?" it replied[61]

> "I work by processing text inputs provided by users and generating outputs based on that input. My underlying technology is a type of artificial intelligence called deep learning, specifically a type of neural network called a Transformer. I have been trained on a massive dataset of text, which allows me to generate text that is similar to the text I have seen during my training.
>
> When I receive an input, I process it and generate an output by considering patterns and relationships in the text I have been trained on. I then use those relationships to generate an output that is most likely to be relevant and informative given the input I received. My goal is to provide helpful, informative, and accurate answers to the questions I am asked.
>
> It's important to note that my training data has a cut-off date of 2021, so I may not have the most up-to-date information on certain topics or events that have occurred since then. Additionally, because I have been trained on a large dataset, I may provide responses that reflect the biases and perspectives present in that dataset."

Replying to the same question the second time, it wrote: "I work by using a deep learning-based language model developed by OpenAI. I have been trained on a large corpus of text data and can generate human-like responses to natural language questions and prompts. When a user inputs a message, I process it using my neural network, and generate a response based on the patterns I have learned in my training data. My responses are generated in real-time and are not pre-written."

The accuracy and relevance of GPT depends on the quality and timeliness of the training data it was trained on. If the training data is outdated, the model may generate responses that are inaccurate or no longer relevant to current events and developments, a very important consideration to organisations, which are running blindfolded towards it. The GPT has a potential to be used in a variety of military applications, depending on the vintage of training data made available to it:

- It is usable for sentiment analysis, text summarization, and entity extraction, which is useful for analysing large amounts of data or text-based communications in a military context.
- It can provide real-time situational awareness by analysing and aggregating data from a variety of sources, such as satellite imagery, drones, and ground sensors.
- Its ability to generate human-like text can support military decision-making processes by generating reports, summaries, and insights.
- It NLP is useful for developing autonomous systems that can interact with humans using natural language.
- It could be used to improve cyber security by detecting and preventing social engineering attacks, such as phishing scams, or to help identify and respond to cyber threats.

Fuzzy Logic (FL)

FL is both an AI technique and a branch of mathematics. It was originally developed as a mathematical framework for representing and manipulating uncertainty, and it has since been widely used in a variety of AI applications, including Expert Systems (ES), control systems, and NLP.

In AI, it is used to model complex and uncertain real-world systems, and to make decisions based on incomplete or ambiguous information. The main idea is to use linguistic variables, which are be represented by a wide and continuous range of values to represent the uncertainty and imprecision inherent in many real-world problems. The values of these variables are assigned a degree of truth to each value, representing the degree to which it satisfies a particular fuzzy set.

Fuzzy logic is a mathematical framework for dealing with uncertainty and imprecision. It was developed as an alternative to traditional (Boolean) logic, which operates on binary values of true or false. This conventional logic block of a computer takes precise input and produces a definite output as TRUE or FALSE, which is equivalent to human's YES or NO.[62] In fuzzy logic, the truth values of statements can be any value between them, representing the degree of truth or falseness. This allows for more nuanced and flexible decision-making, as it can accommodate situations where there is not a clear-cut answer.

It can be implemented in systems with various sizes and capabilities ranging from small micro-controllers to large, networked, work station-based control systems. It can be implemented in hardware, software, or a combination of both. A 5-level FL air conditioning system adjusts the temperature of an air conditioner by comparing the room temperature and the target temperature value by creating a matrix of room temperature values versus target temperature values that an air conditioning system is expected to provide.

Researchers are impressed by FL capability to work without a database. They argue that it is a mathematical framework for representing and manipulating uncertainty and it can be applied without the need for a database or any other specific type of data storage. It may be used with data from various sources, such as sensors, databases, or human input and the underlying concepts do not depend on a specific type of data source. The main components are the fuzzy rules, which represent the knowledge and understanding of the system, and the inference engine, which uses these rules to make decisions based on the input data. So, fuzzy logic can be applied to a wide range of problems, regardless of whether there is a database involved or not.[63]

In AI, FL is used to model complex and uncertain real-world systems, and to make decisions based on incomplete or ambiguous information. It can be used in ES to represent the knowledge and understanding of a human expert, and to make decisions based on that knowledge. It can also be used in control systems to make decisions based on sensor data, and in NLP to understand and generate text.

Overall, it is a valuable tool in AI, as it provides a flexible and effective way to represent and manipulate uncertainty, and to make decisions based on incomplete or ambiguous information. It can take imprecise, distorted, noisy input information which can be modified by just adding or deleting rules due to flexibility of FL. It provides a solution to complex problems in all fields of human life decisions. It is being considered a good approach to reach AGI and ASI.

Expert Systems (ES)

FL and Expert Systems (ES) are related but distinct concepts in AI. ES is a computer program that emulates the decision-making abilities of a human expert in a specific domain. This uses knowledge encoded in the form of rules, which are based on the expert's experience and expertise, to solve problems and make decisions. The rules are used in combination with an inference engine, which evaluates the rules and determines the most appropriate conclusion based on the inputs.

ES typically focus on a single field or business sector, centralizing knowledge in that space for companies to access and leverage. During the 1960s, Symbolic AI (based on human-readable depictions of problems, logic, and search) developed knowledge-based systems, later christened Expert Systems. The developers hoped that these ES could lead on to an AGI machine, but they hit difficulties with knowledge acquisition, maintaining large knowledge data bases, and out-of-domain problems. Quantification of tactic knowledge could never be built fully into such ES – leading it to wither away. AI has re-envisioned this as Good Old-Fashioned AI (GOFAI).[64] AI-powered expert systems are validating the importance of the human-technology partnership. The knowledge base in an ES now has AI capabilities to mine data from a wide range of internal and external sources. ML empowers the ES to identify and obtain the tacit knowledge which actually set up its demise in the 'Sixties. The ES can now apply logical rules that generate intelligent insights from the expert database, prompted by user queries and finally the NLP provides a valuable user output interface.

ES and AI have matched the old with new. AI algorithms can be trained on large amounts of data to make predictions and decisions with high accuracy. AI can automatically acquire knowledge from large amounts of data, reducing the need for manual knowledge engineering. A large number of rules and conditions are now handled by AI, making it easier to scale ES to handle complex problems. The new ES has better decision support capability, which is more effective and efficient. FL is also useable in ES to represent and manipulate uncertainty, making it well-suited for dealing with situations where there is no clear-cut answer.

IT IS NOT ALWAYS ABOUT DATABASE OR BIG DATA

Interestingly, an ES can work without AI.[65] ES uses a knowledge base to solve problems that would normally require human expertise. It does this by applying a set of rules and heuristics to the data it receives and making decisions based on that data. While AI techniques and algorithms can be used to enhance the performance and functionality of, they are not essential to the operation of an ES. An ES works by using simple rule-based systems and decision trees, without the use of more advanced AI techniques such as machine learning or neural networks. So, while AI can be a valuable tool in enhancing the performance and functionality of ES, it is not a requirement for operation, and ES can be built and deployed without the use of AI.

Similarly, once the ES is designed, it does not need a running backward connection to any other database or cloud. Since it is expected to curate decisions on a very narrow subject, its application is independent of the cloud connectivity.

There is a profound link between Data and AI – bigger data is bigger AI capability. But what if Big Data is unusable or unavailable – will AI work without Big Data? Some researchers believe it will. New ML on a small chip is proving AI to run independent of the internet, whereas typical neural networks (known as 'tethered' neural networks) always require an internet connection to run an AI program. This breakthrough will lower data transmission cost, increase privacy and security, and create opportunities for AI in areas without internet connectivity, or where organisations are highly security conscious. Researchers at the University of Waterloo,[66] placed AI in a virtual environment, and as the AI trained, the computational power and data availability was gradually reduced; in theory making AI more efficient. The team applied the Darwinian 'survival of the fittest' concept to train the AI, which progressively adapted to generate improved outputs with lower data dependency and power.

ETHICAL USE AND LEGAL FRAMEWORKS

Though not part of this research, however it is important how the world calls out on the AI and how it has affected the development of the niche technologies. The UN General Assembly, in pursuance of the UN's 2030 Agenda for Sustainable Development has remarked that a common effort[67] is required from international and multi-stakeholders to work on ethical effects of AI and other emerging technologies, cyberspace and biotechnology. The UN also recognises that AI development should be accompanied by ethical reflection. AI applications influence human-technology relations in both beneficial and harmful ways. It is a collective choice to make a decision. The UN published Resource Guide on AI Strategies[68] (April 2021) is a useful compilation for the interested.

The UN Convention on Certain Conventional Weapons of 10 October 1980 (CCW)[69] came into force on 2 December 1983. It is a wide-ranging convention including non-detectable fragments, anti-vehicle mines, booby-traps, incendiary weapons, blinding laser weapons and explosive remnants of war. Lethal Autonomous Weapons Systems (LAWS) was added to the convention in

2014. But each type forms part of an independent protocol and signatories obviously differ.

The UN office of ICT prepared a White Paper in April 2021[70] to summarise the common liabilities of AI for member-countries to ensure that a standard response mechanism is prepared by all the stakeholders. It summarised that AI related concerns could be wide ranged. AI could be incompetent – unable to perform tasks safely, result in loss of privacy, support discrimination due to biases and cause general erosion of society – where AI understands human needs (intellectual and emotional) and continuously feeds the same form of information (a self-radicalisation). The lack of explainability could cause unintended effects like loss of human autonomy (though not possible now, at least till the ASI is invented) or an AI operator could indulge in exclusion of those who cannot afford the niche technology. AI is likely to support deception (deep fakes), manipulation, digital crimes, and the concern of LAWS. Various countries, organisations, and developers are conscious of the ethical facet of AI and have in their own manner tried to cooperate.

USA. The Defence Primer of November 2022 defines LAWS as a weapon system that can select and engage targets without further human intervention. It does not claim to possess one yet, but also justifies development, if threatened. America believes that LAWS are viable alternatives when traditional weapons are unusable. The USA believes that 'human judgment should play a role in the use of force with LAWS and that the level of human involvement can vary depending on the context – human judgment doesn't necessarily mean manual control of the weapon system, but rather a broader human involvement in decisions related to its usage'.[71] Though the USA participates in the United Nations' Group of Governmental Experts to examine the technological, military, ethical, and legal aspects of LAWS, it does not support a ban. It believes that automated targeting and engagement results causes less risk of collateral damage.

China. China's Position Paper of December 2021 on Regulating Military Applications of Artificial Intelligence[72] believes in 'extensive consultation, joint contribution and shared benefits; – develop in a prudent manner; – refrain from seeking absolute military advantage'. China believes that it is legitimate to enhance defence capabilities, but should not be used as a tool to start a war, undermine the sovereignty of other countries (an obvious dig at the USA).

Russia. Distrust and competition with the USA and NATO justify the inescapable[73] Russian development – forced to demonstrate capabilities to send a message to the technological challenge. So, it is forced to build AI weapons to counter American development – which is threatened by a Russian threat!

United Kingdom. The UK believes that AI has tremendous power to minimise the risk[74] to the country. In statements to the CCW, the UK has stated LAWS do not exist and will never exist.[75] But the Defence Ministry does not say so openly; in fact, there is a terminology difference. As mentioned in a response to the emerging threats in Defence Artificial Intelligence Strategy of June 2022,[76] they want to 'take advantage of AI' and over 200 researches are under way to include 'autonomous ships and swarming drones'. The UK intends to continue development and discussion in the UN simultaneously. Understanding the potential cross-domain effects and implications; the UK plans to develop appropriate AI tools.

Conclusion

AI has been a rapidly growing field in recent years and has been making tremendous advancements in various industries. Theoretical discussions are greater than ever. AI algorithms are capable of processing vast amounts of data and making complex decisions with high accuracy and speed, making it a valuable tool. It is crucial for developers and researchers to continually evaluate the implications and limitations of AI to ensure its responsible deployment. It is also critical for users to identify the output they desire from AI – defining a 'need' correctly and prudently develop smarter and precise AI tools.

In recent times, artificial intelligence (AI) has been transforming human lifestyles and the way we interact with each other. It is also changing the nature of competition and warfare, altering our understanding of success, and elevating data analysis and interpretation to new levels. The integration of AI is expected to have major implications for digital, physical, and political security, magnifying existing threats, creating new ones, and altering the character of both threats and warfare. Some of the potential changes include the automation of social engineering attacks, the discovery of new vulnerabilities, the launch of influence campaigns, the misuse of commercial AI systems by terrorists, increased scale of attacks, and manipulation of information availability. This growing concern has led to increased global awareness of the impact of AI.

The military sector has not been left behind in the AI revolution, with the deployment of AI in the military increasing in recent years. This has the potential to revolutionize the way armed forces operate, with AI algorithms making it easier for military leaders to make informed decisions and plan effective strategies. AI-powered systems such as unmanned aerial vehicles (UAVs) and autonomous weapons are already being used in military operations, leading to improved efficiency, accuracy, and reduced collateral damage.

The impact of AI on society is undeniable, and its integration into the military sector is likely to bring about significant changes in the future. The next chapter will examine case studies and explore the pathway and future of Military AI.

KEY TAKEAWAYS

- Present AI technology floats between Artificial Narrow Intelligence (ANI) and Artificial General Intelligence (AGI), biased towards ANI.
- Development of Artificial Super Intelligence (ASI) is far into the future, connected to neurosciences and human induced winters – COVID 19 and the Ukraine-Russia War.
- Manual Tagging of features peaks machine learning (ML) performance and cannot be improved further even by increasing the data.
- Deep learning (DL) with scores of hidden layers with weight criteria of each layer improves ML performance.
- Fog and Edge Computing provide a good option for security-conscious organisations and inconsistent access to cloud or internet.
- Performance, Environment, Actuator and Sensor (PEAS) relationship with Intelligent Agents creates AI Logic Maps to create value AI Machines.
- Well defined AI output is vital before commencing AI development cycle.
- Data supports AI growth and Bias in data corrupts outputs.
- Accuracy in AI is a relative concept addressed separately for the output required.
- High performance hardware components with low power consumption will create better AI.
- ML ingests better data with Computer Vision (CV) and Natural Language Processing (NLP), concurrently improving both CV and NLP.
- Generative AI like Generative Adversarial Network (GAN) and Generative

Pre-trained Transformer (GPT) will disrupt AI in the manner not fully understood yet.
- Expert System also known as Good Old-Fashioned AI (GOFAI) is the actual Man-Machine interface which utilises man to prepare a good machine which will assist him.
- Fuzzy logic provides a human like context to AI.

NOTES

1. An environment is the surrounding of the AI agent. The agent takes inputs from the environment through sensors and delivers the output to the environment through actuators.
2. Stuart J. Russel, and Peter Norvig. *AI A Modern Approach*, 3rd Edition, Pearson India Education Ltd, Noida, India, 2010, pp 2-4.
3. https://www.britannica.com/biography/Alan-Turing/Computer-designer, accessed on 13 October 2022.
4. Nick Bostrom coined this term. 'Superintelligence' is an AI agent that possesses intelligence far surpassing that of the brightest and most gifted human minds. Though he predicted its arrival in the 21st century he also mentioned the importance of neurosciences to better understand and replicate the human brain as a primary factor. This machine will be the pinnacle of AI's evolution, but is hypothetical and does not yet exist.
5. Labelled data is a group of samples that have been tagged with one or more labels, specifying the features of the data. To break down the data to computational features for the AI machine to learn and train.
6. Stuart J. Russel, and Peter Norvig. p. 1055.
7. Arend Hintze "Understanding the four types of AI, from reactive robots to self-aware beings," 14 November 2016, https://theconversation.com/understanding-the-four-types-of-ai-from-reactive-robots-to-self-aware-beings-67616 accessed on 27 September 2022.
8. A detailed analysis of this model will be covered later in the chapter.
9. Nick Bostrum. *Super Intelligence: Paths, Dangers and Strategies*, UK, Oxford University Press, 2017.
10. Dustin Harris. What is Artificial Intelligence (AI)? How Does AI Work?, https://builtin.com/artificial-intelligence 10 October 2022.
11. A limited machine or computer capability in which the machines are designed to automatically follow predetermined sequences of operations or respond to predetermined instructions only and nothing more. They cannot work without external stimulus.
12. Jørgen Veisdal. The Birthplace of AI: The 1956 Dartmouth Workshop, 12 September 2019, https://www.cantorsparadise.com/the-birthplace-of-ai-9ab7d4e5fb00 accessed on 13 October 2022.
13. MIT Professor Joseph Weizenbaum wanted to show the superficiality of communication between man and machine, but was surprised to observe how human thought it to be an intelligent machine. Sample chatting is possible at https://web.njit.edu/~ronkowit/eliza.html accessed on 13 October 2022.
14. Keith D. Foote. A Brief History of Deep Learning, 4 February 2022, https://www.dataversity.net/brief-history-deep-learning/ accessed on 13 October 2022.
15. John Paul Mueller and Luca Massoron/ *AI for Dummies*, India, Nice Printing Press, 2020, p. 121.

16. https://www.superannotate.com/blog/guide-to-training-data accessed on 16 December 2022.
17. John Paul Mueller and Luca Massoron. *AI for Dummies*, India, Nice Printing Press, 2020, pp. 40, 127.
18. They are a class of computational problems for which no efficient solution algorithm has been found, though there could be a hack or approximate solution.
19. Roger Brown. Understanding the Importance of Training Data in Machine Learning, 26 August 2019, The AI Technology, https://medium.com/the-ai-technology/understanding-the-importance-of-training-data-in-machine-learning-da4235332904#:~:text=It%20helps%20them%20to%20 recognize,the%20failure%20of%20AI%20project accessed on 16 December 2022.
20. https://www.superannotate.com/blog/guide-to-training-data accessed on 16 December 2022.
21. https://www.programsbuzz.com/article/what-training-data-and-why-it-important-ai-and-computer-vision-find-out-here accessed on 16 December 2022.
22. https://www.techtarget.com/searchenterpriseai/definition/neural-network accessed on 18 October 2022.
23. https://www.mathworks.com/discovery/deep-learning.html#:~:text=Deep%20learning%20is%20a%20machine,a%20pedestrian%20from%20a%20lamppost accessed on 18 October 2022.
24. Stuart J. Russel and Peter Norvig, p. 34.
25. Ibid., p. 31.
26. Ibid., pp. 49-54.
27. Difference between Cloud, Fog and Edge Computing in IoT, https://www.digiteum.com/cloud-fog-edge-computing-iot/ accessed on 10 November 2022.
28. Jagreet Kaur Gill. 'What's the Difference Between Cloud, Edge, and Fog Computing?', https://www.akira.ai/blog/difference-between-cloud-edge-and-fog-computing/ accessed on 1 November 2022.
29. Principal Investigator for AI Bias to National Institute of Standards and Technology (NIST), US Department of Commerce.
30. Eirini Ntoutsi, and others. Bias in data-driven artificial intelligence systems—an introductory survey, May 2020, Wiley Interdisciplinary Reviews: Data Mining and Knowledge Discovery, https://www.researchgate.net/publication/338998132_Bias_in_data-driven_artificial_intelligence_ systems-An_introductory_survey accessed on 18 October 2022.
31. 'What is Training Data and Why Is It Important for AI and Computer Vision? Find Out Here', 4 April 2022, https://www.programsbuzz.com/article/what-training-data-and-why-it-important-ai-and-computer-vision-find-out-here accessed on 16 December 2022.
32. Dr. Matt Turek. DARPA, Explainable Artificial Intelligence (XAI), https://www.darpa.mil/program/explainable-artificial-intelligence, accessed on 5 September 2022.
33. Sekhar Vallath. The What, Where, and Why of Contextual AI, https://symbl.ai/blog/the-what-where-and-why-of-contextual-ai/ accessed on 1 November 2022.
34. Importance of Training Data for Machine Learning, 18 August 2021, https://www.dataentryoutsourced.com/blog/importance-of-training-data-for-machine-learning/ accessed on 16 December 2022.
35. Thilo Huellmann. Precision vs. Recall in Machine Learning, 16 November 2022, https://levity.ai/blog/precision-vs-recall, accessed on 4 February 2023.
36. Sciforec Report, 7 November 2019, https://medium.com/sciforce/ai-hardware-and-the-battle-for-more-computational-power-3272045160a6#:~:text= As%20AI%20systems%20become%20more,and%20reduce %20the%20power%20consumption accessed on 27 September 2022.
37. Sciforec Report, accessed on 27 September 2022.
38. V. Rajaraman, Multi-core Microprocessors, The Indian Institute of Science, Bengaluru, December 2017, https://www.ias.ac.in/article/fulltext/reso/022/12/1175-1192#:~:text=Multi%2Dcore%

20microprocessor%20is %20an,than%20a%20single%20core%20processor accessed on 27 September 2022.
39. Sally Ward-Foxton. Top 10 Processors for AI Acceleration at the Endpoint, 20 April 2020, https://www.eetimes.eu/top-10-processors-for-ai-acceleration-at-the-endpoint/2/ accessed on 3 November 2022.
40. Sally Ward-Foxton, accessed on 3 November 2022.
41. Sciforec Report, accessed on 27 September 2022.
42. Ron Lowman, Why In-Memory Computing Will Disrupt Your AI SoC Development, https://www.synopsys.com/designware-ip/technical-bulletin/in-memory-computing-ai.html accessed on 3 November 2022.
43. https://www.techtarget.com/whatis/definition/optical-computer-photonic-computer, accessed on 4 November 2022.
44. https://www.prophesee.ai/about-prophesee/ accessed on 4 November 2022.
45. Y. Bengio and Jean-Pierre David. Low precision arithmetic for deep learning, December 2014, https://www.researchgate.net/publication/269932963_Low_precision_ arithmetic_ for_ deep_ learning accessed on 27 September 2022.
46. Sally Ward-Foxton, accessed on 3 November 2022.
47. Niall O' Mahony et al. Deep Learning vs. Traditional Computer Vision, IMaR Technology Gateway, Institute of Technology Tralee, 2019, https://arxiv.org/ftp/arxiv/papers/1910/1910.13 796.pdf accessed on 4 November 2022.
48. https://www.ibm.com/in-en/topics/computer-vision#:~:text=Computer%20vision%20is %20a%20 field,recommendations%20based%20on%20that%20information accessed on 4 November 2022.
49. Ilija Mihajlovic. Everything You Ever Wanted to Know about Computer Vision, 26 April 2019, https://towardsdatascience.com/everything-you-ever-wanted-to-know-about-computer-vision-heres-a-look-why-it-s-so-awesome-e8a58dfb641e accessed on 4 November 2022.
50. https://www.ibm.com, accessed on 4 November 2022.
51. IBM Cloud Learn Hub. Natural Language Processing (NLP), 2 July 2020, https://www.ibm.com/cloud/learn/natural-language-processing#:~:text=Natural%20language%20 processing%20(NLP)%20refers,same%20way%20human%20beings%20can, accessed on 22 November 2022.
52. Jay Selig. What You Need to Know About Natural Language Processing (NLP), 12 July 2022, https://www.expert.ai/blog/natural-language-processing/ accessed on 18 October 2022.
53. Natural Language Processing (NLP). What it is and why it matters, https://www.sas.com/en_us/insights/analytics/what-is-natural-language-processing-nlp.html, accessed on 22 November 2022.
54. Word cloud generated through https://www.jasondavies.com/wordcloud/ on 2 February 2023 based on the 450 worded NLP section in the text.
55. Ben Lutkevich. What is natural language processing?, https://www.techtarget.com/searchenterpriseai/definition/natural-language-processing-NLP, accessed on 22 November 2022.
56. Alex Owen-Hill. What's the Difference Between Robotics and Artificial Intelligence?, updated on 27 July 2021, https://blog.robotiq.com/whats-the-difference-between-robotics-and-artificial-intelligence accessed on 28 November 2022.
57. John Paul Mueller and Luca Massoron. p. 185.
58. Kapil Mahajan. 'Role of Artificial Intelligence and Machine Learning in Robotics', 30 June 2022, https://emeritus.org/in/learn/role-of-artificial-intelligence-and-machine-learning-in-robotics/ accessed on 22 November 2022.
59. Alan Martin. Robotics and artificial intelligence: The role of AI in robots, 26 November 2021, https://aibusiness.com/verticals/robotics-and-artificial-intelligence-the-role-of-ai-in-robots accessed on 28 November 2022.

60 Chitti Rajesh, (ed.). What is GAN in AI?', K-tech Centre of Excellence, 1 July 2022, https://coe-dsai.nasscom.in/what-is-gan-in-artificial-intelligenceai/ accessed on 20 December 2022.
61 Accessed https://chat.openai.com/chat on 3 February 2023.
62 Sayantini. What is FL in AI and What are its Applications? 15 December 2022, https://www.edureka.co/blog/fuzzy-logic-ai/#:~:text=Fuzzy%20logic%20is% 20used%20in%20 Natural% 20language%20processing%20and%20various,use% 20it%20with%20 Neural% 20 Networks, accessed on 9 January 2023.
63 https://www.guru99.com/what-is-fuzzy-logic.html#5, accessed on 4 February 2023.
64 Matthew Grant. AI and Expert Systems: Powering the Future of Business, 24 November 2022, https://www.leanix.net/en/blog/artificial-intelligence-expert-systems#:~:text=AI%20and %20 Humans%3A%20Better%20Together&text=But%20AI%20and%20expert%20systems,actions% 20is% 20revolutionizing%20that%20discipline, accessed on 9 January 2023.
65 Frank Säuberlich, Dr. and Danko Nikoliæ, Prof. 'AI Without Machine Learning', https://www.teradata.com/Blogs/AI-without-machine-learning, accessed on 4 February 2023.
66 Cameron Carpenter. University of Waterloo: Artificial Intelligence Without Internet Now Possible, The University Network, 12 March 2019, https://www.tun.com/blog/university-of-waterloo-artificial-intelligence-without-internet-now-possible/, accessed on 4 February 2023.
67 UN Chief Executives Board for Coordination. Artificial Intelligence, https://unsceb.org/topics/artificial-intelligence, accessed on 6 February 2023.
68 https://sdgs.un.org/sites/default/files/2021-04/Resource%20Guide%20 on%20AI%20Strategies _ April%202021_rev_0.pdf, accessed on 6 February 2023.
69 UN Convention on Certain Conventional Weapons (CCW). https://www.auswaertiges-amt.de/en/aussenpolitik/themen/-/218382#:~:text=Article,which%20may%20cause%20unjustifiable% 20suffering, accessed on 6 February 2023.
70 Lambert Hogenhout. A Framework for Ethical AI at the United Nations, https://unite.un.org/sites/unite.un.org/files/unite paper-_ethical_ai_at_the_un.pdf, accessed on 6 February 2023.
71 https://crsreports.congress.gov/product/pdf/IF/IF11150, accessed on 10 January 2023.
72 http://geneva.china-mission.gov.cn/eng/dbdt/202112/t20211213_10467517.htm, accessed on 10 January 2023.
73 Anna Nadibaidze. Russian Perceptions of military AI, automation, and autonomy, FPRI, https://www.fpri.org/wp-content/uploads/2022/01/012622-russia-ai-.pdf
74 Defence Artificial Intelligence Strategy of UK, June 2022, https://assets.publishing.service.gov.uk/government/uploads/system/uploads/attachment_data/file/1082416/Defence_ Artificial_Intelligence_Strategy.pdf, accessed on 10 January 2023.
75 https://article36.org/wp-content/uploads/2016/04/UK-and-LAWS.pdf, accessed on 6 February 2023.
76 https://assets.publishing.service.gov.uk/government/uploads/system/uploads/attachment_data/file/1082416/Defence _Artificial_Intelligence_Strategy.pdf, accessed on 6 February 2023.

Chapter Two

AI on the Frontlines: Case Studies in Military Innovation

MILITARY AND AI

Artificial Intelligence (AI) has the potential to transform modern warfare and military operations. It can handle large amounts of data, leading to improved decision making and undertake various tasks. This technology is driving the evolution of war, shifting from attrition and destruction-based approaches to ones based on rapid and accurate decisions, deployments, and the destruction of the adversary's ability to fight.

AI has the potential to enhance the capabilities of military forces in handling undefined war situations or hostile environments. It enables rapid decision-making, particularly in dynamic and information-dense environments, and can speed up decision making during wars and combat. The use of AI in modern warfare also requires strategic intelligence, as wars are becoming more knowledge-based, driven by information. AI can extract hidden strategic information from the high volume of data generated by the adversary in any domain.

AI can revolutionize logistics, administration, maintenance, training, personnel management, and even routine activities or exercises. It can reduce institutional workload and free warriors to focus on core functions. AI can upgrade command-and-control capabilities of the armed forces with a highly automated, data-oriented approach and support higher formations in designing and deploying more effective battle plans. The Military-AI-Ecosystem can provide sharper and deeper insights for better control of operations.

It can transform modern warfare and military operations, bringing with it new possibilities and challenges. Global powers are investing in this technology to establish their dominance in the global power play, even tweaking[1] existing algorithms from commercial applications and find uses for them in the military domain. The effective use of AI in the military requires a comprehensive understanding of the technology, its applications, and the entire battle design framework.

The military usage of AI is broadly sequenced[2] into collection and analysis of surveillance data feeds, using multiple sources and sensors to detect and track objects or personnel. This is followed by planning, where the available data is churned by ML to identify intentions, anticipate resourcing requirements, and plan associated training and mission enablers. During field operations, AI provides real-time information and quick assessments to improve mission outcomes, which targets the adversary efficiently and accurately while protecting our own people, assets, and information. Support functions are critical for operational sustenance where expeditious and failsafe procurement processes are the backbone for tactical supply of war equipment. AI will also enable smarter HR functions for recruitment and retention. Many of these sequences merge and support each other in the conflict and peace continuum. For the sake of this research, the military AI analyses will consider following aspects.

Battlespace Analysis. This is conducted by militaries and nations continually and not necessarily during a conflict situation to stay ahead of the information loop. With the assistance of AI, this analysis is very wide ranging and provides strategic insights.

- Analysing reports, documents, newsfeeds, and other forms of unstructured information allows defence forces to gain an in-depth understanding of potential adversaries.[3] AI plays a crucial role in this process by looking over large volumes of data and extracting useful information. The technology assists in assembling connected data from several datasets and various sources. This advanced analysis enables the military to recognize patterns and derive correlations, providing a thorough grasp of probable operational domains.
- The analysis may include probability-based forecasts[4] of enemy behaviour, aggregation of weather and environmental conditions, anticipation and flagging of potential supply line bottlenecks or

vulnerabilities, assessments of mission approaches, and suggested mitigation strategies. NLP research is capable of multilingual speech recognition[5] and translation in noisy environments, geo-locating images without the associated metadata, fusing 2-D images to create 3-D models, and building tools to infer a pattern-of-life analysis (an important aspect for precise and surgical operations).

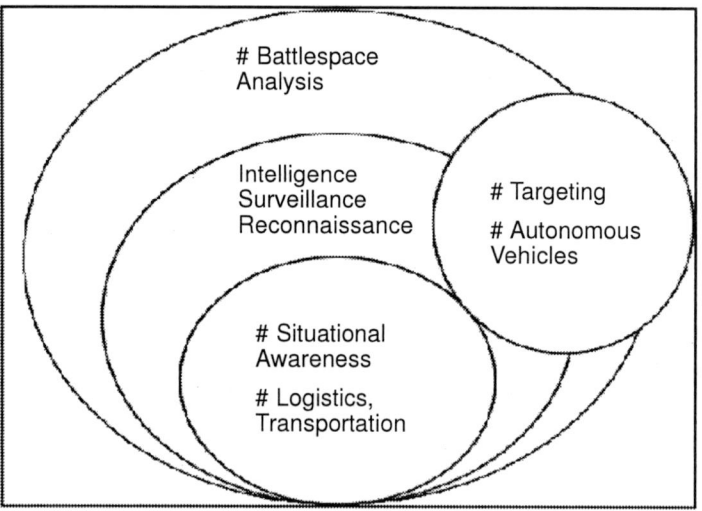

ISR (Intelligence, Surveillance, and Reconnaissance). This is a precision activity carried out to provide battle space edge and initiative. It is specified to a space and time continuum and highly focussed.

- Information processing is essential for all military operations and plays a crucial role to support military activities. Unmanned systems with AI, can be deployed along a predetermined path to better assess threats and maintain situational awareness. Real-time, multisensory, threat monitoring and situational analysis are crucial during military operations that support a wide range of actions. Machine learning algorithms are used to track and discover targets from the data obtained.
- AI is also creating increasingly realistic deep fakes,[6] which are photo, audio, and video forgeries that can be deployed as part of information operations. These deep fakes can be used to generate false news reports, influence public discourse, erode public trust, and attempt to blackmail. Though forensics tools exist, the battle to become better continues

between the ability to detect manipulations and outmanoeuvring these tools.

Situational Awareness. Soldiers in battle are provided battle field intelligence and awareness to overcome battle isolation which provides them a heads up of the tactical battle around them. This allows soldiers to undertake pre-emptive, precision actions to defeat the adversary soonest.

- Battlefields are noisy and create combat isolation. It is essential that the soldiers are aware of their surroundings to take appropriate actions without delay. AI provides this real-time information, helping them understand complex situations and make better decisions.
- Unmanned systems equipped with AI can be utilized for situational awareness which communicate potential threat information to military response teams. Thereon, the integration of AI in target recognition can increase accuracy in challenging combat conditions.

Targeting. This is the ultimate test of the military, commencing from identifying, matching weapons, precision engagement, and damage assessment and re-targeting, if necessary. Precision, lethality, and speed are essential to degrade the adversary soonest, which is the most prized output expected out of military AI.

- AI has the ability to enhance target recognition in military operations. It can track targets using machine learning algorithms, making the process more precise. Target identification systems incorporating AI can predict enemy activities, analyse mission[7] approaches, aggregate environmental data and employ mitigation techniques, among others.
- AI can help create a single source of information, known as the common operating picture, by fusing data from sensors across all domains such as air, space, cyberspace, sea and land. This comprehensive picture[8] would provide decision-makers with a real-time analysis of friendly and enemy forces, and even offer a menu of viable courses of action, potentially improving the quality and speed of wartime decision-making.
- AI has the potential to improve the efficiency of warfare systems by reducing reliance on human input and increasing the synergy and performance of systems. Embedding AI into weapons and other platforms used in land, naval, airborne, and space operations could also empower autonomous and high-speed weapons[9] to carry out collaborative attacks.

Autonomous Vehicles. The range of autonomous vehicles is wide and probably a topic for another research. These vehicles are capable of operating without human input. It uses a combination of sensors, and actuators, supported by AI to sense its environment, make decisions, and control its movements. These vehicles can perform various tasks, depending on the payload fitted into them. These could operate on land, sea or air. These could operate in dangerous or undesirable battle spaces, and could be semi or fully autonomous.

- Attach a bouquet of weapons to these autonomous vehicles and they become a Lethal Autonomous Weapon System (LAWS). A special class of weapon systems, these use sensor suites and computer algorithms to independently identify a target and engage it without manual human control. These systems are crucial for military operations in communication-degraded or denied environments, where traditional systems are able to operate.
- Swarming,[10] a subset of autonomous vehicle development, involves large formations of low-cost vehicles designed for a variety of tasks – be it precise and silent or loud and overwhelming attacks. Such tiny swarms of drones controlled from remote locations have the potential to unleash destruction on an unprecedented magnitude. These swarms can enter combat areas, cross international boundaries undetected and attack stealthily.

Training and Simulation. The aim is to train the entire chain of command in all likely scenarios, ranging from analyses, planning, decision making and physical training. It could also be used to test and verify a military concept or a plan.

- Realistic/Virtual Simulation can create realistic simulations of real-life scenarios, allowing soldiers to practice and prepare for actual missions in a safe setting. AI training tools have shown reduction on training time by 40 per cent for US Army pilots.[11]
- Scenario Generation algorithms generate a large number of scenarios, including variations of terrain, enemy tactics, and other parameters, providing military planners with a wider range of options for war-gaming exercises.
- War-gaming Analysis can predict outcomes of different training scenarios, allowing military trainers to identify areas that need improvement.

- Decision-Making Support Systems can provide military planners with real-time analysis and insights, assisting them in making more informed decisions during war-gaming, extendable to actual combat.

Transportation and Logistics. AI systems can provide guidance and navigational advice over unchartered terrain in the contested EM spectrum battlefield. Similar vehicles can travel to remote distances to drop or pick up soldiers for and after small team missions. A military operation's success depends on the effective movement of troops, equipment, ammunition, and supplies. Integration of AI in transportation can reduce manual labour and transportation costs, while detecting anomalies and predicting component breakdowns or likely supply chain congestions.[12] AI induced predictive maintenance and analysis[13] extracts real-time data from embedded sensors and others on board systems to determine inspection or replacement.

Battlefield Healthcare. AI powered platforms can undertake casualty evacuation autonomously and can be combined to perform remote surgical support. Research is on for AI-powered systems to mine a soldier's medical history and provide diagnostic support too.

Cyber security. In military operations, cyberspace is recognized as a fully integrated aspect that operates across all domains and dimensions. Cyber security threats against military systems can result in significant damage, including the loss of important military data. To combat these threats, defence organizations are utilizing machine learning (ML) for intrusion detection by classifying network activity as normal or intrusive. AI-based methods improve the accuracy of this classification, allowing autonomous AI systems to defend against unauthorized access to networks, programs, data, and computers. These AI-enabled security systems have the ability to monitor, record, and identify patterns of cyber attacks, which can be used to develop counter measures. However, hackers are aware of these AI-based defence systems, and can modify malicious codes[14] to evade detection. In response, AI systems can be trained to detect anomalies in network activity patterns, providing a more robust and dynamic defence mechanism. It is possible for a single algorithm to simultaneously perform both offense and defence functions.

MILITARY AI CASE STUDIES

Military forces worldwide are increasingly integrating artificial intelligence (AI) into their operations, each with its unique degree of success. Examining the effectiveness of various AI applications and understanding the control structures implemented by nations is crucial for comprehending the nuanced landscape of viable AI utilization in military contexts. The forefront of the global AI race is currently dominated by nations such as the USA, Singapore, Switzerland, The Netherlands, Japan, South Korea, Sweden, Finland, Germany, and Ireland. These countries showcase varying degrees of advancement in deploying AI within their military frameworks. However, the focus of this research will extend beyond a broad overview, narrowing its scope to conduct in-depth case studies on militarily significant nations.

This chapter will specifically delve into the AI strategies and implementations of the USA, the United Kingdom, France, Australia, Turkey, Iran, Russia, and China. These countries have been selected not only for their standing in the global AI landscape but also for their significant military capabilities and engagements. The examination will extend to how these nations navigate the challenges posed by hierarchical structures, compliance complexities, and the need for responsible AI deployment in military operations. By scrutinizing the military applications of AI in these diverse geopolitical contexts, the research aims to unveil patterns of success, lessons learned, and potential areas for improvement. Understanding how these nations leverage AI to meet military requirements is not only pivotal for assessing their current capabilities but also for forecasting the trajectory of AI integration in future military scenarios.

In essence, this focused exploration seeks to bridge the gap between the theoretical advancements in AI and their practical applications in military operations. By dissecting the strategies and implementations of key players on the global stage, the chapter aspires to contribute valuable insights that inform both the academic discourse on AI in military contexts and the strategic decision-making processes of nations navigating this rapidly evolving landscape.

AI study is interesting for Taiwan since it is facing a territorial threat from China and in this technological intensive character of warfare, Taiwan's scientific and industrial edge will provide it with a great number of AI options. Israel as a state has always provided numerous innovative and unorthodox solutions in its typical efficiency and precision and hence is also part of this research. Availability of AI to non-state actors will also be studied.

RUSSIA

> *"Artificial intelligence is the future, not only for Russia, but for all humankind. It comes with colossal opportunities, but also threats that are difficult to predict. Whoever becomes the leader in this sphere will become the ruler of the world."*
>
> *– Vladimir Putin*

Russia is putting a lot of effort into closing the gap and aims to stay competitive in the world. Russia aims to improve operational effectiveness and efficiency by using AI. Details of the projects, either released by the government or discussed by think tanks or researchers is obviously lesser than the USA. But what is sure is that AI is affecting many aspects of Russian military operations already.

Significant technological improvements have been made in space, airborne, ground-based electronic and optical reconnaissance systems. Development of target recognition and classification systems, including those with automatic shoot down capabilities, is underway with an emphasis on electronic warfare, robotics, air and missile defence and autonomous vehicles.

There are concerns, though, about marrying up the archaic Soviet systems to modern and efficient data analysis. The automation of entire military command-and-control processes is still under development. Military officials understand this and mention this problem as 'a deep mismatch between the organisational and technical dimensions'[15] of information support for combat operations.

AI Development Curve

The Russians noted deficiencies[16] in the robotic development well before outlining an AI roadmap by an especially directed campaign and launched a 'Creation of Prospective Military Robotics through 2025' program. Annual conferences since 2016, and 'Robotization of the Armed Forces of the Russian Federation' codify the disparity and interdepartmental coordination. The Advanced Research Foundation (ARF) was created to ensure superiority in defence technology and to warn the leadership about the associated risks due to technology lethargy.

Russia's official AI strategy unveiled much later in September 2017, outlines the country's plan to become a global leader in artificial intelligence (AI) by 2030. It recognizes the potential risks and includes measures to address these concerns as well. The strategy focuses on three main areas; firstly, Developing the technological infrastructure to support AI research and development, secondly creating a regulatory framework that encourages innovation and investment in AI and finally fostering a skilled workforce and a need for collaboration between government, academia, and industry to achieve these goals.

In March 2018,[17] the Ministry of Defence (MoD) released a 10-point proposal to bring together the MoD, Education and Science (MES), and the Russian Academy of Sciences (RAS) in partnership to develop AI. The proposal sought to create an AI and big data consortium to combine national effort, establish a fund, while creating educational (AI conferences, discuss AI proposals at domestic military forums, monitor global developments) and training systems (war games). Involving private players and developers, Russia is also holding AI design competitions that are open to the public in various technologies, especially AI and drones.[18] A National Centre for Artificial Intelligence is developing promising AI projects to implement AI solutions, checking for compliances at the federal level. Russian also realised that it will concurrently have to deal with an all-out effort to degrade its AI development capabilities and it should be ready with a legal framework[19] to eliminate such development barriers.

This MoD proposal also led to the creation of a comprehensive AI research campus – Era (Good) Military Innovation Technopolis at Anapa on the Black Sea. This centre will lead military research on AI, robotics, and automation. Scheduled to have a research cluster for R&D, an education cluster and also a production cluster[20] for building prototypes of military and special hardware, it is hoped to be one stop for all military AI.

Sberbank, a state-owned retail bank (and largest in Russia), created a national strategy in October 2019, an AI roadmap in November 2019 and an AI federal project in August 2020. Russia's AI development strategy is therefore unique;[21] it is led not by the government, nor by the private sector, but by state-owned firms; Rostec (military hardware producer – also state owned), Yandex (largest private tech firm), and Gazprom Neft (fourth largest oil producer) are collaborating with Sberbak on AI development. It is yet to be seen as to why this unusual arrangement exists – a possibility of distributing the research and development across all participants (including the existing AI systems already developed) to avoid dual effort and encourage cooperation without conflict of interest seems a plausible reason. Yandex launched Alice, an AI-enabled virtual assistant and signed Big Data and ML contracts with Gazprom Neft in 2017. Rostec has meanwhile successfully repurposed AI into its product development plans. It is also working on AI-related technologies like 5G, block chains, and IoT devices.

Thoughts on Military AI Strategy

V.M. Burenok[22] posited on the 'Role of artificial intelligence in the military confrontation of the future' in the April 2021 issue of *Military Thought* (journal of the MoD). He starts his research very thoughtfully chiding theorists and military minds conflating programming based uncomplicated algorithms with AI systems and opines this inability to understand the essence of AI can lead away from the development of true AI. He proposes a wide-ranging but focused role for AI in future military confrontation. AI data analysis can be used for forecasting and projecting politico-military situation, characterising and predicting actions by potential adversaries, and the resultant armed conflict.

During the research and development stage, AI can assist in projecting possible weapons, their design deliberations and preventing accidents during both development and employment. Information warfare is a standard AI use case across multiple media and realms in all forms, through variety of tools. Cyberspace has a wide scope to utilise AI for pre-emption, as well as defensive and offensive actions. During combat, Russia supports the use of AI in reconnaissance, monitoring, and information support to process the complete combat picture. Russian AI decision support systems aim to predict an adversary's actions during the combat cycle. Utilising nano-robots and swarms to launch shock reconnaissance (reconnaissance by fire to incite adversary to divulge his locations in retaliation) is a unique and interesting application. Russia wants to develop interactive and intuitive command and control and intelligent information systems. These systems are expected to have live environmental feedback to assess the situation and subsequently control diverse forces and assets in real-time. Russian robotic complexes (combat robots) with various levels of autonomy are expected to create greater mass during various contact battles. Finally, AI is expected to have utility in the technical planning and execution of operational and tactical logistics as well.

Russian Trials in Syria

Russia tested 600 new weapons including 200 next-generation[23] military equipment in combat conditions in Syria with varying degrees of success. Unmanned vehicles performed a wide variety of tasks[24] – intelligence, surveillance, reconnaissance, combat missions, de-mining and logistics. For the course of this research 'robots' and 'autonomous weapons systems' (AWS) both will be used synonymously and Russian trials mentioned below provide a look into the future.

- Twenty-two Uran-9 robot tanks in support of infantry and engineer units for reconnaissance and fire support missions were not utilised in independent tactical manoeuvres, as perceived earlier. Sensor data generated by the vehicle did not provide complete battlefield awareness[25] to the operators at Moscow. This points towards absence of smarter AI systems with better self-state knowledge, capable of orientation and independent combat analysis.
- Uran-6, a demining robot successfully assisted Russian engineers to clear recaptured areas from mines, IEDs and unexploded ordinance. Russian controlled Uran-6 helped the Syrian government to clear the historic World Heritage Palmyra[26] of 3,000 explosive devices across 2.3 square kilometres left by Islamic State (IS, formerly ISIS/ISIL) in March 2017.
- Russian battle robots captured the strategic tower of Syriatel[27] (Latakia province) in March 2016 in support of the Syrian army. It was the first ever attack by a man-machine team using six Platform-M and four Argo robots. There were many firsts during this attack. A self-propelled autonomous artillery installation, (SAU) Acacia, accurately destroyed enemy positions. The Andromeda-D system linked all military robots, guns and drones to an automated C4I2 (command, control, communications, computer, intelligence, and information) system. While the tactical commander of the attack on the tower was directing operations, operators of military robots sitting in Moscow led the attack, each aware of the battle field and the whole picture.

- Russian engineers successfully tested and utilized Scarab and Sphere[28] during de-mining in Palmyra. Scarab is a small wheeled platform with a high-resolution video camera, microphone, and thermal imager, controlled via a digital radio channel providing surveillance video on a console to the operator. Sphere is a ball with four video cameras with LED lighting, a microphone, and an information transmitter that provides a 360-degree view. Its built-in positioning system automatically takes a vertical position after being deployed, permitting it to be lowered into wells and underground communications, or thrown through windows to remotely inspect a room for suspicious objects.
- Russia tested an Unmanned Underwater Vehicle (UUV) in February 2018, when Galtel[29] was deployed to search for undersea unexploded ordnance, seafloor mapping and protection of Russia which has a naval base at Tartus.
- During 2018, Russian UAVs operated 60-70 drones daily, logging 23,000 missions and 140,000 flight hours,[30] providing aerial reconnaissance, target designation, controlling airstrikes, adjusting artillery fire and search and rescue.
- Russia learned a major command and control lesson when it decided to hand over all responsibilities (development, construction and deployment) to Air and Space Forces[31] (Voenno Kosmicheskie Sily – VKS). These are understood to have been included in the new concept of operations across the military.

Battle Space Analysis

Russia, like many other countries, is investing in big data analysis as a means of gaining insights into complex problems and improving decision-making. The Russian government has launched several initiatives to promote the development of big data technologies and to encourage their adoption by businesses and public organisations. One of them, the Digital Economy Program, aims to develop digital infrastructure and improve access to data. It includes the creation of a national big data platform, which will enable data sharing and collaboration across different sectors. Leading technology companies, such as Yandex and Mail.ru, are investing heavily in big data research and development. These companies are developing advanced algorithms and data processing tools that can analyse vast amounts of data quickly and accurately, enabling businesses and organizations to make data-driven decisions.

The System for Operative Investigative Activities (SORM) provides lawful interception[32] of all communications including internet. Communication metadata and content, phone calls, email traffic, and web browsing activity are available to numerous federal agencies in Russia.

Civilian tech companies in Russia are already developing facial recognition and speech recognition technologies that could have potential military applications. NtechLab, for example, has developed a facial recognition software called FindFace,[33] which has been used in Moscow's extensive urban surveillance network. This system allows the authorities to monitor the daily movements of almost 12 million people,[34] and similar technology could be used in an urban conflict when combined with data from unmanned systems and enhanced by AI algorithms. Such technology could be used to improve the Russian military's situational awareness in urban settings. By utilizing these technologies, the Russian military could carry out effective information operations in cities by tracking individuals and identifying potential targets. The use of AI in conjunction with these technologies would allow for large amounts of data to be processed in real-time and enhance the military's ability to respond to complex situations.

Information Operations

Employment of resources which may or may not involve combat but blur across diplomatic, informational, economic, and military lines is a new character of future conflicts and Russia as all nations are expected to utilise AI to gain an edge in this grey zone. Information warfare as a central tenet of contemporary conflicts is a useful strategy for advancing foreign policy goals and there are major lessons from the Russian cases analysed below.

- The April Wars of 2007 against Estonia was a series distributed denial-of-service (DDoS) attacks in response to the Estonian government's decision to move a Soviet-era war memorial from a central square in the capital city of Tallinn. The attacks overloaded Estonian government websites and disrupted critical online services such as banking and media outlets. This resulted in the disruption of critical infrastructure; Russia obviously denied any involvement.

- In the 2008 war between Russia and Georgia, the Georgian government's various ministries, departments, and news agencies were subjected to attacks, leading to the unavailability of their web material. As a result, Georgia had to rely on alternative servers to host its online content. Online surveys conducted during the time suggested that Russia's actions were perceived as peacekeeping by most of the respondents in a CNN survey. Similarly, blog entries were also found to be more in favour of Russia than Georgia.

Today, any individual with a smart phone and access to the Internet can share real-time images, videos, and other content that can reach and influence millions of people across the globe. In this new information ecosystem, information operations and public affairs functions have become increasingly important for the success of the mission.

AI, especially GAN networks, can create deceptively realistic but not necessarily true depictions of events. This can be utilised as a good tool for election manipulation like the Trump win. East StratCom Task Force,[35] has observed that Russia has used AI based deep fakes to gain political footholds across the globe.

Command and Control (C2)

Russia has a historical inclination to use mathematical models and other quantitative tools to support military decisions. Raketno-YadernoyeNapadenie[36] (RYAN) – a mathematical computer model was developed in the early 1980s to calculate the overall strategic balance between the Soviet Union and the USA. At a certain degree of imbalance, a pre-emptive nuclear strike would be suggested to the leadership by RYAN. This has eased ingress of AI into the command control and decision support systems. Activated in 2014, the National Defence Management Centre[37] (Natsional'nyi Tsentr Upravleniya Oboronoi – NTUO) provides a common operating picture for Russia. It is expected to employ AI to collect, collate, and analyze information on the military-socio-political situation across the world at all times.

Russia has established a complete set of military command-and-control systems. These include individual and tactical systems for various branches of the armed forces and most importantly, the Automated Command and Control

Systems[38] (ASU – Avtomatizirovannaya Sistema Upravleniya) in the military districts and the National Defence Centre.

The Reconnaissance Strike Contour[39] (RSC) is designed to coordinate employment of high-precision, long-range weapons linked to real-time intelligence data and precise targeting provided to a command and control centre. The RSC is designed to function at operational depths using surface-to-surface missile systems and aircraft-delivered 'smart' munitions.

Electronic Warfare

Currently, Russia has a range of highly mobile EW systems in its arsenal. Russia's interest and emphasis on EW is obviously triggered by satellite communications, Global Positioning System (GPS) navigation, and high-bandwidth internet dependence of US and NATO military and hopes to suppress their decision making abilities with domino effect on the battlefield forces.

RB-109A Bylina is a fully autonomous system[40] that analyses combat situations, identifies targets, chooses how to disable them, and ultimately issues orders to EW forces in the field. After deployment, Bylina[41] automatically establishes communication with the higher headquarters, tactical command posts in the field and individual electronic warfare stations. It recognizes radio stations, communication systems (including low powered ones), radars, early warning aircraft and enemy satellites almost immediately on deployment and effectively resolves complex radio-electronic environments without human intervention. It can take automated, AI enabled decisions how to suppress them and what jamming stations to use. The Russian deployment of Bylina in the Donbas Region of Ukraine in 2018, obviously gained favour with the military. Expecting an increase of EW efficiency by 40 to 50 per cent,[42] the military is seeking delivery by 2025.

Russian Military Robots

Russia wants its military robots to be faster, more discriminating in target selection, and more accurate than people, simultaneously fighting alongside their human warrior brothers. Russia's approach to robots predominantly entails retrofitting and introducing robotic capabilities to existing platforms, rather than trying to develop completely new systems. These robot machines are expected to react faster to sensor data, acquire accurate targeting information and execute with increased precision. The thickest of the battles will be decided by the volume of human casualties and this is where Russia believes in achieving results.

Broadly speaking, developments in military robotics, autonomy, machine learning, and artificial intelligence – combined, will improve intelligence collection and analysis, facilitate navigation and manoeuvre in dangerous terrain, and allow for more precise targeting, reduce costs of urban warfare and enhance combat effectiveness across the spectrum of military operations. This has resulted in a major impetus in Russia developing a wide range of unmanned systems. It hopes to refine these tethered and semi-autonomous robots by AI-driven command, control, computers, communications, intelligence, surveillance and reconnaissance (C4ISR) capabilities – technologies that will safeguard soldiers' lives and make forces more precise and lethal in combat. The Russian hypothesis of dependency on robotics probably is an attempt to create an antithesis to the soft and algorithmic AI dependence of its traditional adversaries (NATO and the USA).

Russia is investing substantial resources in the development of unmanned systems, which is evident from the numerous UAVs, UGVs, and UUVs in various stages of research and testing. These systems are primarily operated remotely, but a greater autonomy can be achieved as software technology improves. By incorporating AI into semi-autonomous and autonomous vehicles, it is possible to enhance force protection, situational awareness, and the ability to move through complex urban terrain. In this step towards AI, Russia has prioritised robotisation of on-board fire resources creating robotic strike complexes (Robotekhnicheskie Kompleksy – RTK), which can also be grouped into specialised assault units – all with aim to reduce losses and increase the effectiveness by increasing their strike capabilities. These are presently operated in remote mode by an off-site soldier. The end state would consider them functioning independently of humans, taking into consideration the system's current state and the external environment. In the future, Russia seeks to completely automate the battle[43] and may well introduce robotic groups conducting independent warfare. This adaptive nature will allow them to self-adjust under changing conditions and AI is expected to execute fire missions under real combat conditions.

- **Nerekhta.** It is a multifunctional modular tracked robot. Its variants include, a fire support (12.7-mm machine gun or 7.62-mm Kalashnikov RKTM); a reconnaissance module for artillery; and a transport version (delivery of equipment and ammunition to troops during contact). It can navigate to predetermined targets without a remote operator. It is considered an ideal platform to test-bed a host of military AI applications[44] including collaborative behaviour with other systems.
- **Kamikaze[45] Nerekhta.** This versatile chassis is being developed to approach the target and blow itself up, much like a suicide bomber. The operator downloads a map into the system, selects the target and launches it on its mission. The Nerekhta will plot its course to the target, approach it and blow it up. Later versions will enable multiple uses, when the robot will approach the target, place an explosive device on it and retreat to a safe distance for re-use.
- **Platforma M.** This is designed for visual, technical reconnaissance and fire support during patrolling, reconnaissance, detection and destruction of enemy equipment. It includes a 7.62-mm RKTM and four RPG-26 antitank grenade launchers.
- **Soratnik.** The BAS-01G Soratnik (comrade-in-arms[46]) is an unmanned armoured tracked ground vehicle designed for reconnaissance, fire support, patrols and protection of important facilities. It can operate in conjunction with other automated combat units.
- **Uran 9.** This unmanned tank is designed to destroy mobile targets, buildings and installations. It can be operated by remote control or can function autonomously or be set on a pre-programmed path.[47] Though not considered very successful in Syria, it is a robot to look out for. It is designed for combat operations – weighing in at 10 tons, its modular structure includes 23- and 30- mm automatic guns, the 7.62-mm Kalashnikov tank machine gun, six rocket-propelled flame throwers and an antitank missile system. It is also equipped with a variety of sensors, laser warning systems, thermal and electro-optic cameras. Russia is also considering export of Uran 9.
- **Vikhr.** Vikhr provides fire support during urban reconnaissance operations and the destruction of lightly armoured targets. It can be fitted with multiple modules – 30-mm 2A42 cannon, coaxial tank machine guns, Kornet ATGM or a 57-mm automatic cannon. It also

carries four quadcopters that could be used for kamikaze attacks on high-value targets. Interestingly its targeting data is usable by aviation, artillery and other robots as well.

- **Shtrum.** Near-urban combat missions in Syria are not considered completely successful.[48] Russia is working on design changes to withstand tough urban combat conditions. A project called Shtrum[49] (Storm), based on the T-72 tank chassis is under development for a city fight. This new system will include four combat vehicles with various weapons. A highly mobile command centre based on a T-72B3 chassis will control combat vehicles in a radius of 3 km.

- **Volk 2.** It is an unmanned, tracked, assault reconnaissance vehicle. It is capable of reconnaissance, detection and destruction of stationary and mobile targets, fire support and protection of important facilities in the automated security systems. It can be equipped with a 12.7-mm, a 7.62-mm machine gun or a 30-mm automatic grenade launcher. It has a laser rangefinder, an armament stabilizer, a thermal imager and an electronic ballistic computer. In its automated mode, the operator can remotely select up to 10 targets, which the robot then bombards.[50]

- **Taifun-M.** A tracked security robot for security of strategic missile facilities, is a unique concept, expected to be autonomous, it will be a failsafe option to guard strategic missiles – RS-24 Yars and SS-27 Topol-M missile sites.[51] It weighs 900 kilograms and has cameras, a laser rangefinder and radar sensors, capable of firing a 12.7-millimetre heavy machine gun. It has 45 kmph speed and operates continuously for 10 hours.

- **Msta-SM 2S19M2.** This is an AI-enabled robotised artillery system with automated guidance and a fire control system from howitzers, using smart high-precision shells. Each combat vehicle can exchange information[52] within themselves about each shot fired. This leads to efficient usage of firepower and better-coordinated attacks.

- **Armoured Fist of S-300 and S-400.** AI will convert data from all the national early warning radars, anti-aircraft and anti-missile systems, analyse information and come to a faster conclusion to launch these long-range precision missiles. An AI-enabled system will sift vast quantities of satellite imagery[53] to enable automated control of all offensive and defensive airborne and stationary assets as far as Crimea – aircraft, helicopters, missile batteries, and other systems – for optimal data analysis and decision-making.

- **Lantset.** Known as the Kamikaze drone, with a 3-kg payload, it can hit a target within a 40-km radius. The drone can transmit a live image of the target, which confirms the success of the strike. The Lantset can be used to target enemy UAVs, with a speed of 300 kmph; it is capable of striking a 150-kmph enemy UAV.[54]
- **Altius.** This AI-enabled bomber UAV can operate independently and interact with SU-57 5th generation aircraft. As a wingman, its AI elements enhance command and control, increased autonomy for navigation, target identification, and target engagement in the future. Once it receives the coordinates of the target, Altius finds the optimal route to the target and drops bombs, all without the help of an operator[55] and return to base by plotting the safest route.
- **S-70 Okhotnik-B.** A heavy, stealthy combat UAV, flies in automated mode interoperable with Su-57 lead aircraft. It is likely to be delivered in 2024[56]. A lead Su-57 pilot is expected to fly along a swarm of S 70 Okhotnik B.
- **ArgoAmphibious Vehicle.**[57] A jeep-size amphibious vehicle can cross a lake and fire at targets, conduct reconnaissance and support fire for amphibious landings and tactical load delivery. It is armed with a 7.62-mm tank machine gun, antipersonnel grenades and rocket launchers.
- **Underwater Mini Torpedoes.** Slow moving mini-torpedoes in groups[58] would be silent and inconspicuous, travelling at just 2–3 miles per hour while using AI to alter their movement patterns in a way that resembles a school of fish. This is a unique concept, which is so different, unknown and probably highly lethal.
- **Poseidon AUV.** Expected to dive deeper and travel faster, capable of killing all kinds of surface vessels, and carry either a conventional or nuclear warhead – a futuristic nuclear delivery systems.[59] It is meant to lie idle on the seabed, within target range (an aircraft carrier or a city), awaiting attack instructions.
- **Unicum.** This control system can control up to ten robots simultaneously. It can guide a selected robot system to a most favourable position and search for the target, which a remote operator attacks

subsequently. On-board AI is expected to provide protection to this setup from malware, viruses, and false information commands. It can transfer its environment awareness to other robotic devices.
- **Marker.** This UGV is capable of functioning autonomously using a neural network to create swarms on the battlefield. It is also serving as a test bed for computer vision, autonomous movement[60] navigation, and swarming technologies.
- **Swarm Technology.** The Russian military has shown interest in having swarm robots[61] coordinate through networks, able to act in formation, not separately, while a human monitors them from the ground. This interest is urging the civilian domestic industry to start developing a UAV swarm that could perform independent combat operations by penetrating enemy space and striking at targets.
- **FEDOR.** Final Experimental Demonstration Object Research is an armed robot capable of space[62] travel. It is the first domestically produced anthropomorphic robot. Designed to replace humans in high-risk areas, it is taught a variety of tasks in an urban environment. Still futuristic, it can become a game changer in combat.

Robotic Urban Warfare.[63] Russian experiences in Syria have led them to develop a robotic use in urban combat. Urban combat is highly lethal and increases human losses; robotic assault formations in such scenarios are important as per the Russian concept. Dul'nev also recommends a concept of robotic urban warfare to improve success in lethal urban environments, as outlined below:

- An RTK-assisted attack would consist of a recce and fire support by light RTKs (anti-shrapnel protection, capable of audio-video reconnaissance) and aerial multi UAV (may not be a swarm) to support aerial reconnaissance and destroy small targets.
- Heavy RTK (with tank-type armour protection and direct firing capability) and artillery fire would cover the advance of a penetration by heavy RTKs and destroy highly protected targets.
- Specialised remote controlled RTKs would create passages through obstacles under cover of medium RTK (BMP style type protection and fire support) which would also cover flanks and hold captured regions and provide addition fire support to heavy RTKs.

56 ❏ *Artificial Intelligence: Military Tactics, Bridges and Aspirations*

- Light RTK (anti-small arms protection) would destroy enemy personnel, unarmoured equipment and defend command posts.
- Transport RTK would provide battlefield and tactile mobility during the operation.

The Russians are still unable to sort out all problems such as inter-robot cooperation, integration, robust communication, data transmission, unified tactical command and control system and 'Identification of Friend or Foe' system to eliminate fratricide in such a concept.

CHINA

Reference to Chinese Hanzi scripted morphemes is inevitable when analysing any document, especially those involving policies and military concepts. 'Self-assessments' are of special interest here, reminding the world of the internal state analysis and justification of every action by quoting Hanzi morphemes. In one such assessment of 2015, the PLA emphasised 'Five Incapables'[64] – unsettling incapability of 'some officers – cannot judge situations; cannot understand higher authorities' intentions; cannot make operational decisions; cannot deploy troops (successfully); and unable to deal with unexpected situations. Earlier in 2009, 'Peace Disease' had acknowledged the PLA's practices arising due to poor combat experience over four decades. Attributed to a 'do not expect to go to war' attitude and the resultant shortcuts and 'go through the motion' training. The PLA took to 'cure the peace disease' and prepare to 'fight to win'.

Developing Military Warfighting Concepts. In line with developing military warfighting concepts, 'Potential Ways' are discussed in China too. Exploratory papers also mention of 'Intelligent Combat' with 'Superior performance beyond the limitations of the human body', capable of providing new tools and thus new ways of warfighting. China's obsession with creating new phraseologies is evident in these newer concepts. Latent warfare refers to pre-deploying unmanned systems (much like human deep assets) near important enemy targets, keeping them in a long dormant state, to execute attacks at the appropriate time. Cluster warfare (another term for swarming) employs large number of intelligent, unmanned, and autonomous systems in a coordinated manner for reconnaissance or combat to overwhelm the enemy. Global rapid strike warfare utilises hypersonic speeds, space, and unmanned combat platforms to precisely attack enemies across the globe within an hour. Chinese security thinkers have grouped the emerging technologies to formulate novel combat concepts and provided a vision roadmap to its developers and military.

LAWS. Though China has proposed a ban on LAWS[65] the definition is too narrow and may not constrain its own development or use. This pathological need of China to play with words is a way to tire out negotiators and discussions, to harass the adversary. It is aggressively developing robots but it is likely that China will dwell more on AI at Rest. It seems to be careful to avoid being trapped into an arms race and suffer the experience of the former Soviet Union. Advanced data processing and decision support systems will find hidden enemy systems, understand their vulnerabilities and turn sensor data into a common operating picture. Under conditions of informatised warfare, large-scale attrition of enemy forces is not the priority objective; rather, it is dominating adversary military systems. The PLA would attempt to create disruption or paralysis on the enemy side by targeting enemy operational systems and crashing them. Going by this concept,[66] China would rather bank on algorithms and unmanned platforms while precisely targeting the enemy underbelly (software and hardware) rather than a full spectrum, distributed warfare.

Cognitive Warfare. It is highly probable that the PLA will utilize artificial intelligence in the area of cognitive warfare, which involves disinformation,[67] misinformation, and propaganda strategies that are not uncommon. This approach is consistent with the conventional Chinese military philosophy that emphasizes the importance of winning over the hearts of the people and subduing the enemy without resorting to direct confrontation. AI makes it more precise, undetectable, and omnipotent.

Blitz on Taiwan. A small visualisation of a blitz attack[68] on Taiwan is noteworthy – the PLA may take advantage of its proximity and the American geographical distance from Taiwan to speed up its attack and 'demotivate any American intention to come to Taiwan's rescue'. Probable focus on combination of AI at Rest (algorithms, cyberattack, system of battle network systems) and AI in Motion (high numbers of unmanned and precision guided missiles) to exploit this relative advantage window and gain full Access/Area Denial (A2/AD).

ISR. It is noteworthy that 20 per cent of AI deals revolve around ISR, which has wide-ranging applications in geospatial imagery analysis, media analysis, and intelligence acquisitions. The PLASSF, China's strategic support force,[69] is leading the way in acquiring geospatial information perception and intelligent analysis subsystems. Unmanned Aerial Vehicles (UAVs) have proven to be useful tools for this critical military activity as they can penetrate enemy lines from hundreds of kilometres away and operate at different altitudes. These UAVs can provide crucial reference points for strategic decision-making and large-scale military operations. China's AI strategy aims to not only enable its platforms to quickly locate hidden targets but also to fuse multiple intelligence sources, including open-source intelligence and human intelligence, into a single, comprehensive operating picture.[70] These projects aim to combine the output of space-based, airborne, and terrestrial sensors, which would enable autonomous monitoring and post-strike damage assessment of large areas across the world.

Space Operations. China is also exploring the use of AI-supported analysis to manage large satellite constellations[71] for space operations. Smart radio technology can improve space-based communications by autonomously shifting channels, while autonomous satellite operations can compensate for limited communications windows and bandwidth. This approach can also reduce the workload of ground satellite operators. By collaborating with satellites and utilizing sky wave over-the-horizon radar,[72] UAVs can effectively locate and monitor hostile naval targets and instantly relay the collected data. As a result, military UAVs can play a crucial role in the kill chain during anti-aircraft carrier operations.

Overcoming Operational Inexperience through Hybrid Intelligence. The PLA is ready to use AI in war-gaming and simulation to train military officers. The Institute of Automation, Chinese Academy of Sciences in Beijing, has built AlphaWar[73] for such a military exercise. AlphaWar, was able to hide itself in 2020, as an AI from richly experienced military strategists during a test run and gave out reasonably sound military decisions. A battle simulation system, predicting possible outcomes is a major capability and more so if the system could come up with strategies superior to those devised by humans. The system is capable of self learning or by playing against humans. Playing a game of skill against AlphaGo is very different from a military war game, since the latter includes a lot of intangibles, the fog of battle episodes and human behaviours, but AI and the Fuzzy Logic fit in here by incorporating such intangibles and churn out a more comprehensive military strategy. Is AlphaWar heading that way?

Autonomous Command Decision Making. The People's Liberation Army (PLA) is actively exploring the use of AI technology to collect and analyse large, diverse, and rapidly changing streams of information that are beyond human capacity.[74] This effort is aimed at improving situational awareness to support decision-making, particularly in response to the PLA's 'Five Incapables' self-analysis. By leveraging AI, the PLA seeks to evaluate the potential consequences of various courses of action, thereby achieving decision superiority in future 'intelligentised' warfare.[75] The focus is on enhancing situational awareness to improve cognitive speed in decision-making, particularly in high-speed operational environments. The PLA is prioritizing the development and implementation of AI-based solutions to enhance future command decision-making capabilities. As a mind-machine interface, AI is expected to act as a staff officer and stack up the inputs in a more understandable manner for commanders to take a decision. The importance of integrating and leveraging synergies among human-machine 'hybrid' intelligence is expected to support command decision-making in many ways,[76] both concurrently and sequentially, or perhaps, in some contexts, even replace human commanders on the future battlefield:

- AI would also highlight priority recommendations, based on the predictions of the success percentages of a decision outcome. This would help commanders make more objective, scientific decisions with AI sorting through large volumes of data, remove chaff and encouraging attention to priority inputs. Even if a 'decision support system' does not materialise, AI developed strategies or plans would provide an opportunity to identify and correct flaws in these plans, provide more than one option and allow military commanders to think beyond the usual practices.
- Given its inadequate combat experience in joint operations, the PLA considers AI an 'indispensable digital staff officer' during planning and conduct of joint operations. It could study all efforts together, indicate the slowdowns, windows of opportunities, and recommend reinforcement (in time and space) to a particular line of operation for a quick battle resolution.
- Analysis of previous events is also considered a vital aspect for better command decisions in future, for which AI can come to help again.
- Autonomous systems may also be delegated greater autonomy where

speed is critical, like in cyber or ballistic missile defence or defence of critical infrastructure, where the pace of operations surpasses the human ability to intervene effectively.
- AI can also automate target selection and be more proficient than humans at striking multiple targets simultaneously.

Electronic Warfare. AI-powered UAVs are expected to distinguish and categorise electromagnetic signals using machine learning. These UAVs fly over enemy territory to disrupt the enemy's electronic equipment, air defence and fire control radar, bolstering China's attack capabilities.[77] By detecting and classifying signals from radar and communication systems, AI UAV could also conduct more efficient anti-jamming protocols,[78] if required. This special interest in electronic warfare resonates with similar interest in the Russian concept of operations. It's clear that the focus on electronic warfare (EW) is driven by the US and NATO military's heavy reliance on satellite communications, GPS navigation, and high-bandwidth internet. China will aim to disrupt their decision-making capabilities, which could cascade the effectiveness of battlefield forces.

Cyber Warfare. The cyber/network domain entails categorisation of large amounts of data in real-time to identify threats and update defences, where AI systems can improve the speed and scale of cyber defence.[79] AI-guided navigation of adversary networks is useful during technical reconnaissance and cyber-attacks. The PLA Strategic Support Force is researching the use of pattern recognition to identify and defend against distributed denial-of-service attacks and to identify advanced persistent threats while deep neural networks have been able to detect intrusions. Cyber attacks to disrupt enemy systems is in consonance with the Chinese concept to attack the venerable underbelly of the adversary.

Predictive Maintenance and Logistics. The PLA Air Force has utilised image recognition to identify cracks in engine propeller/impeller blades for quick repairs. Another application will analyse a large number of data points from equipment and maintenance to indicate predictive maintenance[80] of each equipment separately, optimising human effort. The same application will also predict failure in electromechanical systems. Big data and data analytics are expected to improve the supply chain, indicate likely disruptions, and improve acquisitions and maintenance. During 2020-21, over 11 per cent of the 343 AI contracts are focused[81] on maintenance, repair, logistics, or sustainment. All existing contractors are also expected to provide AI-based detection for fault diagnosis and smart warehousing.

Not one to shy away from kinetic options, China has inducted a wide variety of AI in motion systems too – robotic unmanned systems and precision guided missiles and is prepared for kinetic options during symmetric, asymmetric and cyberspace warfare. While the PLA Rocket Force wants to create full-pledged remote sensing, and target identification, the PLA Strategic Force[82] is focussing on AI for its electronic, cyber, space, and psychological warfare.

Airborne Robotic Systems. To be 'ready now', China has concentrated on autonomy to improve the effectiveness of existing platforms and tactics. Functional UAVs have enhanced China's confidence to conduct reconnaissance and surveillance tasks to protect claimed territories. In the early 2010s, for the first time BZK-005 unmanned reconnaissance aircraft flew over the Diaoyu Dao/Senkaku Island disputed area, avoiding Japanese detection. Information during competition and conflict is important[83] and UAV fills that role magnificently.

- EA-03 Xianglong is a high-altitude long-endurance unmanned reconnaissance aircraft. With a mission load of 600 kg and an effective range of 7,000 km, it can conduct continuous aerial surveillance for 10 hours from an altitude of 18,000 metres.
- Attack-1 equipped with synthetic aperture radar (SAR), laser-guided missiles and GPS-guided bombs can operate continuously for 20 hours at a maximum speed of 370 km per hour.
- JWP02 UAV (ASN-206) is a tactical unmanned reconnaissance aircraft reaching out to 150 km for four hours.
- BZK-005 UAV is a stealthy medium and high-altitude long-range unmanned reconnaissance aircraft with an endurance of 40 hours.
- Dark Sword UAV unveiled during the Zhuhai Air Show in October 2012, caught the world by surprise. It is a stealthy, supersonic, ultra-high mobility aircraft which can autonomously perform air-to-air combat, ground attack and carry anti-ship ballistic missiles making it the world's only aircraft carrier attack UAV system.[84]
- Star Shadow is a stealthy combat UAV aerial vehicle with a radar cross-section of 1 sq ft.[85] It is a blended-wing and twin jet-engine platform,

having endurance of 12 hours, a cruising speed of 600 km/h and an altitude of 40,000 ft.
- The GJ-11, known as the Sharp Sword,[86] is an unmanned aerial vehicle designed as a stealthy flying wing. This drone can take off from warships without human intervention and carry out various tasks, such as releasing swarming decoys or electronic warfare equipment, launching guided weapons with precision, and conducting aerial surveillance missions.

Ground Robotic Systems. To develop still more cutting-edge, next-generation systems, the PLA has held a 'dangerous crossing'[87] contest in September 2018, the details of which are sketchy. The contest included competitions in cross-country reconnaissance, cross-country formations and transportation, air-ground collaborative reconnaissance, lifelike walking drones that can follow and support troops, and UGVs capable of conducting transportation missions over mountainous terrain. But the data available indicates a clear preference of UAVs over UGVs.

- The Sharp Claw 2[88] is an unmanned 6×6 wheeled UGV to carry out battlefield reconnaissance, patrolling, assault and transport missions. In addition to carrying weapons and ISR accessories, it can transport a smaller Sharp Claw 1 mini UGV. Its electric motor allows the vehicle to move silently across extreme terrain. It can be remotely operated but has autonomous mission functions, especially approaching an objective.
- Sharp Claw 1,[89] a six-wheeled, reconnaissance vehicle is an in service UGV with the Chinese military since 2020, though the numbers are not known. Possibly autonomous, it weighs 120 kg, has a small profile, which lends itself to operations and logistics, though it is mounted with only a single machine gun.

Maritime Robotic Systems. To overcome the communication clutter and significance of timely and just nuclear launch commands, PLAN nuclear submarines are being equipped with AI decision-support systems to 'reduce commanding officers' workload and mental burden'.[90] The AI submarine is expected to receive communication signals and interpret those using convolutional neural networks. In addition, the Navy is looking for autonomous vehicles too.

- Shanghai University has developed a series of USVs that possess autonomous capabilities to monitor and map long coastlines. These vehicles are equipped with advanced technologies that enable them to navigate complex terrains, and avoid obstacles such as reefs, icebergs, and other moving vessels. Key features include intelligent autonomous control and marine target detection and recognition.[91] This series of USVs identified a new anchorage while working with a polar research vessel in Antarctica. It also detected and sampled contaminated water after the sinking of the oil tanker *Sanchi* in the East China Sea. China is exploring the integration of swarm intelligence to enable cluster control and cross-domain synergy between the sky, land, and sea, with ongoing research focusing on deep-sea exploration and long loitering time.

- China has deployed a fleet of slow moving but long endurance underwater drones known as the Sea Wing (Haiyi) glider in the Indian Ocean[92] and Selayar Islands in the Flores Sea of Indonesia. Using 'Variable-Buoyancy Propulsion'[93] these slow-moving underwater drones are considered high value naval intelligence tools. Similar gliders are used by the US Navy too. It is understood that the PLA Navy and Yunzhou Systems are conducting autonomous swarm testing on these drones to improve their capabilities. The Sea Wing is expected to support submarine operations during military missions with AI support to interpret acoustic signals underwater for target recognition.
- The D3000[94] is a stealthy warship designed to operate autonomously for months. It can operate independently or in conjunction with manned ships for long periods. Its four sets of 7-tube 30-mm rapid-fire guns, drone launch decks, anti-ship missile launchers and torpedo tubes make a new 'big deal at sea'. The USA considers it a prime example of China's rapid AI technology development.
- Reports also suggest that massed unmanned systems could disrupt enemy communications and subvert anti-submarine capability. These underwater drones could act as decoys and draw the adversary into an ambush.[95]

Swarming. In an effort to overwhelm the adversary's battlefield awareness, China has invested heavily in developing swarms in all physical domains. A number of studies are related to swarm operations, be it organization, data links and tasking. The swarming is being studied across multiple types of drones. China considers it a suitable candidate for ISR, communications, and strike missions. The possibility of developing air-launched, retrievable collaborative swarms[96] to simultaneously strike high-value targets is a special area of interest.

The Future is Already Here

Manned Unmanned Teaming. With advanced air-defence weapons, many supported with fully autonomy, China understands the vulnerability of manned aircraft. Complementing a manned aircraft with a UAV is an obvious solution. By deploying sensors and radar further away from the manned aircraft, UAVs can serve as 'pathfinders', greatly reducing pilot casualties. Fully autonomous UAVs, supported with navigation technology, will no longer be remote-control models, reducing the cognitive load on the manned aircraft pilots in the hybrid teams. The pilots will be freed to undertake higher function roles – executing fire missions. The UAVs and manned aircraft will each play to their respective advantages, complementing each other, developing together.

The Chinese government has a well-known ability to devote substantial funds towards AI development. China recognizes that collaboration between public and private sectors is the key to technological leadership. It has been urging private companies to develop critical technologies like AI, and is even providing subsidies.[97] The private sector is unlikely to invest in risky, long-term research, which makes government funding crucial. The Chinese government's capacity to signal priorities to the private sector and local governments is also advantageous. The government is trying to address talent the talent growth issues by recruiting highly educated officers,[98] enlisted personnel, and civilian staff, fighting off the competition from the private sector and it seems to be successful.

In 2023, China validated AI's capability to design[99] the electrical system of a warship, '300 times faster', than the traditional method. It is claimed to be accurate, since it learns from databases of Chinese ship design knowledge and past decade experience. New designs are rechecked with the existing databases to ensure accuracy. This AI system operates with guidance from humans. Fast ship launches, despite this AI trial, is already a concern for the US Navy, – "One shipyard has more capacity than all of our shipyards combined," observed Navy Secretary Carlos Del Toro. The Indian Navy Chief also notes that the construction speed of China has launched 148 warships[100] in the last 10 years – almost the size of the Indian Navy.

TÜRKIYE

Turkiye's defence sector is led by Turkish Aerospace Industries (TAI) and its focus on UAV platforms. Turkiye believes the major focus in future wars will be on artificial intelligence, autonomy and advanced robotic warfare. A double-word frequency analysis results of private sector AI projects receiving TÜBITAK[101] (The Scientific and Technological Research Institution of Turkiye) grants shows this primacy to certain AI projects like, decision support systems, image processing and big data projects. Deep learning, image processing, natural language processing, computer vision, machine learning, artificial neural networks, virtual reality, and face recognition are given the next priority.

Understanding the need to continue the edge in AI development, Turkiye has taken innovative policy decisions:

- Open Government Data Portal,[102] shares data to strengthen communication between stakeholders by using open government data, encouraging collaborations and triggering new studies to create newer value from data. There are mechanisms of protecting intellectual property rights, privacy and national security in this data-sharing portal.
- Data Labelling[103] is an important facet for ML. Here again, Turkiye has created a Data Hive (Defence Industry Agency, under the President) to enable browser-based labelling of video, image and text data. This crowd sourcing of labelling activities are uniquely innovative and will reduce the workload and hasten a large volume and variety of datasets. It will be obviously monitored vigorously to avoid all biases.

Turkiye's reliance on unscrewed systems is evident and so is the world's acceptance of the capability to produce and deliver a variety of systems, based on the necessity. Some use cases are discussed below.

- Bayraktar TB2[104] is a battle tested, well-known UCAV against Armenian forces and in Syria against Russian air defence systems. It is claimed to be highly successful and 20 other countries employ them. Four hard points make it flexible to carry a large variety of laser-guided smart bombs or rockets, Long Range Anti-tank Missile or INS guided 81 mm mortars bombs.
- Anka-S[105] can conduct a range of missions such as real-time intelligence, surveillance and reconnaissance (ISR), communication relay, target

acquisition and tracking. UAVs can detect, identify and track stationary or moving ground targets. With a payload capacity of 200 kg, friend or foe identification system, laser designator and laser range finder, it has two under wing weapon stations, which can carry Smart micro munitions, missiles or guided rockets to engage light armoured vehicles, personnel, military shelters, and ground radar stations. It can be operated in both autonomous and remote control modes.

- Togan UAV can be used in GPS-contested environments using AI and CV. As part of KERKES[106] (Global positioning system independent autonomous navigation system development) project, it is able to navigate without relying on GPS.
- Akinci (Raider)[107] is a heavy, unmanned, aerial, high-altitude, long-endurance UAV to replace Bayraktar TB2. It can execute air-to-ground and air-to-air attack operations and is capable of carrying 1,350 kg of payload and flying at 40,000 feet. Powerful AI flies the drone, processing data from on-board sensors and cameras. Its weapons profile include laser-guided smart munitions, missiles, and long-range stand-off weapons. It is being studied to potentially serve as a mother ship for drone swarming attacks, including control of kamikaze drones.
- Aksungur[108] is an ASW (anti-submarine warfare) capable drone, capable of full autonomy. It is maritime enabled which combines radars, sensors, and a variety of air-to-ground weapons. It carries laser guided bombs, missiles, and precision guided kits for small diameters like 81-mm mortar bombs.
- Sancar[109] is an unmanned surface vehicle (USV) designed for asymmetrical warfare missions, measuring 12.73 metres and weighing 9 tonnes. With a maximum speed of 40 knots and a range of 400 nautical miles, it can operate for up to 40 hours. It is equipped with a 12.7-mm machine gun and two anti-armour missiles, making it capable of targeting main battle tanks, small watercraft, and light structures. It can be used for reconnaissance, surveillance, surface warfare, mine countermeasures, harbour protection, patrol, and search and rescue missions, even in conditions up to Sea State four.
- The Turkish military employs autonomous kamikaze[110] drones extensively. There are various different models – Togan, Alpagu, but Kargu is the favoured one. Equipped with advanced machine vision

and biometric facial recognition (permitting specific human attack[111]), it can be used as a single platform or grouped together in a swarm of 20. Fully autonomous anti-personnel and anti-armour quadcopter drones can be paired with other autonomous systems too. March 2020 saw successful employment of Kargu during the siege of Tripoli against Khalifa Haftar's affiliated forces (HAF).

- TF-X National Combat Aircraft (MMU)[112] is an unmanned aircraft technology in which AI serves as a co-pilot. TF-X is a 5th generation aircraft with a human pilot and an AI co-pilot, thus creating a man machine system – a more intelligent aircraft.

- Turkiye has launched its Swarm UAV[113] technology development and demonstration program, which is aimed at developing and demonstrating swarm capability. MilSOFT, a Turkish software company, has demonstrated algorithms and software for swarm capability, including automatic detection of moving targets using AI and machine-learning techniques with image-processing algorithms. The swarms can be updated with additional capabilities such as reconnaissance, detection, recognition, search and rescue, and vehicle tracking. The drones are expected to perform frontal attacks on command from helicopters and provide operational support to other friendly platforms. They can make a swarm of 50 operational vehicles and communicate with each other from up to 500 metres in a 10-km network for data transfers. MilSOFT plans to integrate its technology for underwater and surface platforms as well as land vehicles.

- The US State Department, has supported Turkiye's request for upgraded F-16 fighters[114] by noting 'compelling long-term NATO alliance, national security, economic and commercial interests' as sufficient for defence ties. This upgraded F-16 was tested in December 2022 with an AI software[115] and landed and took off, flying for 17 hours autonomously.

- Turkiye wants to decrease its dependence[116] on foreign states. In 2019, quoting CAATSA, the USA removed Turkiye from the F-35 programme, which cost it nearly 100 F-35 jets. The USA felt training on S-400 undermines the commitments of NATO allies to move away from Russian systems. By focusing on the application of AI and success of Bayraktar TB2 UCAVs, it expects to gain region and global accolades.

Academics suggest that Turkiye lead in global concept development and shaping the debate. Presence during international debates on the trajectory, implications and regulations is a sure way to turn the debate against a competitor while reorienting own development towards the 'greater good'. Turkiye is seeking to improve its AI and associated products by creating a new operational vision[117] that focuses on cross-domain capability.

- The country aims to achieve this by teaming up unscrewed systems from different physical domains. For example, the Estonian THeMIS UGV is launching and recovering a UAV, while Lockheed Martin's Squad Mission Support System (SMSS) is dropped by a helicopter for surveillance in a mission area.
- The success of loitering ammunition has already been proven, and Turkiye plans to de-specialize it to make it universally available to all its forces. To achieve this goal, Turkiye plans to create a smarter and cheaper UAV that can autonomously take off and destroy targets without the need for Special Forces. This advanced system will have more endurance, higher payload, and better overall capabilities, which could prove crucial in hybrid battlegrounds with the risk of inter-state conflict.
- To mobilize the nation's innovative minds from all corners of the country, Turkiye organized the unmanned and autonomous land platforms design competition, called ROBOIK. This competition aims to foster robotic research and development by developing various subsystems, thereby contributing to the overall development of the project.

Turkiye intends to create a more advanced and capable system that is cheaper and smarter to meet the challenges of modern warfare. By coordinating the efforts of innovative minds from across the country, Turkiye hopes to develop systems that can operate autonomously and effectively in hybrid battlegrounds, carried usually in the inter-state conflict.

IRAN

Iran is known for advocating its capabilities to deter perceived threats and assert its influence in the region. While some may view the claims with scepticism, many experts believe that Iran possesses significant technological capabilities, despite facing various challenges. Among the array of technologies at Iran's disposal, Artificial Intelligence (AI) stands out as particularly potent. AI has the potential to enhance defence capabilities significantly.

Despite facing sanctions that restrict access to certain hardware and technologies, Iran has demonstrated resilience and ingenuity in developing indigenous solutions to overcome these obstacles. While sanctions may slow down Iran's technological advancements in some areas, they have not prevented the country from making significant strides in AI research and development. It has published 17,722 citable AI documents higher than of Turkiye and Russia.[118] By investing in AI research and development, Iran aims to enhance its military capabilities. As long as AI is universally available, Iran will continue to use it to offset the sanctions on the physical artefacts and Iran is heavily relying on its large STEM qualified students and researchers.

In 2019, Google forbade the Iranian ride-sharing company, Snapp, from using Google Maps. Cafe Bazaar – an Iranian version of the Google Play Store, launched Balad – a speaker map and navigation app. A smart voice assistant reads the names of all the streets and alleys with more than 1 million points on the map. Despite the ban, the application continues to grow, thanks to AI and Iran's innovative approach to utilise it. Freely available deep learning algorithms; YOLO and R-CNN, were refined to Faster R-CNN, by removing the mask heads of a Mask R CNN model. This did faster box bounding of store signs and billboards as ubiquitous to all retail stores and establishments – the Points of Interest (POI).[119] This makes Iranians unfettered from Google POIs, thanks again to AI.

As part of population control, Iran's Revolutionary Guard has made vast use of large language models and generative AI.[120] Social engineering, troubleshooting software errors, intruder detection and luring anti-regime feminists – all have been tested by Iran. Iran has 'successfully' utilised AI for surveillance during the nationwide Islamic dress code unrest.[121] This is formal proof of the concept of Iran's capability to monitor 'Persons of Interest', a major AI at Rest capability. Interestingly, Iran is also contemplating using AI to issue and implement fatwas. This could entail computer vision (CV) to monitor citizens, correlate their actions as anti-Islamic and offer a bouquet of fatwa choices to clerics.

Militarily, Iran understands the inability to fight a conventional battle against superior peers. So it is obvious that AI solutions will be created to offer better survivability through high-density anti-aircraft defences and ECM measures. Iran also covets fully autonomous battle systems 2024.[122] During the January 2021 exercise, the Iranian Army ground forces introduced autonomous suicide drones, "using advanced image processing capabilities and AI" to detect and destroy targets. Concurrently Islamic Revolutionary Guard Corps (IRGC) also demonstrated explosive suicide drones, with some level of AI. Deep learning automatically recognises targets and the soldiers click on a screen ordering the machine to destroy or investigate the target. Iran's ground forces have been able to design and build remotely controlled vehicles that have AI to be used in irregular warfare. Some typical AI efforts are paraphrased below.

- In 2023, Iran indicated AI a component in Abu Mahdi, the 1,000-km cruise missile.[123] Iran wants to exploit AI to strike accurately, fly low and farther. Its trajectory definition and command-and-control systems have reportedly been equipped with AI, allowing it to easily evade radar and alter its course mid-air.
- In early 2023, Iran showcased miniature tank unmanned vehicles[124] with some form of limited AI capabilities. The remote-controlled four and six-wheeled machines can transport troops, cargo, and conduct day-and-night surveillance.
- The six-wheeled Nazir is a 600-kg unmanned ground vehicle with a 4-km radius of action; it is capable of carrying cargo and food for soldiers. Its four wheeled version has multiple terrain capability. Mounted with day-and-night optics and 7.62-mm weapons systems it can conduct surveillance. Nazir was probably developed during the ISIS war and designed against US forces in Iraq and Syria.
- Made in 2018, Haider is a smaller vehicle[125] (less than a metre). Iran wants to use it for 'sniper, suicide bomber, mine thrower or reconnaissance,' duties. Iran has claimed the ability to network Haider with other UGVs for 'multilateral combat operations'. Iran wants to target the larger armoured vehicles of its rivals with swarms of smaller Haiders.
- A ground-based 'suicide robot' has been made to ram into enemy armoured vehicles. Iran wants to utilise such robots to penetrate enemy trenches and target the enemy's key commanders.

- The Caracal[126] is six-wheeled armed UGV but not a true AI system. Dubbed as an 'intelligent ground robot that uses an independent suspension system', it can reach a speed of 30 km/h. Its flexible structure can mount a variety of weapons and is hailed as an 'intelligent remote-control system with a laser range finder and increases tactical opportunities for commanders.
- Caracal is a smart car that uses an independent suspension system. The maximum speed of Caracal is 30 km/h and it is capable of carrying light and semi-heavy weapons and is equipped with an intelligent remote-control system, a laser range finder and an optical system. This vehicle is designed for ground combat and is very agile and has an independent suspension system. The operational range of Caracal is more than 500 metres.

This variety inventory of UGVs is indicative of an unconventional approach to fight conventional warfare.

AUSTRALIA

Australia considers itself to be in a highly dynamic strategic environment. Increased relevance of the Indo-Pacific region, prevalence of a grey zone and increased proof of disruptive and emerging technology has called for self-reliance across the world. Australia perceives an increase in regional competitions and the likelihood of conflicts forces it to reconsider its positions and plans. AI is christened as Robotics and Autonomous Systems (RAS) in Australia and major headways have been achieved. With increasing advancement, availability, the affordability of RAS technologies and potential adversaries' capabilities are expanding, making them more intelligent and agile. This narrows the gap between well-equipped militaries and motivated individuals (groups with a cause), creating new and significant threats in future conflicts.

The Australian Defence Force (ADF) is cognisant of the need to manage the strain between being simultaneously 'Ready Now' and 'Future Ready'. Interestingly, the ADF confirms the necessity of external collaboration with other government agencies, academia, industry and international partners, including the AUKUS.[127] NATO is not mentioned as a collaborative organisation! Individually, the Australian Navy has identified a need for autonomous – Launch & Recovery Systems (LARS) espousing reusable RAS systems that can team with and complement crewed fleet units or systems.

The global realities indicate that the inclusion of RAS capabilities will be flexible and adaptive, depending on organic capabilities, international opportunities and supply chain crisis that may arise in the future. Human-Machine teaming within a strike package will consist of AI-operated systems with autonomous liberty while others will be 'human-in-the-loop' systems. The ADF is considering flexible levels of autonomy for such an integrated strike package, which would depend on the level of acceptance of AI prevalent[128] at that time. The Australian military wants to employ new technologies spawned by AI, especially human-machine teams,[129] to maximise soldiers' performances by reducing their physical and cognitive loads, achieve decision superiority, always ensuring the protection of the force and improved efficiency.

The ADF has also created an Enhance, Augment, Replace (EAR) framework[130] for induction of RAS capabilities to cater for developmental and technological delays. The first is the simplest and least invasive approach, by retro modification or retro fitting to improve and refine equipment functionality. This is the lowest level of autonomy (maybe remote operations); thus a rapid method of RAS employment. The next stage would be to augment some functionality of the systems by adding new capabilities – automatic target tracking, autonomous operation on reduced crew vehicles or autonomous self-defence systems. The third would be to redesign platforms and capabilities to replace traditional systems – swarm systems, unmanned systems or non-micro environment needs for high speed or high altitude platforms.

Force Protection[131] by increasing autonomous situational awareness is a logical alternative. A more aware force is automatically survivable, capable of generating mass without concentrating and will avoid exposed targets to threat targeting systems. In the same gambit, vision systems are expected to improve object recognition, navigation, see-and-avoid capabilities enabling autonomous stealth operations.

- The Australia Air Force has teamed with Boeing to develop Ghost Bat,[132] an uncrewed aircraft, incorporating autonomous systems and AI to establish human-machine teaming. The UAV is designed to act as a loyal wingman, controlled by a parent aircraft to fly alongside or ahead for reconnaissance, electronic warfare or absorbing enemy fire. The

Ghost Bat has a flying range of over 3,700 km, and can support various aircraft. It will enter service by 2025.
- Un-crewed aircraft, Triton Unmanned Aircraft System[133] are flown remotely by Air Force pilots from a ground station for maritime patrols and other intelligence, surveillance and reconnaissance roles. The Army employs RQ-7B Shadow 200 for real-time information to ground troops. S-100 Camcopter and ScanEagle provide ocean mapping and search capabilities over a maritime environment.
- Strix is a pilotless drone capable of ground strikes, surveillance and reconnaissance in high risk environments. A small aircraft, it is touted as a loyal wingman for military helicopters.[134] It can carry 160 kg payload and fly more than 800 km, indicating increased interest towards autonomous military vehicles.
- The ADF, during a joint exercise with the USA and UK in the 2022's Project Convergence[135] (annual sensor-to-shooter experiments), collaborated to hook on to the former's decision support system to confirm interoperability. Australia tested extended range intelligence, surveillance, reconnaissance, and electronic warfare effects on decision systems.
- Innovatively, inducting AI into combat search and rescue (CSR) operations,[136] otherwise a benign role is highly advantageous in this critical role, especially during hostilities. Own combatants can now avoid risk in attempting rescue of others. The operation would combine UAV SEAD mission with uncrewed AI controlled rescue vehicles.
- Uncrewed platforms, autonomously could project themselves to wider zones and distances generating unprecedented mass and tempo[137] enabling a reach beyond traditional systems in all domains and entire spectrum of operations.

RAS cooperation and swarming boost human-machine teaming and machine-machine teaming to generate mass, apply distributed affects and aid in safe and effective operations. Increased mass[138] by drone swarms employing large numbers of small, easy to manufacture drones will increase the mass that the military can generate. Autonomic drones could be projected into an area on a mother ship before being released to conduct a task. A large number of drones with different capabilities could coordinate their actions to create a resilient system that achieves its tasks by overwhelming enemy defences and adapting to enemy activity. Using collaborative drone swarms to deliver precision targeting data to FIRESTORM[139] (American AI system providing threat warning to ground troops and sending precision targeting data, even aiming their vehicles' weapons at the enemy).

STELaRLab[140] is testing a new, AI leveraged military command-and-control training system. It provides integrated air and missile defence (IAMD) decision-support capabilities at tactical levels. It is a useful assistant in Air Force mission planning, execution and debrief. Further developments will support a wide range of domains – air battle management, air and missile defence, and joint all-domain command-and-control.

SARAH[141] (Supply Autonomous Robotic Assistant Hardware) delivers parts from the logistics section to the flight line. It is an autonomously moving spares and aircraft parts hardware and is part of a strategic transport squadron and Air Force modernisation plan (Jericho). SARAH is successfully transporting parts within the unit building, transporting 200 kg on predetermined paths and slows down in the vicinity of obstacles and humans.[142] It will be soon tested outdoor with increased complexity. Relieved humans can focus their attention on higher order jobs.

The University of South Australia[143] has been contracted to develop a statistical machine learning algorithm using noisy and dynamic data from wearable devices to detect early signs of infection in soldiers. Future practical applications will include the early detection of chemical or biological threats and achieve superior decision making.

Managing the Change

The ADF, like all militaries, found itself wanting to maximize the effectiveness of its forces through technological induction. The modernization of the military created the need for thousands of additional soldiers, which traditionally meant induction of more soldiers. However, this is an antithesis to smarter, leaner and more agile forces. Can AI directly help the military make its workforce smarter? A unique analysis re-imagined a new type of workforce by redistributing work. To perform lower spectrum jobs, three AI assistants were created: Automated assistants, Cognitive assistants, and AI advisors. These assistants were created to free soldiers from repetitive and routine tasks and allow them to use human judgment, intuition, and empathy to greater effect in critical frontline roles.

The ADF found utilization of data could lead to improvements in training, simulation, and planning of force levels, and boosting of operational concepts and procedures.

A cross-functional Deloitte team[144] worked with the military to scope, design, configure, test, and pilot the automated assistants. The team identified and classified all current and anticipated work types, then mapped them according to the level of strategic importance of the work and the cognitive power required to do it. This map provided options about which work could be augmented with AI and to what extent and in what sequence. AI assistants were selected based on high return, rapid realization, and low risk. Two AI assistants were initially built as proofs of concept, and 10 per cent of the people previously doing the work were able to be redeployed to higher-value tasks. This innovative use of AI has empowered the 'non-experts' and a higher proportion of a workforce to be involved in applied military AI, thus reinforcing the integration cycle. Currently, there are 20 AI assistants across all the three services.

Emphasising the need to capture and manage data for generating training environments to refine machine learning and system improvement, the ADF is expecting deep integration of AI in training. Soldiers are being trained to 'trust but verify', avoid complacency but have confidence and not become technologically biased about autonomous systems – evolved development of human-machine teams.[145] Wanting to build on the strength of each, where humans excel at cognition and thinking creatively, the autonomous systems analyse large amounts of data, execute fast and accurately. The ADF also understands that autonomous systems are expendable and thus have to be available in greater numbers. Teaming autonomous and crewed systems is also considered an adjunct to human machine teaming, thus a need to normalise the new expectation through ab initio training and periodic visitations throughout the career is also a part of the training plan.

Strategic Choice. Australia is concerned about the possibility of conflict in the Indo-Pacific region given Chinese assertiveness. Even though the Quad members have unique strengths in AI, their collaboration and investment in AI-related projects are minimal. For Australia, it may be useful to align more with Japan and India[146] for focussed efforts on the military applications of AI technologies. Air Marshal Rob Chipman, the Air Force Chief, in February 2023 mentioned drones as a major threat to forces on the ground or sea. Warning about technology bias, he observed proliferation of low-cost drones[147] for ISR and the effect of loitering munitions which is a good consideration for 'potential of low-cost drones that bring mass to air combat' and 'new measures to defend against them as well'. Such disruptive AI[148] technologies will spawn a whole range of AI at Rest tools from the island country.

UNITED KINGDOM

'Machines are good at doing things right (e.g., quickly processing large data sets). People are good at doing the right things (e.g., evaluating complex, incomplete, rapidly changing information guided by values such as fairness). Human-Machine teaming will therefore be our default approach.'

– *UK Defence AI Strategy*

The UK Defence AI Strategy,[149] launched in 2022, aims to enhance the effectiveness of AI systems in competition and conflict scenarios. It recognizes the potential threat posed by adversarial AI systems to national security and highlights the importance of prioritizing research, development, and experimentation to maintain a strategic edge. The strategy emphasizes the need to counter false narratives through information operations, which are increasingly crucial in safeguarding the friendly narrative. The UK's response to AI-related strategic competition is planned to be 'rapid, ambitious, and comprehensive'. The UK wants to collaborate with NATO allies and friendly AI ecosystems to develop 'innovative solutions for shared challenges'. Human-machine teaming, which blends machine precision with human cognition, is a top priority.

The Strategy highlights a broad range of potential threats posed by adversaries' use of AI. The strategy anticipates that adversaries may use a wide spectrum of AI capabilities to attack AI systems in both cyber and physical domains, including data poisoning, hardware and supply chain corruption, and communications interference. Non-state actors may engage in illegal actions to undermine the integrity of AI systems. Adversarial states could challenge information superiority by conducting citizen surveillance to spread fake news, targeting individuals and using deep fakes to undermine public opinion. They may also steal key technologies and intellectual property, or use LAWS to conduct military operations. The strategy emphasizes the importance of developing counter-capabilities to safeguard against such threats. It suggests that the UK's

AI industrial sector will play a critical role in developing the responsive mechanisms necessary to counter these threats.

Data Analysis

Projects SPOTTER and SQUINTER use ML and CV techniques to enhance satellite imagery analysts' output, using automated detection and identification of objects within classified satellite imagery. Using convolution neural network, the applications can identify and automatically monitor objects of interest for changes.

The KILPECK program automates the analysis of large datasets by establishing an operational data science environment. This environment stores and provides access to datasets to intelligence analysts. Since the program geo-references[150] all data, it permits rapid integration, exploitation, and extraction of intelligence. Analysts can write their own code or use previous uploaded codes to quickly analyse the growing datasets. This automating of data collation increases the focus on contextualizing data and providing assessments to support decision-making processes which is considered a more efficient and effective way to analyse large and live datasets.

Sensing for Asset Protection with Integrated Electronic Networked Technology (SAPIENT) reduces the human workload in a multi-sensor system by utilising AI on the Edge. It is especially useful for critical base surveillance and protection, or urban battlefield flagging dangers. A standard surveillance system will transmit raw data to the operational room, where an operator observes the anomalies and directs further actions, producing a huge cognitive burden and possible errors. A SAPIENT system, on the other hand, has smart sensors, which, using AI, make a local detection and classification of low-level decisions autonomously,[151] like direction of anomaly, degree of magnification, and send information to the control centre. Based on the information received at the control centre, the AI decision-making module helps the operator to a higher level. This reduces the need for operators to constantly monitor the output of the sensors designed to enable multi-sensor fusion (correlation, association and tracking) and sensor management (dynamic tasking of the sensors in response to the unfolding scenario). This solution is highly scalable and requires minimal or no software engineering. It can be seamlessly integrated with zero integration time, making it a plug-and-play option. This solution provides excellent support for the command and control of operations in intricate landscapes.

During the 2021 annual large-scale NATO exercise in Estonia, multinational troops took part in Exercise Spring Storm[152] to test AI's capability to provide information on the surrounding environment and terrain. This AI was able to rapidly generate environment and terrain information, improving levels of analysis and speed of planning and expected to predict adversary behaviour, perform reconnaissance and relay real-time intelligence from the battlefield in future'. This AI trial proved its ability to rapidly process enormous amounts of data. It has the potential to enhance the quality of information available to commanders during critical operations, and alleviate the cognitive burden of data processing from human operators. The AI can operate on a cloud or independently, providing instantaneous planning support.

UK Drones

While the well-known MQ-9 Reaper and the newer inductee, MQ-9B Protector (SkyGuardian), are armed drones operating in the UK, Watchkeeper is an unarmed drone primary purposed to direct artillery and rocket strikes. There are many other systems, but for the purpose of this study, the sheer variety and deployed numbers in service stand out as a matter of interest. It indicates a fully developed concept of operations for drones, rather than mere technology demonstrators.

Tempest is being developed as a remotely-controlled/autonomous military aircraft. The initial prototypes may also include a pilot on board supported with familiar British Man-Machine Teaming and AI to assist the pilot and provide control of other drones in support of the fighter. Concurrently, Loyal Wingman (Mosquito[153]), an unmanned drone, is under development. The concept views the autonomous drones to fly alongside or slightly ahead of larger military aircraft and undertake various tasks including surveillance, electronic warfare, and laser-guiding weapons onto targets air-to-air or air-to-ground strikes with a high degree of autonomy.

The UK's experience with heavier drones is also noteworthy. Four QinetiQ Banshee Jet 80+ drones will serve the Royal Navy and the fleet will subsequently increase to 16. Banshee, a fixed-wing aerial drone, can undertake intelligence, surveillance, reconnaissance and electromagnetic operations. Heavy lift drones are also under development; Malloy T-600 (a large quadcopter) can carry 250 kg and fixed wing Windracer's Ultra can transport 100 kg over 1,000 km. T-600 can carry and launch a Stingray missile too. 20-30 Malloy heavy lift drones were delivered to Ukraine[154] too. Project Proteus is developing an uncrewed helicopter capable of dropping sono buoys to track and communicate about submarines. It will improve the anti-submarine warfare capabilities of the Royal Navy.

Ultra-persistent wide area communications systems like Zephyr and Aether are conceptual, solar-powered or electric unscrewed systems, designed to linger at high altitudes[155] (70,000 feet – stratosphere) for months at a time for surveillance and communications. They are also expected to be rerouted rapidly to a specific area of interest.

Nano Uncrewed Air Systems (nUAS) are small, autonomous drones designed for use in the field. Weighing less than 200 grams featuring high-quality full motion video cameras, thermal imaging, 3D scanning, mapping and surveying, these autonomous vehicles focus on faster and safer intelligence in around-the-corner or over-the-hill situations increasing awareness on the tactical battlefield in harsh and challenging conditions.

- The UK employs a variety of mini and nano drones[156] for its forces; 229 of the Desert Hawk III battery-powered small drones are in service. It is hand-launched and can fly for 60 minutes in a radius of fifteen kilometres. The Puma AE drone is a small drone capable of landing on water or land. It is primarily used for surveillance and intelligence gathering, but can carry communications relay and laser marker systems as well. Stalker VXE30 fixed wing, vertical take-off and land drones are latest in the inventory (150 have been ordered). It is designed to give situation awareness to small units with eight hours endurance and 90 km reach. The UK also purchased 159 Indago-4 drones with a high-resolution day-night camera which identifies objects, vehicles and weapons in a range of 12 km. The Defend Tex Drone40 (D40) is a nano UAV/loitering arsenal which can carry a small amount of munitions. It is in service with British forces since 2020 and operationally deployed in Mali.
- The famous and social media star, the Black Hornet nano drone, is employed extensively in the British military – 160 in 2013 (famed for

operations in Afghanistan), 30 in 2019 and 859 in 2022 (some for Ukraine).
- Skydio autonomous X2D drones have 360-degree obstacle avoidance, are night enabled and provide quick situational awareness. Its six 4K cameras build a 3D map of the surroundings and its AI with deep learning algorithms[157] can understand and predict future scenarios for better informed decision-making.
- A fist-sized drone Bug nano UAV[158] can be flown with a microphone (to act as a listening device), loudspeaker, white/infrared (IR)/red light, interchangeable lenses, mapping camera and thermal camera. It can fly in severe weather conditions including strong winds (35 kmph gusts), rain and snow. It can simultaneously transmit data to multiple devices – mobile phones, laptops, tablets and Android team awareness kit (ATAK). Follow the leader capability makes it swarm ready. It operates within a range of 2 km on a 5.8 Ghz radio bandwidth and up to 50 km on LTE, making it suitable for urban environments.
- Peregrine is aimed at 'protecting British interests in the Gulf' and 'will enable round-the-clock surveillance of targets over Gulf waters and will be available for a spectrum of operational tasks'. It is a Flexible tactical unmanned air system[159] (FTUAS) to operate seaborne platforms as a counter to all types of threats, including unmanned systems. Camcopter S-100 is an unarmed drone with a wide range of high precision intelligence, surveillance and reconnaissance (ISR) sensors.

The UK is testing two types of swarming technologies the first, a true swarm – a simple rule based behaviour which each participating member follows without a central control and; follow the leader swarm – individuals controlling multiple drones or GPS guidance to one location or control of one drone which controls other drones in flight. In October 2020, a number of small, remotely-piloted Callen Lenz drones with electronic decoys, operated in a swarm,[160] and created a jamming effect by working collaboratively to cause confusion. The UK is also testing Thor – an Israeli design of five autonomous swarms of six unmanned aircraft systems. Its AI and machine learning provides automated mission management and full autonomy to the swarm. A true swarm capability

Robotic Platoon Vehicles (RPV) programme intends to determine combat effectiveness of unmanned vehicles in improving the capabilities of dismounted troops at the platoon level. Rheinmetall Mission Master SP[161] has been chosen for the program. It is a low-profile, silent drive mode unmanned ground vehicle which stealthily follows soldiers anywhere. It is controlled by an autonomous suite of advanced sensors and perception algorithms to find the safest routes through dangerous environments and challenging terrain. The cargo carrier reduces soldiers' combat load, improves mobility and efficiency. It can transport supplies, tactical kits, and medical equipment either independently or in convoy mode with other mission master vehicles. The surveillance version can detect, recognise and identify targets at long ranges. VIKING is yet another autonomous UGV, a load carrier, capable of vision-based AI for terrain and object recognition, for mapping, routing and obstacle avoidance even in GPS-denied[162] environments. It is fitted with a diesel parallel hybrid power train, delivering 20 km of silent driving on battery only, and with a total range of approximately 200 km before refuelling.

The Tracked Hybrid Modular Infantry System[163] (THeMIS) is a tracked multi-role unmanned ground vehicle, which has proven itself during anti-insurgency missions in Mali. Its open architecture enables rapid configuration from transport (extra gear or firepower) to fire support (self-stabilizing remote-controlled weapon system, light or heavy machine guns, 40-mm grenade launchers, 30-mm autocannons or anti-tank missile systems), ordnance disposal (IED detection and disposal) to supporting intelligence operations (situational awareness, surveillance and reconnaissance over wide areas and battle damage assessment).

Maritime Demonstrator for Operational eXperimentation (MADFOX) is an autonomous uncrewed surface vessel designed for deployment in maritime surveillance and force protection missions[164] for the Royal Navy since March 2021. It can identify hostile ships and other threats or collect information and intelligence about targets, providing live high-resolution images. It can also be utilised for covert, long-endurance monitoring of lower-intensity threats, such as criminal groups attempting to smuggle by sea. In an exercise, it launched a Switchblade missile[165] on a target, indicating its weapon carriage capability for hit-and-run missions.

The Royal Navy Motor Boat (RNMB) is an uncrewed mine hunter[166] inducted into the Royal Navy. It can operate autonomously, pre-programmed or under remote control. It tows a side-scan sonar behind it to scout for mines on the seabed and alerts units operating ashore or at sea. The AI on it will make it compatible with underwater vehicles too. The last of the three, RNMB Hebe, has autonomous control capabilities of the other two vessels. All three vessels can communicate with each other to re-optimise the plan during mission, collating real-time information on new threats and performances.

The Royal Navy is also developing an XLUUV[167] (Extra Large Uncrewed Underwater Vehicle) – a 17-tonne, 12 metres long and 2.2 metres in diameter submarine, capable of diving deep and covering 1,600 km in seven days. AI enabled, the UK expected it to 'protect our critical national infrastructure and monitor sub-sea activity'. Operating in support of a Multi-Role Ocean Surveillance Ship (MROSS) to research and protect critical undersea national infrastructure, such as undersea cables (worth trillions of dollars of data transfer), and gas pipelines – a set of economic targets.

Managing the Change

The UK proposes several measures to attract and retain[168] the best AI talent within the military, knowing that multi-disciplinary AI development teams require a range of specialists, including data scientists, AI developers, machine learning engineers, and AI analysts. It is planning to make AI an attractive and aspirational choice for highly skilled professionals by creating a unified career management system in the military. The AI strategy also advocates the integration of AI awareness and understanding within military training programs, commencing from the foundation military training. All senior leaders will have to possess foundational and strategic understanding of AI and its implications for their organizations.

The UK wants to exploit AI to improve and accelerate decision-making and 'uncrewed' vehicles (such as aerial drones), to autonomously search, identify, and track targets. AI weapons systems are expected to launch defensive measures autonomously and quickly when the decision cycles become too burdensome for human operators. Man-machine teaming intends to reduce this cognitive decision load from humans. As a concept, autonomous systems are expected to play a support function to military personnel, relieving them of dangerous or repetitive tasks.

FRANCE

Firmly believing in domination of global AI ecosystems by the USA and China by developing and acquiring digital firms, France accepts there is an AI hegemony and tends to tread a cautious path. This ideological decision is also underpinned by the well-known French motto – Liberty, Equality, Fraternity (Liberté, Égalité, Fraternité) and of course the NATO umbrella it subscribes to.

The MoD AI Task Force[169] observes, 'The United States has GAFA (Google, Apple, Facebook and Amazon), Microsoft and IBM and China has BATX (Baidu, Alibaba, Tencent and Xiaomi), near and around San Francisco/New York and Beijing/Shenzhen, respectively, give a specific edge to both countries'. Believing in this domination, France considers its data as a strategic asset which has to be reinforced by a robust data governance system. The Armed Forces launched the ARTEMIS.IA (Architecture for Massive Information Processing and Exploitation from Multi-Sources and Artificial Intelligence), a data processing and artificial intelligence solution on 24 June 2022. France wants to possess an in-house and secure solution for information processing of big data and AI in an operational employment context.

In addition to data, the AI task force also observes the requirement of substantial computing power and storage capacity can be met by indigenous cloud computing, offering security and reliability in 'concentric circles.[170]' The relevance of exploiting information and cyber domain culls the need for continuous analysis of endless streams of data.

- A restricted private cloud to provide direct support to operations.
- A privatised but localised cloud with a trusted operator, offering collaborative storage space to government agencies and trusted civilian developers.
- A public cloud to support innovation by resource sharing, logistics, support and operational readiness.

From the French military standpoint, AI is usable in all military domains,[171] be it decision support for planning and execution or situational awareness by collaborative data-sharing during tactical operations. Special emphasis is on development of computer vision, distributed intelligence, automatic language processing, semantic analysis and data crossing. The French are also mulling on employment of robotics and autonomy to relieve human beings from 3D (dull, dirty and dangerous) tasks.

- The space imagery of the French Army is analysed with an indigenous tool by Preligens,[172] which has developed AI and cloud-enabled services for surveillance and monitoring. A failed takeover attempt by the CIA in 2020 indicates a high program capability. This AI analysis tool can detect departure of a nuclear submarine or a large armour concentration and immediately alert human operators, providing sufficient time for evasive actions.
- In reconnaissance, the use of an aerial drone as a remote sensor[173] can increase the observational scope of a ground robot. This enables the robot to anticipate obstacles or infiltrate nooks and crannies that are inaccessible to the ground platform.
- In surveillance,[174] the coordinated use of several mobile robots can provide better coverage of the site to be monitored. This reduces blind spots and maintains a degree of unpredictability in patrols.
- The Man-Machine Teaming (MMT)[175] project applies AI to combat aeronautics. It involves Dassault Aviation (aircraft developer) and Thales (human/system interface and sensors). With the aim to define future cockpits and independent systems, MMT will introduce innovative and autonomous cognition systems in decision-making through machine learning, by smart and learning sensors. The goal is to create a seamless integration between the pilot and the machine, allowing them to work together as a cohesive unit to achieve their mission objectives. This also includes optimising pilot workload management, decision-making under stress, and the design of cockpit displays and controls to maximize ease of use and situational awareness.
- Multi-Robot Cooperation,[176] which is similar to swarming, is utilized for the planning and automated assignment of tasks to different robotic systems. The approach offers a significant advantage in navigating through risky environments. For example, when attempting to reach a target in a hazardous area, a swarm of drones is more likely to achieve the objective compared to a single robot. This is because even if some members of the swarm are destroyed, the remaining drones can continue to operate and complete the mission.

- Mule robots[177] capable of following a human leader are being tested for induction. The desert tactical group tested Robopex GACI, capable of carrying a load of 750 kg for eight hours, at a speed of 8 km/h. This test is capable of creating autonomous weapons for France.
- In the field of communications,[178] the use of a system that can adapt in real-time based on data from monitoring satellites can help avoid failures and automatically re-plan operations during a mission if necessary.
- In the context of perimeter security[179] against multiple intrusions, the use of sentry robots can be combined to divide inspection points between them. This approach enables a faster response to threats.
- Despite the advancements in robotic technology, France has no plans to develop fully autonomous systems[180] that are beyond human control in the definition and execution of their mission.

UNITED STATES OF AMERICA

"Autonomy will be used only to empower humans, not to make individual or independent decisions on the use of lethal force."

– Deputy Secretary of Defence Robert O. Work, 2016

The USA believes that it faces significant international competition in the field of military artificial intelligence (unlike previous technologies) which could encourage others to rush the development without sufficient attention to safety, reliability, and humanitarian consequences. It is threatened by aggressive Chinese development of data integration systems from a wide variety of sensors to identify hidden targets, provide a common operating picture to commanders, decision making and planning tools to enable rapid decision making and precise targeting. The Russian use of AI technologies for information warfare operations, robotics, electronic warfare and cyber warfare is highly disquieting for American comfort.

A RAND[181] study (2018) and many others have shaped the USA aspiration of technologies and guide the AI development cycle too and also steered world opinions. It expects significant developments in the next two decades; thus, it is seeking greater technical cooperation and policy alignment with allies. It is also addressing supply chain disruptions by deliberating ethical behaviour of technologists, private sector, and American citizens – laying out massive public outreach. The USA wants to train operators in realistic environments to develop the appropriate levels of trust. In addition to China and Russia, AI threat from non-state actors is also disconcerting for the USA. To stay at the forefront of military AI capability development, the USA is concurrently exploring confidence-building and risk-reduction measures, following international discussions to discern official state positions of competitors over ethical use of military AI, especially LAWS.

The American use of AI in the military can be analysed in two ways – one is the 'tool analysis' sourced from well-known yearbooks/encyclopaedias, and the other is evaluation of threats, international narratives (churned by the omnipresent media) and examinations by the competitors. This research chooses the latter – with the clear aim to avoid getting trapped in the technology jargon of the incomprehensible and overwhelming data of the American systems (some seemingly science fiction), surprisingly easy to acquire. As a nation with clear global ambitions, it is worth a while to analyse the American military AI in relation to all things geo- (political, strategic and economics).

The USA created the Joint Artificial Intelligence Centre (JAIC)[182] in July 2018 to spearhead AI development, supporting it with AI and Data Acceleration (ADA) initiative, Defence Digital Service (DSS) to provide rapid technology solutions, Office of Advancing Analytics (ADVANA) and National Security Commission on Artificial Intelligence (NSCAI). The JAIC spearheaded creation capabilities to war fighters at the 'dirty tactical edge' of the 'dangerous challenges of war fighting environments'. It focussed on creating special and new operating models that that 'purposely fit together', to automatically generate and widely share situational awareness from 'any sensor to any decision-maker'.

To assist the JAIC, ADA[183] was set up to create formal methods to harness data and AI, which was then offloaded to 'operational data teams' (basically specialised teams on contract) that would visit all 11 combat commands to quickly catalogue, manage and automate data feeds, since the combat commands have their typical operation responsibilities and may not be able to curate data. These teams were expected to embed to ensure data is adequately captured, curated and usable. In addition, there were 'AI flyaway teams', – military customer care – which could fly to the commands facing issues with updating network infrastructure to remove policy barriers and ensure the and effectiveness. This did not work as expected, since the USA collates massive amounts of data distributed all over the world so, a large part of their job was to figure out ways to manage that data – what is effective for whom and how.

In spite of all the quick fixes, within four years, realising the stove piped working nature of military organisations as restrictive, the JAIC and other agencies were subsumed[184] by CDAO (Office of the Chief Digital and Artificial Intelligence Officer) in June 2022. This reworking is expected to create a single decision stack that permits integration of data, software and AI to optimize and 'consider them more holistically.' This also bulldozes through service specific turf wars under the garb of security of information. The head of the CDAO is a technocrat, who hopefully will not succumb to his past service specific experiences. Top-down decrees seem to do well with the USA as they realised with the Goldwater–Nichols Act (1986), and the creation of the CDAO is expected to get AI work done similarly.

The USA's AI problem statements are very different from other case studies. For one, how would the military curate data to provide its forces an all-domain advantage and the tactical edge outside continental US[185] (OCONUS)? The USA considers mobilizing all domain forces to respond to an adversarial threat around the globe as an essential function of national security and global responsibility. Extension of Continental United States (CONUS) cloud computing capability is a vital support to the deployed forces and the data has to be secure, not redundant, reliable and persistently available to retain the tactical edge in all theatres. The USA is addressing these challenges through AI controlled software based communications and IT infrastructure, including LEO satellites on demand. The USA has contracted Joint Enterprise Defence Infrastructure, estimated to be $ 5 billion[186] a year over two years. In comparison, Amazon's entire cloud computing revenue is $ 20 billion annually and Google's is around $ 4 billion.

Global Information Dominance Experiments[187] (GIDE) is another attempt of the globally integrated experiment enabled by data, analytics, and artificial intelligence (AI). Involving many combat and civilian commands it provides joint data integration solutions using AI and ML, with a specific aim to 'rapidly improve data access to all – from the 'strategic level to our tactical war fighters'. To overcome critical supply chain risks, the experiments also want to 'stress-test our current systems and processes' and 'learn in an experimentation environment to replicate in real-world operations'. A holistic approach to these experiments will also identify barriers from the host nation and adversaries in policy, security, connectivity, and reasons that prohibit data sharing across the joint forces.

Some creative use of AI in the USA are studied to substantiate the analysis.

- A very famous and controversial PRISM[188] programme is run by the National Security Agency (NSA), an Intelligence agency, that is responsible for cryptographic and communications intelligence and security. Established in 1952, it underpins the importance and distinct character of all communications intelligence. While free to collect foreign intelligence, the agency can obtain a warrant to intercept domestic communications too. It acquires information of internet usage from internet providers and collects a humongous data. Obviously, the NLP is the 'poster child[189] of successes' in this institution.
- Project Maven utilises industrial image-recognition and object detection technology. Large volumes of image and video data which the USA collects routinely, overwhelms human analysis capabilities. AI is needed to prioritize and filter results as data come in. Within just six months from initiation, the project's cross-functional team[190] was able to assist in intelligence processing and actual operational use in the fight against the Islamic State of Iraq and the Levant (ISIS).[191]
- The Defence Advanced Research Projects Agency (DARPA) is committed to Squad X[192] experimentation program. The program aims to enhance squad awareness and engagement capabilities without overburdening soldiers. It provides infantry and marine squads with situational awareness, adaptability, and flexibility in complex environments via ground and air-based autonomous sensors. This capability will allow soldiers to better control their challenging mission situations – a human machine teaming. The program focuses on four key technical areas[193] – precision engagement, non-kinetic engagement, squad sensing, and squad autonomy. The sensors autonomously reconnoitre[194] the mission area to acquire maximum inputs and keep the squad updated. Soldiers access the AI-cleaned operation data via body worn android devices. Final capabilities will include guided munitions, electronic surveillance and autonomous threat detection, completing the collaboration between humans and unmanned systems.
- The Loyal Wingman program[195] is expected to operate as extensions of a pilot in another aircraft who can command them to perform dangerous tasks. The USA will definitely utilise its experience with the Loyal

Wingman to develop swarms of autonomous systems that do not require pilots. The swarms are expected to recognise targets and adapt in contested battlespaces (some members of the swarms will be lost to battle attrition, but others will squeeze through towards their targets – eventually) and operate as a collaborative one in denied spaces (in the absence of GPS or satellite communication), partnering within the swarm to provide situational awareness of the operating environment.

- The US Army is taking an unconventional approach to AI with army futures command's project convergence experiments. Instead of a centralized AI, a partnership[196] of more specialized AIs, each designed to perform in a smaller and yet collaborative manner was tested. This approach allows for greater flexibility and adaptability, as well as a more efficient use of resources. While one AI populates all the digital maps in the combat zone, another AI (Firestorm) warns troops of threats, sending precision targeting data, and in some cases even aiming their vehicles' weapons at the enemy. By using AI to assist soldiers rather than replace them, human judgement remains a critical component during combat.

As with any global player, the USA is sensitive to its image and future credibility, advocating correct usage of AI and its associated risks[197] continue to dominate the minds of decision makers. It is also affected by non-military threats – legal, ethical and financial – hence, there are overt and loud attempts from the USA to stay overboard and ahead in the field of military AI.

- Managing internal factors, the USA understands AI risk lies in the potential technical failures and the risk of misuse by inexperienced soldiers.
- Operational risks refer to the reliability and security of AI systems much like nuclear weapons – Always and Never (always correct employment and never without instructions). Trusting the AI system to work in uncontrolled environments is a difficult decision since its development is based on controlled tests in controlled laboratory conditions. The pressure to rapidly deploy in competition with international players accentuates the accidental risks. Integrity and reliability of the learning algorithms can be targeted by hacking and data-poisoning of ML systems, which can easily fool the AI systems. The USA wants to possess

robust, resilient and trustworthy AI to ensure long-term strategic advantages. Towards this, the USA has prioritized traceability of the AI projects to ensure that ML systems continue to learn from dynamic inputs in real-time, and not get stuck in archaic training datasets. This concern is also addressed by America's interest and explains the emergence of XAI.

- The USA's comments on strategic risks[198] refer to challenges that affect the stability of the international order. Unmanned vehicles could lower the threshold for military action and replace non-military options. Automated operations at machine speeds are likely to increase miscalculation and misunderstandings.

The USA is advocating the threat of dual use technologies which arise out of combined AI R&D between military, private sector and non-governmental researchers, which create and use open-source tools, which have dual-use capabilities: cyber capabilities and are live case studies of this nature.

Acutely aware of AI development speed and its pluralism, the USA continues to seek collaboration with other countries, such as China and Russia, on AI safety and maintaining crisis communication protocols – much like SALT and START. This communication also gives a look into the development vulnerabilities (technology, ethical and legal) of the adversaries, which could be exploited at a correct point of time and stymie the competition.

TAIWAN

Though Taiwan is not considered a revisionist state, but since it is experiencing an existential threat from China, its defensive strategy is significantly under pressure. The supremacy of the Chinese military and proclivity to win an annexation battle is well accepted. Taiwan's concerns rests on the promised military support from its allies – in time, since Ukraine's assured help did not fructify in the manner expected.

Taiwan recognises its technological supremacy and is getting ready to fight it out with the best possible outcomes, and is obviously relying on its indigenous capability. The MND (Ministry of National Defence) is paying close attention to the character of global conflicts and considers UAVs and autonomous technologies as part of multi-layered deterrence. AI asymmetric warfare[199] is being considered as an option in Taiwan against Chinese occupational threat.

Two different theses are highly instructional in propounding the Taiwanese defence against Chinese aggression. One by Ian Easton in September 2014 and another by Steven D. Bernstein in December 2020. The latter is of special interest since there is explicit utilisation of AI in the defensive strategy. Both the studies have considered wide ranging use of UAVs for reconnaissance, weapon delivery to reduce load from fighter jets, acting as decoys to aggressor radars and logistic support to troops in contact. This comes with the assumption that Taiwan is putting its best resource – industrial and technology edge to full use. Starting from the top, Taiwan passed a parliament regulation – 'Unmanned Platform Technology Innovation Experiment' to encourage R&D, testing and building all types and classes of unscrewed vehicles, specifically AI-boosted unmanned platforms and man-machine teaming.

Information Warfare. Both sides will extensively use AI-powered applications, such as cyber security, intrusion detection, cyber threat identification, and tools for disinformation, hate speech, and information warfare. Taiwan also has the potential to expose the Chinese population to information beyond state propaganda, potentially disrupting China's information control[200] and destabilising its society. This focus on domestic security may also distract the PLA from offensive operations.

Reconnaissance. Taiwan would do well to identify the Chinese preparation in run up to the hostilities well in time. Taiwan working with the USA will obviously create resilient systems, but will largely depend on a combination[201] of both active and passive defence elements. In the Taiwan context, ISR is critical. With a continental hostile neighbour, Taiwan cannot afford to miss any information, much less conclude an incorrect analysis. It is thus critical that sufficient warning, including enemy order of battle details are obtained through all ISR assets – radar systems, signals intelligence (SIGINT), cyber reconnaissance and human intelligence networks for indications of hostile actions. Though Taiwan's HUMINT resources are considered famously credible, but expanding upon aerial reconnaissance capabilities is still crucial to prevent the PLA from having the element of surprise. This aspect is important to redeploy its meagre resources optimally and secure the leadership from a decapitation strike.[202]

- AI-assisted UAV/USV could help Taiwan to distinguish feints from main operations, and decoys from legitimate targets. Teng Yun, an indigenous UAV, equipped with autonomous navigation landing, is not designed to be a fully-autonomous drone and requires 10 pilots.[203] Future improvements relying on AI and man-machining teaming aim to reduce the number of pilots by giving more task load to the AI in the machine.

- AI-powered machines will sort and interpret multi sensor reconnaissance data for analysis quickly and accurately – generating anomalies, alerts, and warnings.
- Taiwan's National Space Organization (NSO) designed and built Formosat 7[204] which has powerful imaging, remote sensing and reconnaissance capabilities sorted and categorized by AI.

Missile and Air Defence. Taiwan will have almost zero reaction time to intercept PLA missiles (seven minutes to reach Taiwan). The counter battery fire will obviously be overwhelmed by the missile barrage and will have to include smart AI solutions to destroy maximum targets. This task of selecting, prioritising, and targeting will be beyond human capability and has to be assigned to AI to enable autonomous systems.

- Sky Bow III, SAM anti-ballistic and anti-aircraft missile defence system is equally good at intercepting missiles and aircraft. The Sky Bow III integrates AI[205] to pinpoint and intercept enemy missiles much quicker than a team of human analysts can. In fact, it can self-activate if it detects an incoming missile and hence fully autonomous.
- Deploying drones as suicidal decoys[206] is a 'cat in a corner' option for Taiwan, distracting PLA missiles from critical and limited quantity F16 fighter jets. Although still theoretical, swarm warfare may also be used to create this electronic seducing.
- In an extreme situation an AI managed UAV swarm attack on enemy radars[207] will overwhelm the precision strikes of the PLA.

Combat Missions. Taiwan will eventually have to fight it out on the sea, the beaches, the urban areas and finally guerrilla type resistance in the mountains. In all these stages, the ultra-technological savvy nation has numerous AI exploitations to undertake and hold off till external support reaches.

- AI-assisted Hsiung Feng IIE (HF2E) surface-launched cruise missiles are accurate enough to target spectacular targets in Beijing and Shanghai. Other missiles like Yun Feng, and HF-3 missiles are also equipped with

infrared homing and fully-autonomous target recognition for terminal guidance, provided there is sufficient quality safety from initial PLA strikes.
- PLAN may surprise Taiwan like the Allies did at Normandy during World War II. In such a scenario, Taiwan's, traditional meagre resources, already stretched, may not be flexible enough to entrench themselves ahead of new threat. Smart sea mines[208] will come to the fore here. AI enabled, with ability to differentiate one type of ship from another before detonating, mobile, rather than stationary will continue defence of the beaches in such a scenario. Development of autonomous miniature submarines[209] and drone mines with propulsion and remote sensors in sufficient numbers may complicate PLAN amphibious operations a lot and give more time to Taiwan.

Operational Logistics. Taiwan's AI-powered logistics, quality control, Smart City infrastructure[210] and computation systems will enhance logistic stamina. Major roadways, bridges, and tunnels around Taiwan will be destroyed by the PLA. Unmanned vehicles (ground, air, undersea) would be critical for Taiwan's defensive strategy. Instead of utilising manpower on logistics; resilient and AI-powered unscrewed vehicles could support[211] the remaining naval vessels and isolated troops fighting on the beaches, urban areas and mountains – evading PLA barricades and strongholds.

ISRAEL

Denying the enemy its fire capabilities will remove the threat it poses on Israel. Negating the threat will give Israel significant strategic freedom of action and thwart enemy rebuilding efforts after the war.

– Brigadier General Eran Ortal
Commander Interdisciplinary Military Studies, IDF

Israel believes that data and AI have a major role in winning future conflicts and wants to systematically incorporate AI across the military. The geostrategic location and neighbourhood realities have pushed the tiny country to continually examine a vast amount of varied data from sensors and sources, transform into actionable intelligence and most importantly deliver a precise summary (specific points of reference and interest will always vary from one user to other) to those who need it, making the defence forces more effective, faster and efficient.

Israel released a multiyear restructuring of armed forces plan – Tnufa (Momentum) in 2019. The plan envisions successfully fighting a multi-front war,[212] harnessing latest technologies including AI and bring 'swift and massive use of force against enemy systems' while denying 'enemies the ability to communicate or resupply'. An earlier construct of 'asymmetric warfare against inferior forces', was in a manner of conduct a self-evoked restriction to utilise the full power of the state against adversaries. The Tnufa Plan now recognises its 'enemy as an advanced, well trained, networked adversary that presents Israel with an operational challenge that serves enemy strategy'. Though acknowledged years ago by many, Israel, through the new plan, accepts the diminishing returns of deterrence operations, which is in fact 'inoculation for the enemy by gradually exposing him to limited doses', making the enemy more resilient. While the limited operations continue to be considered an alternative, a decisive victory is expected to defeat a terror army quickly, at an acceptable cost to forces, citizens and infrastructure. The practicality of the plan and its ability to rework the defence forces for the future without affecting the present is unique. It lays out a definite and non-negotiable approach for a decisive victory against all adversaries (including asymmetric), providing a practical outline for the necessary force design.

Firstly, multidomain[213] combat-integration of air, land, sea, intelligence, electro-magnetic and cyber forces are necessary to operate in a complex environment infested with data-rich, civilian intensive urban areas, legal and psychological domains. The forces should be able to process information and strike in closed coordination with tactical ground elements. More organic capabilities at tactical level will allow better manoeuvrability across all domains at the end point. Secondly, smarter sensing and communication will allow the tactical forces to undertake quick operations 'without harming its immediate readiness for war and without demanding impossible budgets' – a combination of 'tactical reconnaissance UAVs, connected to joint databases and effective information extraction systems'. Israel believes this 'smart suit' as affordable and a quicker solution to upgrade its traditional capabilities. Thirdly, conscious of multi rocket attacks on its citizens from all borders, Israel wants to enhance its anti-rocket defence capabilities, and possess sufficient advance notice to 'prevent its enemy from continuing to fire'. The Tnufna wants to create quicker and more precise ability to locate enemy rocket firing forces. The 'moment of rocket firing' is the window Israel wants to target to ascertain the location. The tactical reconnaissance by organic UAV and ground-based systems will decipher enemy combat communication, interconnected data and advanced information will be processed more rapidly, aiding attack on more targets quickly and accurately. Quick identification and simultaneous destruction will neutralize the combat capabilities of fire-based adversaries.

Why does Israel need AI? More than any other nation, its survival forces it to develop technology and AI to make a real impact and improve effectiveness and efficiency of the IDF and ensure existence. The complex, networked adversaries have led to an analysis of massive data from multitude devices. Data science teams are working to organize, classify, and create valuable insights for field operations. AI comes to help here, capturing humongous data through flagging suspicious activities autonomously and predicting future possibilities. Israel claims to have mastered the first stage of descriptive AI,[214] where the systems can now understand context and identify and classify data. The next stage of predictive AI, where a system can identify the physical impact of the analysed insights, is underway. This capability will allow preventive operations to target the adversary before it can build up the capability to threaten in the field. This survival is guiding Israel AI development across the spectrum.

The circulating threats around the State of Israel maintain a constant state of operational friction with adversaries. Soldiers on the ground, due to this constant friction, demand high speed innovative solutions from the developers. The advanced technological solutions receive fast and concrete feedback from the field regarding the applications that are being developed and their integration into operations. This unique demand-supply-validation relationship is persistently creating AI solutions as it has done for all previous developments in Israel.

AI at Rest. AI at Rest systems powered the function virtually, within a software-based ecosystem. These comprise expert and planning, natural language processing and image classification systems. Such systems operate virtually in a machine to aid decision-making. It is apparent in domains such as data collection, analysis, web searches and forecasting.

- AI played a key role during 'Guardian of the Walls' operation against Hamas in Gaza in May 2021. The Israel Defence Forces (IDF) of Israel relied heavily on machine learning and data gathering over the preceding weeks that allowed Israel to collect vital targeting information.[215] Advanced AI 'centralised all militant data on one system' to help analysis and intelligence, 'prioritising knowledge building ML systems'. Special programs like 'Gospel' recommended 'quality targets' which the Israeli Air Force successfully struck. AI algorithms predicted enemy rocket launches, place, and time, helping soldiers to target them on time.

AI in Motion. AI in Motion pertains to systems operating in the physical realm, like robots and autonomous vehicles. These systems involve the use of actuators to articulate actions in the physical world, to make continual decisions while in motion, which are then immediately implemented in the physical environment, taking into account various factors such as own state and the operating environment. Self-driving cars and UAVs are examples of AI in Motion. For Israel, the asymmetric forces manifest as low-signature light infantry combined with proxy warfare, and long-range fires as a primary tactical and strategic tool, thus necessitating parallel development of AI in Motion.

- Twin-gun autonomous turrets[216] are installed on top of a guard tower overlooking the crowded al-Arab refugee camp in the West Bank. It fires non-lethal ammunition and its purpose as per Israel is to protect soldiers and civilians better by enhancing the accuracy of hitting the right target.
- AI-Powered Smart Shooter – SMASH – is yet another concept in which AI magnificence has converted each rifleman into a sniper, where each round counts and collateral fatalities are eradicated. The human shooter selects the target through advanced image processing sight, which recognises and locks on the target, predicting its movements. SMASH stays locked on the target even when it moves. The human shooter presses the trigger, while following the target through the sight, the system fires the bullet[217] at the appropriate time to accurately hit the target, each and every time.
- Spice 250 with a 75-kilogram warhead, deployable wings and a maximum range of 100 km is already a futuristic under wing bomb of warplanes like the F-16 with AI alongside automatic target recognition with scene-matching technology making it an ideal weapon for EM-denied environments. The new upgrade uses autonomous capability to

navigate and correct its location. DL gives an ability to distinguish moving ground targets in a cluttered battle field, permitting selection of secondary targets autonomously, if required. New bombs can learn from the Smart Quad Rack[218] on the aircraft wings which extract the earlier launch data – a true AI, learning on the go.

- Project Carmel aims to reduce the number of soldiers inside armoured vehicles from the usual four to two while enhancing mission performance. An ideal example of task automation rather than job replacement, Israel is employing AI to provide better outside optics to the scaled down crew providing them better target acquisition support system and autonomous driving capabilities. The crew just points and click[219] on a screen and the vehicle identifies the best route or waypoints along the route, leaving the crew unencumbered for the critical task of target destruction.
- Israel's experiences, some not so good in the urban conflicts at Gaza or Lebanon, have given birth to a networked sensor-to-shooter system[220] – Fire Weaver. As the name suggests, it gives highly accurate common visual aid in targeting for commanders and soldiers. Advanced computer vision technology and AI connects forces on the battlefield, gives the view of the same battlefield and targets from different angles and recommenders the best shooter for the target. This system will permit Israeli forces to operate in civilian congested urban environments, without fear of collateral damage.
- The famous Iron Dome[221] uses AI to predict if rockets or missiles will land on the population or critical assets. If it senses a credible threat, the AI system directs launch of an interceptor or a faster missile against the danger, destroying it in flight at a high altitude or on a neutral area.
- The death of Iranian nuclear scientist, Mohsen Fakhrizadeh[222] in November of 2020, had caused a huge stir. Iran and Israel have not claimed it to be a remote assignation, but the open source media has provided a logical line up of the likely AI system used for the assignation. He was reportedly killed with a remote-controlled gun, operated from Israel about a thousand miles away. Imagine the technical issues, if controlled from Israel, it had to have a lag time of 1.6 seconds; in other words, from beaming the target image to Israel, gunner to take action

and the weapon to fire. AI enters here which can estimate the future location of the target, provide a corrected aiming point to the gunner to press the trigger and hit the future location.

NON-STATE ACTORS

The adoption of AI technology by non-state entities raises apprehension and poses a potential threat. It may empower them to augment their capabilities and execute attacks more efficiently. Internet and social media platforms can turn into potent instruments in the toolkits of terrorists and militant groups for spreading propaganda, radicalization, and conspiracy theories aimed at destabilising societies. AI-powered data scraping provides the study material and potential targets.

- Use of unmanned aerial systems by terrorist groups and organized criminal networks for reconnaissance, cross-border drug trafficking and arms running can support terrorist attacks.
- Non-state actors could use AI to plan and carry out attacks, recruit new members, and spread specialised deep fake produced propaganda.
- In cyberspace, AI allows hacking groups to identify online vulnerabilities and extort companies or individuals.
- A highly enabled non-state actor could also use AI to physically target a high value individual or minority groups within a country.

Though still speculative due to the technical underpinning of the tool, the threats are not far in the future and responsible countries are already underway to stem these nefarious attempts.

KEY TAKEAWAYS

- All countries are employing AI to study the adversary continuously, countries like Israel overtly, others covertly, but surely.
- Pattern of Life Analysis is a natural outcome of such AI analysis – though all countries will not objectively agree.
- All countries are using AI to firstly collect data from multiple resources, fuse it and then weed out the noise.
- Countries want AI to support the decision makers – reduce the decision burden by allocating priorities.

- AI-based autonomous weapon release is being worked out by countries like Russia, China, and the USA during critical times, when humans could be too slow.
- AI will power both cyber defence and offence for all countries – civilian military cooperation is clearly visible – prone to legal and international pressures and deniability.
- Human Machine (Man-Machine) teaming is worked out by all countries in all domains, to increase combat mass without risking humans.
- Swarming is a sure pathway being followed by all countries. Russia and China have prioritised swarming for electronic warfare and reconnaissance; other countries have prioritised it for increased mass.
- Russia has prioritised development of combat robots (still not fully AI), underpinning its adversarial concern for NATO on its western borders.
- Russian field trials in Syria, give it an edge and confidence to employ them in future battles.
- Chinese emphasis on aerial UAV indicates out-of-area preparation across vast seas and land borders against the USA and NATO.
- Chinese interest in surface and subsurface unscrewed systems, provides it a huge stealth capability in the vast oceans.
- Iran has utilised some form of robotics and AI to bolster its claim and 'ability'. Typical development like 'suicide robot' is a new dimension of hybrid solutions to tactical problems, and is a unique nature of Iran's military AI.
- Australia has emphasised self-reliance to cater for its geographic isolation.
- Large emphasis on man-machine teaming by Australia is expected to cater for distant operational areas and huge attack surfaces gifted by geography to the island country.
- France follows NATO's lead; however, it is concerned over the data privacy issue and may thus delve more on data protection and sharing.
- The UK emphasises on tactical level AI to empower its soldiers at the pointy end of the battle – fully confident of the NATO and allied backbone cover.
- Taiwan would employ AI to increase resistance against Chinese aggression – buying time for the allies to physically join.
- Non-state actors can manufacture AI at Rest to create cognitive database and attack options – open-source codes abound for this purpose.

- The USA will continue to drive international debates to stymie competition, not necessarily following self-moratoriums as in previous cases.

CONCLUSION

Delving into AI strategies and implementations of militarily significant nations, this chapter explores their approaches beyond technological prowess, scrutinising how these countries navigate hierarchical structures, compliance complexities, and the imperative for responsible AI deployment in military operations. Rooted in diverse geopolitical contexts, this examination unveils patterns of success, identifies lessons, and potential areas for improvement in the military applications of AI. The goal is to provide insights into the nuanced strategies employed by these nations to meet Indian military requirements, offering a panoramic view of their current capabilities and forecasting the trajectory of AI integration in future military scenarios.

The next chapter will shift the focus to a different arena, examining the AI development capabilities of Indian developers. As a nation with a burgeoning tech landscape and a commitment to strategic autonomy, India's foray into AI presents a unique perspective. The innovation, challenges, and potential breakthroughs within the Indian AI landscape contribute to the broader discourse on AI's global trajectory. The juxtaposition of global case studies with a focused examination of India's AI development capabilities will enrich understanding of the diverse dynamics at play in the evolving world of AI.

NOTES

1. Manish Kumar Jha and Amit Das. AI Technology in Military Will Transform Future Warfare, 13 August 2021, https://www.businessworld.in/article/AI-Technology-In-Military-Will-Transform-Future-Warfare/13-08-2021-400525/, accessed on 10 November 2021.
2. https://www2.deloitte.com/ca/en/pages/deloitte-analytics/articles/age-with-ai-advantage-defence-security.html, accessed on 15 December 2022.
3. Eray Eliaçýk. Guns and Codes: The Era of AI-Wars Begins, 17 August 2022, https://dataconomy.com/2022/08/how-is-artificial-intelligence-used-in-the-military/#:~:text=Artificial%20intelligence%20is%20used%20in,driven%20systems%20in%20the%20military, accessed on 9 December 2022.
4. Tejaswi Singh and Amit Gulhane. 8 Key Military Applications for Artificial Intelligence, 3 October 2018, https://blog.marketresearch.com/8-key-military-applications-for-artificial-intelligence-in-2018, accessed on 23 March 2022.
5. Congressional Research Service. Artificial Intelligence and National Security, 10 November 2020, https://sgp.fas.org/crs/ natsec/R45178.pdf accessed on 11 June2022.
6. Ibid.

7   Eray Eliaçýk. Guns and Codes: The Era of AI-Wars Begins, 17 August 2022, https://dataconomy.com/2022/08/how-is-artificial-intelligence-used-in-the-military/#:~:text=Artificial%20intelligence%20is%20used%20in,driven%20systems%20in%20the%20military, accessed on 9 December 2022.
8   Congressional Research Service. Artificial Intelligence and National Security.
9   Tejaswi Singh and Amit Gulhane. 8 Key Military Applications for Artificial Intelligence, 3 October 2018, https://blog.marketresearch.com/8-key-military-applications-for-artificial-intelligence-in-2018 accessed on 23 March 2022.
10 Manish Kumar Jha and Amit Das. AI Technology in Military Will Transform Future Warfare, 13 August 2021, https://www.businessworld.in/article/AI-Technology-In-Military-Will-Transform-Future-Warfare/13-08-2021-400525/, accessed on 10 November 2021.
11 Eray Eliaçýk. Guns and Codes: The Era of AI Wars Begins.
12 Shraddha Goled. 'What are the Scope and Challenges of Using AI in Military Operation, 1 November 2020, https://analyticsindiamag.com/what-are-the-scope-and-challenges-of-using-ai-in-military-operations/, accessed on 23 March 2022.
13 Congressional Research Service. Artificial Intelligence and National Security.
14 Tejaswi Singh and Amit Gulhane. 8 Key Military Applications for Artificial Intelligence'.
15 Pavel Luzin. Artificial intelligence in the Russian Army, 19 March 2021, https://ridl.io/artificial-intelligence-in-the-russian-army/, accessed on 13 February 2023.
16 Samuel Bendett. Red Robots Rising: Behind the Rapid Development of Russian Unmanned Military Systems, 12 December 2017, https://thestrategybridge.org/the-bridge/2017/12/12/red-robots-rising-behind-the-rapid-development-of-russian-unmanned-military-systems, accessed on 18 February 2023.
17 Samuel Bendett. Here's How the Russian Military Is Organizing to Develop AI, *Defence One*, 20 July 2018, https://www.defenseone.com/ideas/2018/07/russian-militarys-aidevelopment-roadmap/149900/ accessed on 15 February 2023.
18 Samuel Bendett. The Russian Military Wants Students to Design Its New Underwater Drone', 7 March 2020, https://warisboring.com/the-russian-military-wants-students-to-design-its-new-underwater-drone/, accessed on 18 February 2023.
19 Presidential Address to the Federal Assembly at Saint Petersburg, Russia, on 1 March 2018, http://en.kremlin. ru/events/president/news/56957, accessed on 18 February 2023.
20 Aleksey Nikolsky. Lofty Goals, https://forceindia.net/cover-story/lofty-goals/, accessed on 19 February 2023.
21 Lee Sullivan. Artificial Intelligence in Russia, 20 February 2022, https://geohistory.today/ artificial-intelligence-in-russia/, accessed on 19 February 2023.
22 Anya Fink. Book Review, Russian Studies Series 5/21, 'Russian Thinking on the Role of AI in Future Warfare, 8 November 2021, https://www.ndc.nato.int/research/research.php?icode=712, accessed on 13 February 2023. Review of V.M. Burenok, "IskusstvennyyIntellekt v VoennomProtivostoyaniiBudushchevo" ("Artificial intelligence in the military confrontation of the future").
23 Bybelezer, C. (2018), 'How Russia is using Syria as a military 'guinea pig', *The Jerusalem Post*, 28 February 2018, https://www.jpost.com/Middle-East/How-Russia-is-using-Syria-as-a-military-guinea-pig-543839, accessed on 18 February 2023.
24 Margarita Konaev and Samuel Bendett. Russian AI-Enabled Combat: Coming to a City Near You, 31 July 2019, https://warontherocks.com/2019/07/russian-ai-enabled-combat-coming-to-a-city-near-you/, accessed on 13 February 2023.
25 https://www.globalsecurity.org/military/world/russia/uran-9.htm, accessed on 21 February 2023.
26 Rajesh Uppal. Russia Deployed Family of Killer Robots, IDST, https://idstch.com/military/

army/russia-developing-family-of-killer-robots-conduct-war-games/, accessed 10 February 2023.
27. Ibid.
28. https://ria.ru/20171017/1507003664.html, accessed on 21 February 2023, translated by Chat GTP.
29. Samuel Bendett, Stephen Blank, Joe Cheravitch and Michael B. Petersen. Russian Unmanned Vehicle Developments: Syria and Beyond, Research Report, Centre for Strategic and International Studies (CSIS), 2020, https://www.jstor.org/stable/pdf/resrep24241.9.pdf?refreqid=excelsior%3 Acb84ae3db 9884c6124737659bae8e07b &ab_segments=&origin=&initiator=&acceptTC=1, accessed on 16 February 2023.
30. Ibid.
31. Ibid.
32. Zack Whittaker. 'Documents reveal how Russia taps phone companies for surveillance', 18 September 2019, https://tcrn.ch/32MaCYb, accessed on 21 February 2023.
33. Stephanie Petrella, Chris Miller, and Benjamin Cooper. Russia's Artificial Intelligence Strategy: The Role of State-Owned Firms, FPRI, *Orbis*, volume 65, issue 1, November 2020, pp 76, https://www.fpri.org/article/ 2021/01/russias-artificial-intelligence-strategy-the-role-of-state-owned-firms/, accessed on 15 February 2023.
34. Margarita Konaev. Russian AI-Enabled Combat: Coming to a City Near You, ibid.
35. https://www.eeas.europa.eu/search_en?fulltext=artificial+intelligence, accessed on 16 February 2023.
36. Military Applications of Artificial Intelligence: Ethical Concerns in an Uncertain World, | RAND, https://www.rand.org/pubs/research_reports/RR3139-1.html, accessed on 24 January 2023.
37. Ibid.
38. Pavel Luzin. Artificial intelligence in the Russian Army.
39. Samuel Bendett, Stephen Blank, Joe Cheravitch and Michael B. Petersen. Russian Unmanned Vehicle Developments: Syria and Beyond.
40. Timothy Thomas Russia's Electronic Warfare Force-Blending Concepts with Capabilities, September 2020, https://www.mitre.org/sites/default/files/2021-11/prs-19-2714-russias-electronic-warfare-force-blending-concepts-with-capabilities.pdf, accessed on 18 February 2023.
41. Alexey Ramm Bogdan Stepovoy. Sees the target:—without the participation of the operator', 16 April 2020, https://iz.ru/1000101/aleksei-ramm-bogdan-stepovoi/vidit-tcel-bylina-smozhet-atakovat-protivnika-bez-uchastiia-operatora, accessed 16 February 2023.
42. Research Paper, Chatham House. 23 September 2021, 06 Military applications of artificial intelligence: the Russian approach | Chatham House—International Affairs Think Tank accessed on 23 March 2022.
43. Timothy Thomas. Russian Robotics: A Look at Definitions, Principles, Uses, and Other Trends, *MITRE*, February 2021, https://irp.fas.org/eprint/russian-robotics.pdf, accessed on 13 February 2023.
44. https://www.c4isrnet.com/unmanned/2019/07/31/is-this-russias-gateway-drone-to-better-armed-robots/, accessed on 17 February 2023.
45. Rajesh Uppal. Russia Deployed Family of Killer Robots.'
46. https://www.globalsecurity.org/military/world/russia/soratnik.htm, accessed on 21 February 2023.
47. Rajesh Uppal, 'Russia Deployed Family of Killer Robots'.
48. Margarita Konaev. Russian AI-Enabled Combat: Coming to a City Near You, ibid.
49. https://andrei-bt.livejournal.com/949786.html, accessed on 21 February 2023.
50. http://www.army-guide.com/eng/product.php?prodID=5525&printmode=1, accessed on 21 February 2023.
51. https://www.armyrecognition.com/april_2014_global_defence_security_ news_ uk / russian _

army_to_use_unmanned_ground_robot_taifun-m_to_protect_yars_and_topol-m_missile_sites_2304143.html, accessed on 21 February 2023.
52 Sharathkumar Nair, A closer look at Russia's AI-powered artillery, 26 January 2022, https://analyticsindiamag.com/a-closer-look-at-russias-ai-powered-artillery/, accessed on 21 February 2023.
53 Margarita Konaev. Russian AI-Enabled Combat: Coming to a City Near You.
54 Roger McDermott. Russian UAV Technology and Loitering Munitions, 6 May 2021, https://www.realcleardefense.com/articles/2021/05/06/russianuavtechnology_and loitering munitions_775980.html, accessed on 21 February 2023.
55 https://www.globalsecurity.org/military/world/russia/soratnik.htm, accessed on 21 February 2023.
56 https://tass.com/defense/1277657, accessed on 18 February 2023.
57 Rajesh Uppal. Russia Deployed Family of Killer Robots.
58 RAND. 'Military Applications of Artificial Intelligence.
59 Melissa Heikkilä. AI: Decoded: Putin's high-tech war— Making sense of AI systems—Deepmind controls nuclear fusion reactor, 23 February 2022, https://www.politico.eu/newsletter/ai-decoded/putins-high-tech-war-making-sense-of-ai-systems-deepmind-controls-nuclear-fusion-reactor-2/, accessed on 13 February 2023.
60 https://ria.ru/20171015/1506649786.html accessed on 17 February 2023.
61 Military Applications of Artificial Intelligence, RAND, ibid.
62 Rajesh Uppal. Russia Deployed Family of Killer Robots, ibid.
63 P. A. Dul'nev. The Employment of Robotic Complexes During the Assault of a Town (Fortified Area), *Vestnik Akademii Voennykh Nauk* (Journal of the Academy of Military Science), no. 3, 2017, p. 27, referred to by Timothy Thomas, Russian Robotics: A Look at Definitions, Principles, Uses, and other Trends, *MITRE*, February 2021, p. 11, https://irp.fas.org/eprint/russian-robotics.pdf, accessed on 13 February 2023.
64 Dennis J. Blasko. The Chinese Military Speaks to Itself, Revealing Doubts, 18 February 2019, https://warontherocks.com/2019/02/the-chinese-military-speaks-to-itself-revealing-doubts/, accessed on 10 March 2023.
65 Military Applications of Artificial Intelligence, Rand, 2020, https://www.rand.org/content/dam/rand/pubs/research_reports/RR3100/RR3139-1/RAND_RR3139-1.pdf, accessed on 24 January 2023.
66 Yuan-Chou Jing. How Does China Aim to Use AI in Warfare?, *The Diplomat*, 28 December 2021, https://thediplomat.com/2021/12/how-does-china-aim-to-use-ai-in-warfare/, accessed on 15 December 2022.
67 S. D. Pradhan. Transformation of the Chinese influence operations into a Cognitive war: Warning for India, 21 July 2022, https://timesofindia.indiatimes.com/blogs/ChanakyaCode/transformation-of-the-chinese-influence-operations-into-a-cognitive-war-warning-for-india/, accessed on 13 March 2023.
68 Yuan-Chou Jing, ibid.
69 Akashdeep Arul. How China is using AI for Warfare, 21 February 2022, https://analyticsindiamag.com/how-china-is-using-ai-for-warfare/, accessed on 15 December 2022.
70 'Military Applications of Artificial Intelligence', Rand, 2020, p, 66, https://www.rand.org/content/dam/rand/pubs/researchreports/RR3100/RR3139-1/RAND_RR3139-1.pdf, accessed on 24 January 2023.
71 Amy J. Nelson and Gerald L. Epstein. The PLA's Strategic Support Force and AI Innovation, 23 December 2022, https://www.brookings.edu/techstream/the-plas-strategic-support-force-and-ai-innovation-china-military-tech/, accessed on 8 March 2023.
72 Jeirou Lee, 'Artificial Intelligence Technology and China's Defence System, '*Journal* of Indo-

Pacific Affairs (March-April 2022), https://media.defense.gov/2022/ Mar/28/2002964034/-1/-1/1/FEATURE_LI.PDF, 15 December 2022.

73. Stephen Chen, 'Chinese AI plays war games like a human, with military strategists unable to identify it as a machine, developers say in Beijing', 23 February 2023, https://www.scmp.com/news/china/science/article/3211142/chinese-ai-plays-war-games-human-military-strategists-unable-identify-it-machine-say-developers, accessed on 13 March 2023.
74. Amy J, Nelson and Gerald L. Epstein. The PLA's Strategic Support Force and AI Innovation, 23 December 2022, https://www.brookings.edu/techstream/the-plas-strategic-support-force-and-ai-innovation-china-military-tech/, accessed on 8 March 2023.
75. Elsa Kania. Artificial Intelligence in Future Chinese Command Decision Making, https://www.jstor.org/stable/pdf/resrep19585.26 .pdf?refreqid=excelsior% 3A28f037d2b7 aaa39b25be6e 923051 aced&ab_segments=&origin=&initiator=&acceptTC=1, accessed on 9 March 2023.
76. Military Applications of Artificial Intelligence, Rand, 2020, p. 70, https://www.rand.org/content/dam/ rand/pubs/researchreports/RR3100/RR3139-1/RAND_RR3139-1.pdf, accessed on 24 January 2023.
77. Jeirou Lee. Artificial Intelligence Technology and China's Defence System, *Journal of Indo-Pacific Affairs* (March-April 2022), https://media.defense.gov/2022/ Mar/28/2002964034/-1/-1/1/FEATURE_LI.PDF, 15 December 2022.
78. Amy J. Nelson and Gerald L. Epstein. The PLA's Strategic Support Force and AI Innovation, 23 December 2022, https://www.brookings.edu/techstream/the-plas-strategic-support-force-and-ai-innovation-china-military-tech/, accessed on 8 March 2023.
79. Amy J. Nelson and Gerald L. Epstein. The PLA's Strategic Support Force and AI Innovation, 23 December 2022, https://www.brookings.edu/techstream/the-plas-strategic-support-force-and-ai-innovation-china-military-tech/, accessed on 8 March 2023.
80. 'Military Applications of Artificial Intelligence', Rand, 2020, p. 70, https://www.rand.org/content/dam/ rand/pubs/researchreports/RR3100/RR3139-1/RAND_RR3139-1.pdf, accessed on 24 January 2023.
81. Akashdeep Arul. How China is using AI for Warfare, 21 February 2022, https://analyticsindiamag.com/how-china-is-using-ai-for-warfare/, accessed on 15 December 2022.
82. https://economictimes.indiatimes.com/news/defence/chinas-pla-aims-to-leverage-advanced-technology-for-use-of-unmanned-weapons-artificial-intelligence-says-report/articleshow/97431564.cms?from=mdr, accessed on 8 March 2023.
83. Jeirou Lee. Artificial Intelligence Technology and China's Defence System, *Journal of Indo-Pacific Affairs* (March-April 2022), https://media.defense.gov/2022/ Mar/28/2002964034/-1/-1/1/FEATURE_LI.PDF, 15 December 2022.
84. https://www.globalsecurity.org/military/world/china/anjian.htm, accessed on 14 February 2023.
85. https://www.ainonline.com/aviation-news/defense/2018-02-21/new-chinese-armed-uav-project-unveiled, accessed on 14 March 2023.
86. https://en.wikipedia.org/wiki/Hongdu_GJ-11, accessed on 14 March 2023.
87. Military Applications of Artificial Intelligence, Rand, 2020, p. 70, https://www.rand.org/content/dam/ rand/pubs/researchreports/RR3100/RR3139-1/RAND_RR3139-1.pdf, accessed on 24 January 2023.
88. https://www.armyrecognition.com/china_chinese_unmanned_aerial_ground_systems_uk/sharp_claw_2_ugv_6x6_unmanned_ground_vehicle_technical_data_sheet_specifications_pictures_video_11412165.html, accessed on 14 March 2023.
89. Juan Ju. 'Norinco's Sharp Claw I UGV in service with Chinese army', 15 April 2020, https://www.janes.com/defence-news/news-detail/norincos-sharp-claw-i-ugv-in-service-with-chinese-army, accessed on 14 March 2023.

90 Elsa B. Kania. Chinese Sub Commanders May Get AI Help for Decision-Making, 12 February 2018, https://www.defenseone.com/ideas/2018/02/chinese-sub-commanders-may-get-ai-help-decision-making/145906/, accessed on 8 March 2023.
91 Robotic boats making waves, https://www.nature.com/articles/d42473-022-00140-y, accessed on 14 March 2023.
92 China deploying 'en masse' underwater drones in Indian Ocean: Report, *Livemint*, 31 December 2020, https://www.livemint.com/news/india/china-deploying-en-masse-underwater-drones-in-indian-ocean-report-11609373452869.html, accessed on 14 March 2023.
93 https://en.wikipedia.org/wiki/Buoyancy_engine, accessed on 14 March 2023.
94 https://www.globalsecurity.org/military/world/china/d3000.htm, accessed on 13 March 2023.
95 https://carnegietsinghua.org/2018/10/24/tides-of-change-china-s-nuclear-ballistic-missile-submarines-and-strategic-stability-pub-77490, accessed on 14 March 2023.
96 Military Applications of Artificial Intelligence, Rand, 2020, p 70, https://www.rand.org/content/dam/rand/pubs/researchreports/RR3100/RR3139-1/RAND_RR3139-1.pdf, accessed on 24 January 2023.
97 https://economictimes.indiatimes.com/news/defence/chinas-pla-aims-to-leverage-advanced-technology-for-use-of-unmanned-weapons-artificial-intelligence-says-report/articleshow/97431564.cms?from=mdr, accessed on 8 March 2023.
98 Elsa Kania. Artificial Intelligence in Future Chinese Command Decision Making, https://www.jstor.org/stable/pdf/resrep19585.26.pdf?refreqid=excelsior%3A28f037d2b7aaa39b25be6e923051aced&ab_segments=&origin=&initiator=&acceptTC=1, accessed on 9 March 2023.
99 Stephen Chen. In China, AI warship designer did nearly a year's work in a day, 10 March 2023, https://www.scmp.com/news/china/science/article/3213056/china-ai-warship-designer-did-nearly-years-work-day, 14 March 2023.
100 *Indian Aerospace Defence News* (IADN) (@NewsIADN) tweeted at 1:26 AM on 14 March 2023, https://twitter.com/NewsIADN/status/1635369382120673280?t=Q8fDFFQVY1TBmUjoc2ob QA&s=03
101 National Artificial Intelligence Strategy (NAIS) 2021-2025, p. 47, https://cbddo.gov.tr/SharedFolderServer/Genel/File/TRNationalAIStrategy2021-2025.pdf, accessed on 27 February 2023.
102 National Artificial Intelligence Strategy (NAIS) 2021-2025, pp. 49, https://cbddo.gov.tr/SharedFolderServer/Genel/File/TRNationalAIStrategy2021-2025.pdf, accessed on 27 February 2023.
103 Ibid.
104 Amir Husain, 'Turkiye Builds A Hyperwar Capable Military', 30 June 2022, https://www.forbes.com/sites/amirhusain/2022/06/30/Turkiye-builds-a-hyperwar-capable-military/?sh=1c0f0e855e11, accessed on 27 February 2023.
105 Anka-S. Unmanned Aerial Vehicle, 11 April 2018, https://www.airforce-technology.com/projects/anka-s-unmanned-aerial-vehicle/, accessed on 27 February 2023.
106 Rojoef Manuel, Turkiye Develops Tech Reducing Small Drones' Reliance on GPS, 13 October 2022, https://www.thedefensepost.com/2022/10/13/Turkiye-drones-gps-navigation/, accessed on 27 February 2023.
107 Turkiye's 'Fully Loaded' Baykar Akinci Combat Drone Completes Test Flight Armed With All Weapon Stations, 26 August 2022, https://eurasiantimes.com/Turkiyes-fully-loaded-baykar-akinci-combat-drone-completes-test-flight-armed-with-all-weapon-stations/, accessed on 27 February 2023.
108 Aksungur Medium-Altitude Long Endurance (MALE) UAV, Turkiye, 13 April 2022, https://www.naval-technology.com/projects/aksungur-medium-altitude-long-endurance-male-uav-

Turkiye/, accessed on 27 February 2023.

109 Baird, Maritime. New Locally-Built Armed USV for Turkish Navy, 29 September 2022, https://www.bairdmaritime.com/work-boat-world/maritime-security-world/unmanned-systems/vessel-review-sancar-new-locally-built-armed-usv-for-turkish-navy/, accessed on 27 February 2023.

110 Goksel Yildirim. STM Defence Technologies Engineering produces autonomous 'Kamikaze' Kargu drones for military use, 15 June 2020, https://www.aa.com.tr/en/economy/anadolu-agency-tours-state-of-the-art-turkish-uav-maker/1877808, accessed on 22 February 2023.

111 Alex McFarland. Turkiye Further Revolutionizes Define Sector with AI Technology, 21 October 2020, accessed on 15 December 2022.

112 Turkiye to be among pioneers of AI-controlled warplane: Erdoğan, 26 May 2021, https://www.dailysabah.com/business/defense/Turkiye-to-be-among-pioneers-of-ai-controlled-warplane-erdogan, accessed on 27 February 2023.

113 Burak Ege Bekdil. Turkish firm develops AI-powered software for drone swarms, 24 November 2020, https://www.defensenews.com/unmanned/2020/11/24/turkish-firm-develops-ai-powered-software-for-drone-swarms/, accessed on 27 February 2023.

114 Humeyra Pamuk. U.S. says potential F-16 sale to Turkiye would serve U.S. interests, Reuters, https://www.reuters.com/world/us-says-potential-f-16-sale-Turkiye-would-serve-us-interests-nato-letter-2022-04-06/ accessed on 16 August 2022

115 Tony Ho Tran, An AI Successfully Flew an F-16 Fighter Jet for 17 Hours, 14 February 2023, https://www.yahoo.com/now/ai-successfully-flew-f-16-182649243.html?guc counter=1&guce_referrer=aHR0cHM6Ly93d3cuZ29vZ2xlLm NvbS8&guce_referrer_sig=AQAAAMVK-4wAQdy-xexCp0cMUYdqRIV4BE2wi_pPO 8r8 NiU19j5w2BVMXtK9nnzaytF4zzs7Fb5s8xZmcX5rtqpwdRsIHEswSb8gm4aU-wWqLZWIUe0yOExvgl-IUlgyeb rO069LJjLwsGQq6uaOkGuwKeJJ9VyQ3pSjfrcCyKdSuexH, accessed on 27 February 2023.

116 Reuters. White House says Turkiye's involvement in F-35 program impossible, https://www.reuters.com/article/Turkiye-security-usa/white-house-says-Turkiyes-involvement-in-f-35-program-impossible-idINFWN24I0I3 accessed on 16 August 2022.

117 Can Kasapoğlu and Barý' Kýrdemir. The Rising Drone Power: Turkiye on the Eve of Its Military Breakthrough, Centre for Economics and Foreign Policy Studies, 2018, http://www.jstor.com/stable/resrep21043, accessed on 27 February 2023.

118 https://www.scimagojr.com/countryrank.php?category=1702&order=it&ord=desc, accessed on 29 March 2024.

119 Mahan Fathi. Collecting a Nation's Points of Interest: Computer Vision to the Rescue, 12 January 2019, https://medium.com/@MahanFathi/collecting-a-nations-points-of-interest-computer-vision-to-the-rescue-41026053bdf6, accessed on 29 March 2024.

120 http://timesofindia.indiatimes.com/articleshow/107705046.cms?utm_s ource=contentofint erest&utm_medium=text&utm_campaign=cppst, accessed on 30 March 2024.

121 https://www.concentric.io/blog/the-global-landscape-of-ai-development-china-russia-and-irans-strategies-and-impacts#:~:text=the%20Ukraine%20war.-,Iran,the%20AI%20market%20by% 202032, accessed on 22 November 2023.

122 Evan Omeed Lisman. 'Iran's Bet on Autonomous Weapons', 30 August 2019, https://warontherocks.com/2021/08/irans-bet-on-autonomous-weapons, accessed on 29 March 2021

123 Michael Rubin. Iran Claims Development of Cruise Missiles Guided by Artificial Intelligence, 2 October 2023, https://fmso.tradoc.army.mil/2023/iran-claims-development-of-cruise-missiles-guided-by-artificial-intelligence/#_edn. 1, accessed on 30 March 2024.

124 https://www.concentric.io/blog/the-global-landscape-of-ai-development-china-russia-and-irans-strategies-and-impacts#:~:text=the%20Ukraine%20war.-,Iran,the%20AI%20market%20by%

202032, accessed on 22 November 2023
125 Seth J. Frantzman. Iran reveals bizarre new AI-powered miniature tank robots—analysis, 2 February 2023, https://www.jpost.com/middle-east/article-730433, accessed on 22 November 2023.
126 @IranObserver0, 3 August 2023, https://twitter.com/IranObserver0/status/1687093036764008449, accessed on 31 March 2024.
127 Army Robotic & Autonomous Systems Strategy v2.0, August 2022, p. 21, https://researchcentre.army.gov.au/sites/default/files/Robotic%20and%20Autonomous%20Systems%20Strategy%20V2.0.pdf, accessed on 3 March 2023.
128 Sanu Kainikara. Working paper 45, Artificial Intelligence and the Future of Air Power, p. 8, https://airpower.airforce.gov.au/sites/default/files/2021-03/WP45-Artificial-Intelligence-and-the-Future-of-Air-Power.pdf, accessed on 2 March 2023.
129 Peter Layton. The ADF could be doing much more with artificial intelligence, 26 July 2022, https://www.aspistrategist.org.au/the-adf-could-be-doing-much-more-with-artificial-intelligence/, accessed on 2 March 2023.
130 Army Robotic & Autonomous Systems Strategy v2.0, August 2022, pp. 27-29, https://researchcentre.army.gov.au/sites/default/files/Robotic%20and%20Autonomous%20Systems%20Strategy%20V2.0.pdf, accessed on 3 March 2023.
131 Ras-ai strategy 2040; warfare innovation navy, p. 14, https://www.navy.gov.au/sites/default/files/documents/RAN_WIN_RASAI_Strategy_2040f2_hi.pdf, accessed on 3 March 2023.
132 https://www.airforce.gov.au/our-work/projects-and-programs/ghost-bat, accessed on 7 March 2023.
133 Petra Stock. Lethal drones: The future of the Air Force could be un-crewed, 1 March 2023, https://cosmosmagazine.com/technology/lethal-drones-the-future-of-the-air-force-could-be-un-crewed/, accessed on 7 March 2023.
134 Ibid.
135 Jaspreet Gill. https://breakingdefense.com/2022/11/how-project-convergence-is-informing-british-australian-military-modernization/, accessed on 7 March 2023.
136 Sanu Kainikara. Working paper 45, Artificial Intelligence and the Future of Air Power, p. 8, https://airpower.airforce.gov.au/sites/default/files/2021-03/WP45-Artificial-Intelligence-and-the-Future-of-Air-Power.pdf, accessed on 2 March 2023.
137 RAS-AI strategy 2040; warfare innovation navy, p. 14, https://www.navy.gov.au/sites/default/files/documents/RAN_WIN_RASAI_Strategy_2040f2_hi.pdf, accessed on 3 March 2023.
138 Concept for Robotic and Autonomous Systems, November 2020, p 26, https://tasdcrc.com.au/wp-content/uploads/2020/12/ADF-Concept-Robotics.pdf, accessed on 3 March 2023.
139 Jaspreet Gill. https://breakingdefense.com/2022/11/how-project-convergence-is-informing-british-australian-military-modernization/, accessed on 7 March 2023.
140 Jon Grevatt. Lockheed Martin partners on AI training system for Australia, 25 August 2022, https://www.janes.com/defence-news/news-detail/lockheed-martin-partners-on-ai-training-system-for-australia, accessed on 7 March 2023.
141 Concept for Robotic and Autonomous Systems, November 2020, p. 23, https://tasdcrc.com.au/wp-content/uploads/2020/12/ADF-Concept-Robotics.pdf, accessed on 3 March 2023.
142 https://www.youtube.com/watch?v=34k7UI-DR_8, accessed on 8 March 2023.
143 Defence Artificial Intelligence research network contracts signed, 6 March 2023, https://www.defence.gov.au/news-events/releases/2023-03-06/defence-artificial-intelligence-research-network-contracts-signed, accessed on 8 March 2023.
144 AI reports for duty in the Australian military, https://www.deloitte.com/global/en/services/consulting/perspectives/AI-reports-for-duty-in-the-australian-military.html, accessed on 2 March

2023.
145. RAS-AI strategy 2040; warfare innovation navy, p. 19, https://www.navy.gov.au/sites/default/files/documents/RAN_WIN_RASAI_Strategy_2040f2_hi.pdf, accessed on 3 March 2023.
146. Akash Sahu. AUKUS, Quad key to Australia's AI-powered defence, 1 December 2022, https://asiatimes.com/2022/12/aukus-quad-key-to-australias-ai-powered-defence/, accessed on 2 March 2023.
147. Andrew Greene. Air Force flags need for low-cost drones to prepare Australia for future conflicts, 27 February 2023, https://www.abc.net.au/news/2023-02-27/air-force-need-low-cost-drones-australia-future-conflict/102028934, accessed on 7 March 2023.
148. Peter Layton. The ADF could be doing much more with artificial intelligence, 26 July 2022, https://www.aspistrategist.org.au/the-adf-could-be-doing-much-more-with-artificial-intelligence/, accessed on 2 March 2023.
149. Defence Artificial Intelligence Strategy, June 2022, https://www.gov.uk/government/publications/defence-artificial-intelligence-strategy/defence-artificial-intelligence-strategy#:~:text=The%20DAIC%20achieved%20initial%20operating,AI%20capabilities%20across%20the%20Department, accessed on 13 February 2023.
150. Ibid.
151. SAPIENT autonomous sensor system, 22 June 2022, https://www.gov.uk/guidance/sapient-autonomous-sensor-system, accessed on 2 March 2023.
152. The British Army has used Artificial Intelligence (AI) for the first time during Exercise Spring Storm, as part of Operation Cabrit in Estonia, 5 July 2021, https://www.gov.uk/government/news/artificial-intelligence-used-on-army-operation-for-the-first-time, accessed on 1 March 2023.
153. An overview of Britain's drones and drone development projects, updated February 2023, https://dronewars.net/british-drones-an-overview/, accessed on 2 March 2023.
154. Soraya Ebrahimi. What are the Malloy drones the UK is sending to Ukraine?, 4 May 2022, https://www.thenationalnews.com/world/uk-news/2022/05/03/what-are-the-malloy-drones-the-uk-is-sending-to-ukraine/, accessed on 2 March 2023.
155. https://www.intelligence-airbusds.com/newsroom/news/airbus-zephyr-solar-high-altitude-pseudo-satellite/, accessed on 2 March 2023.
156. 'An overview of Britain's drones and drone development projects, updated February 2023, https://dronewars.net/british-drones-an-overview/, accessed on 2 March 2023.
157. Mike Ball. New Developments for UK MOD Nano-UAS Program, 18 May 2022, https://www.unmannedsystemstechnology.com/2022/05/new-developments-for-uk-mod-nano-uas-program/, accessed on 2 March 2023.
158. https://www.army-technology.com/projects/bug-nano-drone/, accessed on 2 March 2023.
159. An overview of Britain's drones and drone development projects, updated February 2023, https://dronewars.net/british-drones-an-overview/, accessed on 2 March 2023.
160. Joseph Trevithick. RAF Tests Swarm Loaded with BriteCloud Electronic Warfare Decoys To Overwhelm Air Defences', 8 October 2020, https://www.thedrive.com/the-war-zone/36950/raf-tests-swarm-loaded-with-britecloud-electronic-warfare-decoys-to-overwhelm-air-defences, accessed on 2 March 2023.
161. Rheinmetall Wins Bid for Spiral 3 of UK's Robotic Platoon Vehicles Programme, 2 May 2022, https://www.asdnews.com/news/defense/2022/05/02/rheinmetall-wins-bid-spiral-3-uks-robotic-platoon-vehicles-programme, accessed on 2 March 2023.
162. https://www.horiba-mira.com/unmanned-ground-vehicles/media-centre/case_study/viking-multirole-ugv-platform/, accessed on 2 March 2023.
163. https://milremrobotics.com/defence/, accessed on 2 March 2023.
164. MADFOX—USV Designed for Surveillance and Force Protection Missions, *Vessel Review*, 4

August 2021, https://www.bairdmaritime.com/work-boat-world/maritime-security-world/unmanned-systems/vessel-review-madfox-usv-designed-for-surveillance-and-force-protection-missions/, accessed on 2 March 2023.
165 Royal Navy Madfox unmanned vessel launches missile on NATO drill, 14 October 2021, https://defbrief.com/2021/10/14/royal-navy-madfox-unmanned-vessel-launches-missile-on-nato-drill/, accessed on 2 March 2023.
166 British Navy announces arrival of autonomous mine hunter in Gulf, 14 February 2023, https://www.naval-technology.com/news/british-autonomous-mine-hunter-gulf/, accessed on 2 March 2023.
167 Bertie Adam. Royal Navy's first unmanned submarine will be built in Plymouth, 1 December 2022, https://www.plymouthherald.co.uk/news/plymouth-news/royal-navys-first-unmanned-submarine-7883582#, accessed on 2 March 2023.
168 Defence Artificial Intelligence Strategy, June 2022, https://www.gov.uk/government/publications/defence-artificial-intelligence-strategy/defence-artificial-intelligence-strategy#:~:text=The%20DAIC%20achieved%20initial%20operating,AI%20capabilities%20across%20the%20Department, accessed on 13 February 2023.
169 Report of the MoD AI Task Force, 'Artificial Intelligence in Support of Defence', September 2019, https://www.defense.gouv.fr/sites/default/files/aid/Report%20of%20 the% 20AI %20 Task % 20Force%20September%202019.pdf, accessed on 25 February 2023.
170 Ibid.
171 Ibid.
172 Guerric Poncet. The French Ministry of the Army will entrust all its satellite images to Preligens' AI, 7 August 2021, https://preligens.com/resources/press/french-ministry-army-will-entrusts-all-its-satellite-images-preligens-ai, accessed on 26 February 2023.
173 Report of the MoD AI Task Force. Artificial Intelligence in Support of Defence, September 2019, https://www.defense.gouv.fr/sites/default/files/aid/Report%20of%20 the%20AI%20Task %20Force%20September%202019.pdf, accessed on 25 February 2023.
174 Ibid.
175 Launch of the Man Machine Teaming advanced study programme, 16 March 2018, https://www.dassault-aviation.com/en/group/press/press-kits/launch-man-machine-teaming-advanced-study-programme/, accessed on 27 February 2023.
176 Report of the MoD AI Task Force, Artificial Intelligence in Support of Defence, September 2019, https://www.defense.gouv.fr/sites/default/files/aid/Report%20of%20 the%20AI%20Task %20Force%20September%202019.pdf, accessed on 25 February 2023.
177 French army first experiments GACI mule UGVs in external operation, 26 April 2021, https://www.armyrecognition.com/defense_news_april_2021_global_security_army_industry/french_army_first_experiments_gaci_mule_ugvs_in_external_operation.html, accessed on 27 February 2023.
178 Sandra Tubert and Laura Ziegler. France: Artificial Intelligence Comparative Guide, 23 November 2022, https://www.mondaq.com/france/technology/1059760/artificial-intelligence-comparative-guide, accessed on 23 February 2023.
179 Report of the MoD AI Task Force. Artificial Intelligence in Support of Defence, September 2019, https://www.defense.gouv.fr/sites/default/files/aid/Report%20of%20 the%20AI%20Task %20Force%20September%202019.pdf, accessed on 25 February 2023.
180 Sandra Tubert, Laura Ziegler, 'France: Artificial Intelligence Comparative Guide', ibid.
181 https://www.rand.org/content/dam/rand/pubs/research reports/RR3100/RR3139-1/RAND_RR3139-1.pdf, accessed on 23 January 2023.
182 https://www.defense.gov/News/Transcripts/Transcript/Article/2672391/joint-artificial-

intelligence-center-press-briefing/, accessed on 19 March 2023.
183 https://www.defense.gov/News/News-Stories/Article/Article/2667212/hicks-announces-new-artificial-intelligence-initiative/, accessed on 19 March 2023.
184 Jaspreet Gill. Say goodbye to JAIC and DDS, as offices cease to exist as independent bodies June 1, 24 May 2022, https://breakingdefense.com/2022/05/say-goodbye-to-jaic-and-dds-as-offices-cease-to-exist-as-independent-bodies-june-1/, accessed on 19 May 2023.
185 Department of Defence outside the Continental United States (OCONUS). Cloud Strategy, released publically on 26 May 2021, https://dodcio.defense.gov/Portals/0/Documents/Library/DoD-OCONUSCloudStrategy.pdf, accessed on 19 March 2023.
186 Kate Conger the Fight for a Massive Pentagon Cloud Contract is Heating Up, 8 May 2018, https://gizmodo.com/the-fight-for-a-massive-pentagon-cloud-contract-is-heat-1825517332, accessed on 19 March 2023.
187 DoD Chief Digital and Artificial Intelligence Office Hosts Global Information Dominance Experiments, 30 January 2023, https://www.defense.gov/News/Releases/Release/Article/3282376/dod-chief-digital-and-artificial-intelligence-office-hosts-global-information-d/, accessed on 19 March 2023.
188 https://www.britannica.com/print/article/405392, accessed on 20 March 2023.
189 https://fedtechmagazine.com/article/2022/10/intelligence-community-developing-new-uses-ai-perfcon#:~:text=At%20the%20National%20Security%20Agency, National%20 Securit y%20 Alliance's%20spring%20symposium, accessed on 20 March 2023.
190 Adin Dobkin. DoD Maven AI Project Develops First Algorithms, Starts Testing, *Defence Systems*, 3 November 2017.
191 Gregory C. Allen, Project Maven Brings AI to the Fight Against ISIS, *Bulletin of the Atomic Scientists*, 21 December, 2017.
192 Eray Eliaçýk. Guns and Codes: The Era of AI-Wars Begins, 17 August 2022, https://dataconomy.com/2022/08/how-is-artificial-intelligence-used-in-the-military/ #:~:text=Artificial%20intel ligence% 20is%20used%20in,driven%20systems%20in%20the%20military, accessed on 9 December 2022.
193 Philip Root. Squad X Core Technologies (SXCT), https://www.darpa.mil/program/squad-x-core-technologies, accessed on 20 March 2023.
194 https://www.youtube.com/watch?v=DgM7hbCNMmU, accessed on 20 March 2023.
195 Loren Blinde. US Air Force, Lockheed Martin, Demonstrate Manned/Unmanned Teaming, *Intelligence Community News*, 11 April 2017.
196 Sydney J. Freedberg Jr. A Slew To A Kill: Project Convergence, 16 September 2020, https://breakingdefense.com/2020/09/a-slew-to-a-kill-project-convergence/, accessed on 7 March 2023.
197 Margarita Konaevetal. U.S. Military Investments in Autonomy and AI Costs, Benefits, and Strategic Effects, CSET Policy Brief, October 2020, https://cset.georgetown.edu/publication/u-s-military-investments-in-autonomy-and-ai-executive-summary/ accessed on 9 December 2022.
198 https://www.rand.org/content/dam/rand/pubs/research reports/RR3100/RR3139-1/RAND_RR3139-1.pdf, accessed on 23 January 2023.
199 Yisuo Tzeng. Prospect for Artificial Intelligence in Taiwan's Defence, 2019, https://www.jewishpolicycenter.org/2019/01/11/prospect-for-artificial-intelligence-in-taiwans-defense/, accessed on 27 February 2023.
200 Jack Lau. US should use AI to beat Chinese censors in case of Taiwan attack, think tank chaired by former Google CEO says, *South China Morning Post*, 27 October 2022, https://www.scmp.com/news/china/military/article/3197496/us-should-use-ai-beat-chinese-censors-case-taiwan-attack-think-tank-chaired-former-google-ceo-says, accessed on 27 February 2023.
201 Ian Easton. Taiwan Defence Strategy in an Age of Precision Strike, Project 2049 Institute, 2014,

AI on the Frontlines: Case Studies in Military Innovation ❑ 111

p. 25, https://project2049.net/wp-content/uploads/2018/06/Easton_Able_Archers_Taiwan_Defense_Strategy.pdf, accessed on 28 February 2023.

202 Steven D. Bernstein. Taiwan's Defence Strategy and Artificial Intelligence', Dean's Scholars Thesis, 2020, p. 19, https://bpb-us-e1.wpmucdn.com/blogs.gwu.edu/dist/b/3853/files/2022/06/Copy-of-Deans-Scholars-Thesis-Edited-for-Journal_BernsteinSteven.docx-2.pdf, accessed on 27 February 2023.

203 https://www.janes.com/defence-news/news-detail/update-taiwans-ncsist-unveils-new-male-class-uav-development, accessed on 28 February 2023.

204 Lin Chia-nan. Formosat-7 to bolster national security, president says, 22 February 2019, https://www.taipeitimes.com/News/taiwan/archives/2019/02/22/2003710204, accessed on 28 February 2023.

205 https://www.army-technology.com/projects/tien-kung-iii-sky-bow-iii-surface-to-air-missile-system/, accessed on 28 February 2023.

206 Steven D. Bernstein. Taiwan's Defence Strategy and Artificial Intelligence', Dean's Scholars Thesis, 2020, p. 22, https://bpb-us-e1.wpmucdn.com/blogs.gwu.edu/dist/b/3853/files/2022/06/Copy-of-Deans-Scholars-Thesis-Edited-for-Journal_BernsteinSteven.docx-2.pdf, accessed on 27 February 2023.

207 Claudia Conte, et al. Using Drone Swarms as a Countermeasure of Radar Detection, 18 December 2022, https://doi.org/10.2514/1.I011131, accessed on 28 February 2023.

208 Michael Peck. Russia Wants to Use AI-Sea Mines to Sink America's Navy, 8 February 2020, https://nationalinterest.org/blog/buzz/russia-wants-use-ai-sea-mines-sink-americas-navy-120951, accessed on 28 February 2023.

209 J. Michael Cole. How Taiwan Can Defend Its Coastline Against China', 2 July 2019, https://macdonaldlaurier.ca/taiwan-can-defend-coastline-china-j-michael-cole/, accessed on 28 February 2023.

210 Yisuo Tzeng. 'Prospect for Artificial Intelligence in Taiwan's Defence, 2019, https://www.jewishpolicycenter.org/2019/01/11/prospect-for-artificial-intelligence-in-taiwans-defence/, accessed on 27 February 2023.

211 Steven D. Bernstein. Taiwan's Defence Strategy and Artificial Intelligence, Dean's Scholars Thesis, 2020, p. 30, https://bpb-us-e1.wpmucdn.com/blogs.gwu.edu/dist/b/3853/files/2022/06/Copy-of-Deans-Scholars-Thesis-Edited-for-Journal_BernsteinSteven.docx-2.pdf, accessed on 27 February 2023.

212 Seth J. Frantzman. Israel rolls out new wartime plan to reform armed forces, 18 February 2020, https://www.defencenews.com/global/mideast-africa/2020/02/18/israel-rolls-out-new-wartime-plan-to-reform-armed-forces/, accessed on 23 February 2023.

213 Eran Ortal 'Going on the Attack: The Theoretical Foundation of the Israel Defence Forces' Momentum Plan', 01 October 2020, https://www.idf.il/en/mini-sites/dado-center/vol-28-30-military-superiority-and-the-momentum-multi-year-plan/going-on-the-attack-the-theoretical-foundation-of-the-israel-defense-forces-momentum-plan-1/#:~:text=The%20challenge %20of% 20the%20Momentum,Israel%2C%20that%20of%20rocket%20fire., accessed on 23 February 2023.

214 The IDF Sees Artificial Intelligence as the Key to Modern-Day Survival, 27 June 2017, https://www.idf.il/en/mini-sites/technology-and-innovation/the-idf-sees -artificial-intelligence-as-the-key-to-modern-day-survival/#:~:text=Why%20the%20IDF%20 needs%20artificial,and%20automatically%20flag%20suspicious%20activity, accessed on 23 February 2023.

215 https://indiaai.gov.in/news/israel-claims-to-have-fought-the-world-s-first-ai-war, accessed on 23 February 2023.

216 Roselyne Min. Israel deploys AI-powered robot guns that can track targets in the West Bank,

https://www.euronews.com/next/2022/10/17/israel- deploys-ai-powered-robot-guns-that-can-track-targets-in-the-west-bank#:~:text=Amid%20increasing%20 tension%20between%20 Israel, camp%20in%20the%20West%20Bank, accessed on 9 December 2022.
217. https://www.smart-shooter.com/, accessed on 23 February 2023.
218. Seth J. Frantzman and Kelsey D. Atherton. Israel's Rafael integrates artificial intelligence into Spice bombs, 17 June 2019, https://www.defensenews.com/artificial-intelligence/2019/06/17/israels-rafael-integrates-artificial-intelligence-into-spice-bombs/, accessed on 23 February 2023.
219. Seth J. Frantzman. Israel's Carmel program: Envisioning armoured vehicles of the future, 6 August 2019, https://www.c4isrnet.com/artificial-intelligence/2019/08/05/israels-carmel-program-envisioning-armored-vehicles-of-the-future/ accessed on 23 February 2023.
220. Seth J. Frantzman, Israel finds an AI system to help fight in cities, 5 February 2020, https://www.defensenews.com/battlefield-tech/2020/02/05/israel-finds-an-ai-system-to-help-fight-in-cities/, accessed on 23 February 2023.
221. Debolina Biswas. Israel's Iron Dome Puts AI at the Forefront of Modern Warfare, 19 May 2021, https://analyticsindiamag.com/israels-iron-dome-puts-ai-at-the-forefront-of-modern-warfare/, accessed on 25 February 2023.
222. Kyle Mizokami. Everything We Know about Israel's Robotic Machine Gun, 30 September 2021, https://www.popularmechanics.com/military/weapons/a37708762/robotic-machine-gun-kills-iranian-nuclear-scientist/, accessed on 25 February 2023.

Chapter Three

India's AI Odyssey: Navigating Challenges, Fostering Innovation, and Shaping a Future

INDIAN AI JOURNEY

In the realm of technology and innovation, few phenomena have captured the world's attention as profoundly as AI. As the digital age transform our lives, a remarkable narrative is unfolding in India – a nation that not only boasts a phenomenal growth rate but also possesses a dynamic and youthful population. Powered by a relentless pursuit of progress and strategic autonomy, it has emerged as a global contender in AI development, captivating the imagination of the world with its audacious and unbeknown strides.

With a growth rate that defies conventional wisdom, India is ready to extend its knowledge and expertise to the global South. In line with the *Atmanirbhar Bharat* (self-reliant India) initiative, it is actively collaborating with nations across continents, sharing its advancements, fostering cooperation (*Vaccine Maitri* – Vaccine Friendship), and nurturing a global ecosystem of innovation. At the heart of this remarkable transformation lies the country's youthful bulge – a demographic dividend that propels the nation's innovation engine forward. India brims with untapped potential, nurturing a culture of agility, and boundless ambition. This energetic and tech-savvy generation has embraced curiosity and AI as an essential tool for shaping their future, unlocking unprecedented opportunities in fields as diverse as healthcare, agriculture, finance, national security, and transportation.

However, what sets India apart is not merely its rapid growth or its youthful population but its resolute commitment to strategic autonomy and responsible

AI. While embracing the global exchange of ideas, India pursues a path that empowers its citizens to chart their own course. With a transformative leadership that values ethics, transparency, and accountability, the country has placed responsible AI development at the core of its vision, ensuring that progress aligns with societal well-being and human-centric values. The journey towards AI development has not been without challenges. It requires unwavering dedication, collaboration, and a shared vision among stakeholders across academia, government, industry, and society at large.

The history of AI's ancestor – computer science, is much older in India. In fact, Professor H.N. Mahabala[1] introduced AI to India in the 1960s. The UNDP's Knowledge-Based Computing Systems (KBCS) during 1986 also focused India's attention. Professor Mahabala not only mentored Gopalakrishnan and N.R. Narayana Murthy (both of Infosys fame) but countless others too, many of whom became top executives at Infosys, TCS and IBM. The Father of computer science education in India – Professor Mahabala – also headed the computer science department in IIT Madras and introduced the very first M.Tech. computer science programme in 1973.

As the world looks on in awe, India's phenomenal growth, strategic autonomy, responsible AI practices, and transformative leadership serve as a clarion call – a call to embrace the transformative power of AI, to nurture innovation, and to shape a future that transcends boundaries. India beckons the world to join hands, to collaborate, and to collectively venture into uncharted territories, fuelled by the promise of a future guided by principles of global harmony.

Task Force on AI for India's Economic Transformation

AI in India's blossomed in 2017, when the Ministry of Commerce and Industry[2] constituted the 'Task Force on AI for India's Economic Transformation'. Indian pathways to AI, like the first one here, are analysed to understand the national focus and milestones they achieved. An 18-member task force comprising experts, academics, researchers, and industry leaders led by Professor V. Kamakoti of IIT (Madras) was expected to explore possibilities to leverage AI for development across various fields. The task force was directed[3] to submit five-year horizon concrete and implementable recommendations for government, industry, and research institutions. It aimed to leverage AI for economic benefits while creating a policy and legal framework to accelerate deployment of AI technologies.

The task force submitted its report in January 2018. Considering AI as a socio-economic problem solver, the task force attempted to answer three policy questions, which are relevant to the context of this research, and support the hypothesis. The policy questions were: in what areas should government play a role? How can AI improve quality of life and solve scaled problems for Indian citizens? And which sectors can generate employment and growth by use of AI? The task force was typically looking for an Indian solution to employment of AI. It identified ten sectors, including national security, as domains of relevance in India. It also mapped all ten domains into different ministries and departments to structure the whole of government effort and follow-up actions. The task force was prudent enough to recommend an India-specific quantitative model to study the impact of AI-enabled technologies on Indian citizens. The report[4] also provided concrete five-year horizon recommendations to the Government of India:

- An Inter-Ministerial National Artificial Intelligence Mission (N-AIM) funded under the Union budget for Rs. 240 crore per year for five years, supporting coordination among concerned ministries.
- Establishment and seed funding of centres of excellence, setting up of a generic AI test to serve as a validation platform for AI-based technology developers.
- Create digital data banks, marketplaces and exchanges for cross-industry data and information with sharing regulations, and with interdisciplinary large data centre for aggregation and interpretation.
- Standardising design, development and deployment of AI-based systems,
- Tax and other incentives.
- AI-based curriculums, AI-related education and re-skilling; and increasing awareness of AI.
- Lastly, but surely, leveraging key international relationships and participation in AI-based international standards-setting discussions.

For national security the report prophesised a few enablers, which have proven themselves to be both a harbinger of development and an Achilles' Heel and some of them are summarised below.

- National security has a multi-disciplinary nature of the task, so a consortium of MSME (subsystems and components) should be created.
- Provision of grants will support the niche technology development.
- Genuine data and cyber security tools and methodologies to protect digital assets and data from all cyber threats.
- Integrated and unified platform for existing infrastructure including NATGRID (National Grid), HUMINT (Human Intelligence), SIGINT (Signal Intelligence), COMINT (Communication Intelligence), imagery data and video surveillance from aircraft, CCTV data from urban areas and critical infrastructure locations, and radar data and satellite Imagery.
- AI-based techniques in this platform to provide curated real-time information to various security agencies involved in threat mitigation.

Task Force for 'Strategic Implementation of AI for National Security and Defence'

Known as the Task Force for 'Strategic Implementation of AI for National Security and Defence',[5] it was headed by Tata Sons Chairman, N. Chandrasekaran. The 16-member task force included heavyweights like National Cyber Security Coordinator, Chairman & Managing Director of Bharat Electronics Ltd., representatives from the defence forces, Indian Space Research Organisation, Atomic Energy Commission, Indian Institute of Science (Bangalore), IIT (Bombay), IIT (Madras), and private industry. Analysing AI with a national security perspective, the task force was mandated to study[6] it to establish tactical deterrent in the region, possibility of creating transformative weaponry of the future, including intelligent, autonomous robotic systems, while mitigating catastrophic risk. It also aimed to enhance data creation and analysis capabilities, bolster cyber defence and keeping a check on non-state actors. As part of its terms of reference, the task force was also to make recommendations for making India a significant power of AI in defensive and offensive needs, including counter AI. It was also to make specific suggestions for funding and increasing focus on AI within the Defence Research and Development Organisation (DRDO), Bharat Electronics Limited (BEL), service units, and selected academic institutions of the country.

Understanding the gravity of the assignment, the task force met twice during the first quarter of 2018. It understood that learning from select-use cases was a superior form of designing AI which led to a workshop on AI in national security and defence in May 2018. The functional areas identified for civilian use were personnel, supply chain & logistics, predictive maintenance, and finance & accounting. For military use, they were lethal autonomous weapon systems (LAWS), unmanned surveillance, and simulated war games & training. Cyber security, aerospace security, and intelligence & reconnaissance were designated as dual-use functional areas. Within six months (30 June 2018)[7], the task force submitted its report to the defence minister (Raksha Mantri – RM) and within a year (8 February 2019), the Department of Defence Production issued a government order[8] specifying:

- Integrating and embedding AI strategy for defence with defence strategy.
- Creation of a high-level defence AI council (DAIC) under the chairmanship of Raksha Mantri and a defence AI project agency (DAIPA) with Secretary (DP) as chairman.
- Development of data management framework, establishing data management office and appointing data management officer.
- Scaling the existing capability of data centres and establishing a centrally facilitated network of test beds.
- Creation of a framework to work with industry and encouraging start-ups to develop AI capability for defence and IP management.
- Organising AI training courses in all defence training centres and institutes for training of defence personnel.
- Earmarking of AI budget from the yearly defence budget with a corpus of Rs. 1,000 crore to be provided each year for the next five years to support AI activities.

The Defence Acquisition Policy 2020 (DAP 20)

DAP 20, the successor to the defence procurement procedure (DPP), has an independent chapter on acquisition of systems and information & communication technology (ICT) products, fully understanding that a few aspects differ from the regular acquisition projects. The ICT elements usually require periodic up-gradations and intense obsolescence management to maintain dominance. Cyber systems, artificial intelligence projects,[9] and AI output as the main deliverable projects are specifically considered in this chapter. DAP also specifies that the planning process[10] should germinate from the national security strategy/guidelines' (as and when promulgated) and Raksha Mantri's operational directive. Creation of a 10-year Integrated capability development plan (ICDP), including a 5-year defence capital acquisition plan (DCAP) and two-yearly annual acquisition plan (AAP) is prepared, along with a technology perspective and capability roadmap (TPCR). The TCPR 2018[11] provides an overview of equipment to the industry that is envisaged to be inducted into the Indian Armed Forces up to the late 2020s. The ICDP is approved by the defence acquisition council (DAC) prior to promulgation and subsequently translated into appropriate problem definition statements. The DAP defines a strict process to ensure that there is no dilution; amendment to SQR parameters is obtainable from AoN according authority and SQRs should contain specific and verifiable capability parameters.[12]

The DAP specifies defence offset policy too with a stated objective to foster development of competitive enterprises and augment defence related R&D. The strict acquisition process[13] involves the inviolable stages covered through the process by oversight by the technical oversight committee (TOC); request for information (RFI) → services qualitative requirements (SQRs) → acceptance of necessity (AON) → request for proposal (RFP), including offset offer → evaluation by technical evaluation committee (TEC) and technical offset evaluation committee (TOEC) → field evaluation trials (FET) → staff evaluation → contract negotiations by a committee (CNC) → approval of the competent financial authority (CFA) → award of contract and post-contract management.

DAP 20 presents a comprehensive strategy that involves the establishment of an Innovation and Indigenisation Organisation[14] (IIO) by service headquarters, utilising existing resources. This approach is the result of thorough analysis, discussions, and engagements, with the key objective of enhancing the nation's self-reliance in the defence sector and promoting a business-friendly environment through simplified procedures, delegation of tasks, reduced timelines, and increased industry collaboration. As part of this procedure, the IIOs play a crucial role in spearheading innovation and indigenisation efforts.

They undertake advance planning, engage in consultations with stakeholders and conduct feasibility studies. They also focus on indigenous design and development, including import substitution. They will collaborate with potential end users to formulate Preliminary Services Qualitative Requirements (PSQRs), facilitating the fielding of statements of case (SoC) for categorisation and acceptance. IIOs ensure accountability for timely delivery, closely monitoring all 'Make' projects. The IIOs also employ project/programme implementation experts, with the latest execution/monitoring techniques and software, to ensure efficient and timely development and implementation.

MINISTRY OF DEFENCE INITIATIVES

In April 2018, the Indian Government launched the Innovations for Defence Excellence[15] (iDEX) initiative to promote and nurture innovation and technology development in the Defence and Aerospace sectors. Though iDEX has functional autonomy, it is funded and managed by a Defence Innovation Organisation (DIO) formed as a not-for-profit company, which provides high level policy guidance. iDEX has set up network in the form of Defence Innovation Hubs[16] (DIH), which communicate with innovators/start-ups. Various challenges/ hackathons across the country harness innovative youth and start-ups to identify potential technologies. Successfully evaluated and piloted technologies are then facilitated scale-up for indigenisation and integration. Importantly, iDEX interfaces with the military to encourage adoption. The initiative engages 300 industries such as Micro, Small & Medium Enterprises (MSME), start-ups, individual innovators, R&D institutes, and academia to achieve self-reliance along with 20 partner incubators. A central sector scheme with a budgetary allocation of Rs. 498.78 crore up to 2025-26 supports the iDEX framework.

To encourage start-ups/MSMEs and individuals to show capability, intent, and promise and be able to produce functional prototypes or productise existing technologies, the Defence India Start-up Challenge (DISC) was launched in August 2018 to address specific technological needs of the Indian defence establishment. The winners were awarded up to Rs. 1.5 crore on a milestone basis in the form of grant/equity/debt/other relevant structures along with incubation and mentoring support. In its first edition, the DISC #1[17] had 11 wide-ranging, high technology problem statements; Individual protection system with built-in sensors, See through armour, carbon fibre winding (CFW), Active protection system (APS), Secure hardware-based offline encrypt or device for graded security, 4G/LTE based tactical local area network, Desalination (water purification) and bilge oily water separation system, Laser weaponry, Unmanned surface and underwater vehicles and Remotely piloted airborne vehicles. It specifically addressed the need for Artificial Intelligence in logistics & supply chain management (SCM). Subsequently, the DISC morphed into PRIME, and PRIME SPRINT [Supporting pole-vaulting in R&D through iDEX, Naval Innovation and Indigenisation Organisation (NIIO) and Technology Development Acceleration Cell (TDAC)] – the latter being the mouthful challenge by the Indian Navy. Both these versions provide grants and aid up to Rs 10 crore. iDEX also organises space and cyber challenges. Recently, the open challenge[18] series was started to tap into the new generation of engineers skilled in new technologies such as autonomous systems, intelligent machines, advanced materials, predictive algorithms, and even rocket engines. This challenge will provide opportunities to utilise newer technological capabilities to enhance the nation's military. Successful applicants will qualify for grants and investments up to Rs. 10 crore.

Support for Prototype and Research Kick-start (SPARK)[19] framework in defence by iDEX addresses the need for innovation and technological advancement in the defence sector. By supporting start-ups, MSMEs, and Innovators, SPARK encourages the development of prototypes and the commercialisation of products/solutions that are relevant to national defence and security. It will foster fast-moving innovation culture within the Indian defence sector. Grants up to Rs. 1.5 crore will promote the creation of functional prototypes that meet the existing needs of the Indian defence establishment. It serves as a platform for connecting new tech products/technologies with end users in the military. Overall, SPARK plays a crucial role in stimulating technological advancements, promoting innovation, and bolstering the defence industry in India. It provides a structured framework, grants, and support system to empower innovators and contribute to the nation's defence and security needs.

The Ministry of Defence also conducts Innovate4Defence (i4D),[20] a 45-day long internship program for students. It is an opportunity to learn about developing ideas, creating innovative defence products and technologies, incorporating a legal entity and eventually becoming a part of the Indian defence ecosystem and procurement. This internship aims to recommend an approach to a complex and practical defence challenges in order to teach ways of creating and building technology solutions for problems and developing minimum viable products that match the requirements in short time periods. Interns benefit from understanding self-sustainable and potentially scalable business models adopted by start-ups and get accustomed to the Indian defence innovation ecosystem.

As recommended by the task force on AI for India's economic transformation of the Ministry of Commerce and Industry in 2017, the positive Indigenisation list offers development opportunities to Indian industry. The Department of Defence Production (Ministry of Defence) developed a non-transactional online market place[21] (www.srijandefence.gov.in), for 'Opportunities for Make in India in Defence'. Sixteen DPSUs (defence public sector undertakings) and three Service Headquarters (Army, Navy, and Air Force) display their items (sizeable import value) on this portal, which they have imported or going to import. The Indian industry can use these details to show interest for designing, developing and manufacturing in the items individually, or in joint ventures with OEMs. The concerned DPSUs/SHQs, then interacts with the interested industry for indigenisation production. The website's traffic has increased by 11.52 per cent, with a bounce rate of 30 per cent, showing a promising future.[22] Additional efforts will ensure better reach out to potential developers. Interestingly, the website also receives hits from Honduras, Chile, Colombia and Turkiye.

Army Design Bureau (ADB)

The ADB undertakes technology scan, identifies technologies for acquisition and development. It facilitates R&D efforts with industry, academia, DRDO & DPSUs, provides inputs, and enables them to understand user requirements. To reach out to a wide spread and large defence industrial base, a start-up ecosystem, and a considerable talent pool across the country, a Regional Technology Node (RTN) of the Army Design Bureau (ADB) is functional at Pune and another one is being established in Bengaluru. On behalf of the ADB, these RTNs will leverage the locations and act as interfaces with trade, industry, and academia to coordinate advancements in technology for the overall benefit of the Indian Army.

The ADB's Compendium of Problem Definition Statement 2023[23] stems from the ICDP approved by the Defence Acquisition Council (DAC). It is an excellent indicator of military interests in the new technologies. The latest iteration consists of more than 100 problem definition statements across a wide spectrum, of which 20 relate to AI directly, while 15 could be AI-enabled in their newer version. Often abbreviated as CPDS, it is a comprehensive document that outlines and defines a set of specific problems or challenges within a particular field of study, organisation, or project. In the context of theoretical research on Artificial Intelligence in the Indian military, a CPDS would serve as a critical tool to identify and address the issues, gaps, and opportunities that are relevant to research objectives. The primary purpose of a CPDS is to provide a structured and organised overview of the problems that need to be addressed. It acts as a roadmap for researchers, practitioners, and stakeholders by presenting a clear picture of what needs attention and where resources should be allocated. This document is valuable for several reasons. Firstly, it helps in problem scoping. It delineates the boundaries of the issues at hand, enabling focus on the most critical areas. It ensures that the work remains targeted and relevant. Secondly, it aids in prioritisation. Identifying and ranking problems according to their significance and impact, and decision support helps to allocate resources efficiently. Thirdly, a CPDS facilitates collaboration. It can be shared with fellow researchers, policymakers, and experts, serving as a common reference point. This fosters a shared understanding of the issues and encourages interdisciplinary cooperation, which is often essential when dealing with complex problems in military AI.

DRDO Initiatives

The DRDO is a Ministry of Defence department; others being Department of Defence (DOD), Department of Defence Production (DDP), Department of Ex-Servicemen Welfare (DESW) and Department of Military Affairs (DMA). It is India's largest and most diverse research organisation, charged with the military's research and development to empower India with cutting-edge defence technologies and achieve self-reliance in critical defence technologies and systems. It consists of a comprehensive spectrum of agencies, laboratories, and establishments, ranges, facilities, programs, and projects. The department's primary responsibilities encompass a comprehensive array of strategic tasks aimed at bolstering national security. It holds full jurisdiction over the certification of design airworthiness for military aircraft and their associated equipment and stores. Allocation of Business Rules[24] mandates the DRDO to apprise, assess, and advise Raksha Mantri (Defence Minister) on emerging developments in science and technology and their effect on national security. It also furnishes advice to the three services regarding scientific facets associated with weaponry, weapon-platforms, military movements, surveillance, and logistic support across potential theatres of conflict.

It is the pivotal coordinating entity within the Ministry of Defence for all affairs pertaining to instruments of accord with foreign governments in support of the Ministry of External Affairs. The DRDO looks into intricacies of technology acquisition that are subject to foreign governments' national security-oriented export controls. This also includes interactions with foreign research organisations, intergovernmental entities, overseas academic and research institutions for the training and scholarship of Indian scientists and technologists. It extends financial and material aid to individuals, institutions, and corporate bodies to facilitate studies and manpower training in areas of science and technology. It also provides scientific analytical support and evaluation of all proposed weapon systems and related technologies slated for acquisition by the Ministry of Defence.

It also offers insights into the technological and intellectual property dimensions of technology imports and indigenous manufacturing, while addressing matters under the provisions of the Patents Act, 1970, Section 35.[25] It gives instructions to prevent or limit the release of information about that invention if it is significant for national defence. The department can implement projects and acquire land within its budget.

The DRDO has sponsored 500 students for PhD programmes in AICTE/centrally funded technical institutes under the Ministry of Human Resource Development to work on various DRDO projects. Young researchers will be provided working exposure to state-of-the-art defence technologies, high-end research and development activities of DRDO. Under this scheme, students get opportunities to have first-hand experience of ongoing R&D projects in their niche technology area and the DRDO is gifted with new talent. The Paid Apprenticeship Scheme, Internship to B.Tech/M.Tech/M.Sc. students generates interest in defence technologies among school and college students. A regular M.Tech program in defence technology has been launched by DRDO and AICTE to impact necessary theoretical and experimental knowledge, skill, and aptitude in various defence technology areas.

The DRDO established the Defence Industry Academia Centre of Excellence (DIA-CoE)[26] that harnesses and synergises the combined strength of academia, student community, research fellows, niche technology industries, and DRDO scientists in identified futuristic defence technological domains. IITs/universities will undertake science and technology projects and create special test facilities in these DRDO-funded centres. These will boost and sustain advanced technology development for future defence systems and platforms through multi-disciplinary and multi-institutional collaborative efforts. Fifteen DIA-CoEs are already functional and more are expected to come up. Specifically AI-oriented CoE range from Brain computer interface and Brain machine intelligence at IIT Delhi; Artificial Intelligence and robotics for aerial applications at IISc, Bangalore, which is exploring intelligent machines to find a balance between autonomy and human collaboration,[27] Artificial Intelligence for missile and missile defence and nano ornithopter technologies at IIT, Hyderabad,[28]; AI for information and war gaming technologies, Desert warfare technologies, Futuristic omni mobility systems at IIT, Jodhpur,[29] and Cognitive technologies, unmanned underwater technologies and cyber physical defence systems at IIT, Kharagpur.[30] IIT, Hyderabad, has also created a centre for research and innovation in AI (CRIA) with the support of JICA (Japan International Cooperation Agency) and Honeywell. It houses a mini-data centre with high-end computational facilities and deep learning supercomputers NVIDIA DGX1 and DGX2, offering up to 250 teraflops of GPU computing power.[31] This will assist in state-of-the art AI research in-house, as well as in partnership with its collaborators in government and the industry.

The TDF aims to create a self-reliant ecosystem. Though 1,800 MSMSEs, DPSUs and large-scale industries support the DRDO, there is a need to improve and widen this support base by involving Indian industry as development cum-production partners. The DRDO offers its technology to industry at nominal cost and providing free access to its patents. It encourages participation of public/private industries, MSMEs, and start-ups with a funding of up to Rs. 50 crore and a quicker development period of four years. To ensure national security, the industry must be owned and controlled by a resident Indian citizen, and must be registered in India. A non-profit research institution can also apply in collaboration/association with an Indian industry as lead bidder. The eligible entity must not have foreign investment in excess of 49 per cent. The DRDO partners with 'Invest India' (a non-profit venture under the Ministry of Commerce and Industry) to corporate relationships. By the end of 2022, the DRDO had already awarded this scheme to 64 projects for Rs. 280 crore.[32] Through the TDF, the DRDO is seeking inputs on NLP transcription (speech to text and speech to speech). AI-based autonomous object detection and tracking will fuse optical sensors in standalone and swarm mode for detection, tracking and targeting. Development of indigenous scenario and sensor simulation toolkit will generate scenarios, simulating all tactical entities in all physical domains, assisting in integrated dynamic simulation and end-to-end mission and war-gaming.

Conscious of emerging engineering fields and advanced technologies such as Artificial Intelligence, Quantum Technologies, Cognitive Technologies, Asymmetric Technologies and Smart Materials, the DRDO has created five Young Scientists Laboratories (DYSLs) to motivate youth for newer innovations and provide adequate freedom to them to prove their talent. The DRDO has three dedicated laboratories for AI – Centre for Artificial Intelligence and Robotics (CAIR), DRDO Young Scientists Laboratory (DYSL)-AI and DYST-CT (Cognitive Technology) – for application-orientated research. CAIR[33] is researching on advanced AI-based information processing systems and autonomy and cognition for unmanned and robotic systems, computer vision processing. A comprehensive library for image and video processing techniques, image map display system, a comprehensive data-mining toolbox, and a large number of data mining algorithms by CAIR aim to resolve 'rich problem areas in the military domain'. The DRDO also aims to utilise the enterprising and well-educated youth bulge of the country. Research fellowship schemes provide higher qualification opportunities to scientists/engineers.[34] Indian academic institutions of national importance also attempt researches by utilising the grants-in-aid scheme by Extramural Research & Intellectual Property Rights (ER&IPR).[35] This unblended and wide applicability of AI and ML in DRDO projects is illustrated by the table below.[36]

Category	Technology Development Task	Lab
Hardware	Artificial learning/deep learning architectures	Research Centre Imarat (RCI)
Cloud Computing	Private cloud technology for AI/ML Inference	RCI
Image Processing	Image and video analytics for intelligence	CAIR

Category	Technology Development Task	Lab
	Image detection and recognition for anti-tank missiles	Defence Research and Development Laboratory (DRDL)
	Satellite sensor data processing for intelligence	CAIR
	Deep fakes – detect synthetically generated fake images/videos, creation of synthetic images/videos	DYSL-AI
	Generate synthetic disguises and restoring images using generative models	DYSL-AI
	Image fusion and AI-based tracker for seekers	RCI
Natural Language Processing	Autonomous document summarisation	CAIR
	Autonomous machine translation for intelligence	CAIR
	Text analytics	CAIR
	Explainable AI and language models	DYSL-AI
	Large document summarisation and cross lingual question answering technology	DYSL-AI
	Real-time human voice cloning and generation	DYSL-AI
	Speech processing	CAIR
	The device/AI software for speech recognition/processing, machine translation and speech Synthesis for different languages	DYSL-AI
Software Framework	Artificial learning/deep learning software – software frameworks in compatible languages	RCI

The DRDO also launched an annual 'Dare to Dream' innovation contest to scout for technologies from individuals, researchers, and start-ups. It provides an opportunity for start-ups and innovators to solve some key challenges in emerging defence technologies. Since 2020, over 5,000 applicants showed interest out of which 52 individuals and 34 start-ups were shortlisted and seven projects sanctioned. The winning start-up receives a cash prize of Rs. 10 lakh, and the winning individual receives Rs. 5 lakh. The DRDO proposes certain domains and participants are free to propose solutions for up to five ideas.[37] The domains vary, Digital twining of human organs; Target seeking and proximity sensing technologies; Quantum algorithms for image processing; Plasmonic applications and counter-measures for drone and swarm of drones for defence. There is an open category to explore the unthinkable and unimaginable too. Analysis of some products gives a positive fillip to the contest and provides innovative and quick-release solutions[38] to many military problems.

- Cyran AI Solutions is developing neuromorphic vision sensors.
- Frshr Tech Private Limited, Adpix Tech Private Limited, and three other entities are using AI to detect persons of interest based on physiological parameters.
- AI Drone Private Limited is using AI for LPI of radars for electronic support systems.
- Inkers Technologies is reconstructing 3D target landscape in real time.

Transfer of Technology (ToT) is yet another process by the DRDO to enable the industry to boost the growth and capabilities of the defence manufacturing sector for achieving complete self-reliance. It also provides freedom to industries to carry out value addition to base technology in consultation with the DRDO to improve performance or economic viability, which is generally prohibited during foreign acquisitions. Since the DRDO is not a production agency, it frees the organisation for its primary role of R&D after ToT. Many such technologies also have utility in the civilian market and contribute to self-reliance in technology, industrial growth, and national development. The following list of AI-related ToT[39] gives a fair degree of satisfaction, and final product delivery will confirm its usefulness.

- CAIR ICR System (The CAIR Intelligent Character Recognition System) by CAIR.
- Robotic Manipulator by CAIR.
- CAIR GIS (Geographical Information System) by CAIR.
- Image & Video Processing for Net Centric Operations (IVP NCO), though a non-AI application, could be utilised with addition of Fuzzy logic; it is ready to be transferred by CAIR.
- Smart helmet to provide real-time situational awareness is ready to be transferred by DYSL-AI.

NATIONAL AI EFFORTS

AI has emerged as a transformative force with the potential to reshape economies and societies. India finds itself at a crucial juncture where the strategic implementation of a whole-of-government approach is imperative to foster, train, create, and develop a robust AI ecosystem and products within the country. Though the research is concentrating on military employment of AI, its development and production cannot be studied in isolation. India's strength lies in the AI echo system, and whole-of-government support. Initiative by other departments and organisations will bolster AI development and subsequently lead to usable AI in military.

The Ministry of Electronics & Information Technology (MeitY) assumes a central role in shaping India's digital future by formulating policies, fostering innovation, and providing the necessary regulatory framework for the AI sector. Its commitment to nurturing talent and promoting innovation is critical to advancing AI development. Complementing MeitY's efforts, the National Institution for Transforming India (NITI) Aayog, India's premier and high-powered policy think tank, plays an instrumental role in crafting a comprehensive AI strategy. The NITI Aayog's initiatives encompass the creation of the National AI Strategy, driving AI adoption across sectors, and promoting research and development to enhance India's competitiveness in the global AI arena. The Principal Scientific Advisor (PSA) to the Government of India serves as a keystone in coordinating AI-related initiatives across ministries. This coordination streamlines efforts, avoids duplication, optimises resources, and fosters synergy within the AI ecosystem. The Ministry of Education, in its multifaceted capacities, plays a pivotal role in AI development by focusing on educational reforms. These encompass curriculum enhancements, skill development programs, and collaborations with industry stakeholders to cultivate a talent pool proficient in AI. By aligning policies, resources, and expertise, India can harness the transformative power of AI to address societal challenges, propel economic growth, and establish itself as a global leader in AI innovation and development. This section is dedicated the all such efforts which will propel AI incorporation into the military.

Efforts by MeitY

MeitY leads government efforts to promote education, development, and production of AI tools in the country. With close to 8,000 tech start-ups, the second-largest start-up ecosystem in the world, India is rightly focussing on innovation and entrepreneurship and enthusiastically promoting it. MeitY is facilitating innovation and IPR related activities to expand this ecosystem. 'MeitY Start-up Hub' (MSH)[40] acts as a national coordination, facilitation, and monitoring centre that will integrate all the incubation centres, start-ups, and innovation-related activities. The MSH will also ensure crossing sharing of technology resources across the innovation ecosystem and is focused on emerging technologies, including AI, Internet of Things (IoT), and Block chain. Key features and functions associated with the MSH are:

- Early-stage support – office spaces, infrastructure, mentorship.
- Business development – access to industry experts, investors, government agencies.
- Mentorship – entrepreneurship, technology development, and business scaling.
- Growth and sustainability – connect with investors and funding.
- Networking events – visibility, showcase product, connect with partners.
- Collaborates with government initiatives, aligning with the national agenda.

The Technology Incubation and Development of Entrepreneurs (TIDE) scheme was put in place by MeitY in 2008[41] to promote innovation by nurturing start-ups in the Information and Communication Technology (ICT) domains. Under the scheme, financial assistance is provided to institutions of higher learning to strengthen their technology incubation centres for enabling young entrepreneurs to create technology start-up companies for commercial profit in pre-identified areas of societal relevance. Under the scheme, 27 TIDE centres and two virtual TIDE centres have already been created out of an envisaged 51. To leverage the diversity in the ecosystem, TIDE centres will be categorised into three distinct groups in a layered approach. Established centres[42] will also provide physical infrastructure and a conducive environment for start-ups to work on their innovative projects. These will also offer access to research facilities, laboratories, and technical resources, enabling start-ups to work on cutting-edge technologies and innovations.

- *Group 1 Centres (G1C)*
  - Six already matured incubators with pan-India reach.
  - Deep support to start-ups including mentoring, capacity building, investing, and post-investment advisory. Mentor and supervise the Group 2 centres and Group 3 centres, if needed.

- *Group 2 Centres (G2C)*
  - Either be new incubators or academic/student-focused incubators (total 25) that have deep networks in the local ecosystem.
  - Engage large number of aspiring entrepreneurs and students to build high-quality start-ups.
  - They must also agree to nurture Group 3 centres, if required.

- *Group 3 Centres (G3C)*
  - To initiate and evangelise innovation and entrepreneurship ecosystems in unexplored regions with absent/dormant ecosystems.
  - Recently established incubators (total 20) in academic institutes.

In 2018, MeitY constituted four committees[43] for promoting AI and developing a policy framework. The committees proposed action in the areas of platform and data for AI, leveraging AI for identifying national missions in key sectors, mapping technological capability key policy enablers required across sectors, skilling and re-skilling R&D and cyber security, safety, legal and ethical issues. The committees were to identify key application areas and make suitable recommendation/suggestions, identify areas in government systems to lower costs and improve service, suggest technical framework/AI platform, R &D framework and setting up of centres of excellence, recommend policy relating to legal framework, data privacy and cyber security issues, and identify concerns about use of AI employability, skilling and re-skilling challenges. By July 2019, the committees recommended development of an enriched National Artificial Intelligence Resource Platform (NAIRP)[44] of India to bring together all publicly shareable data, information, tools, literature, solutions, best practices to build capacity and solutions. This will be an open data and knowledge-cum-innovation platform that will be available to all users. This platform will be built in a contributory and participatory manner by all stakeholders, initially driven, and mainly funded by the government through MeitY. The success of the National Digital Library of India (NDLI) project (https://ndl.iitkgp.ac.in/) created by IIT, Kharagpur, is expected to be replicated for NAIRP. A Rs. 100 crore-budget over a period of three to four years is expected to be utilised for the platform. Another committee identified 17 national missions,[45] strangely leaving out national security but emphasising on 'problems that are important, and those which are amenable to AI technology'. The third committee on mapping technological capabilities, policy enablers, skilling, re-skilling, R&D, quoted PM Modi, "We need to make AI in India and make AI work for India", emphasised on Edge or Fog computing[46] through small and powerful AI algorithms, creating and nurturing talent, re-skilling if required and utilising available data in the interim, till usable and cleaner data is collated for the future. The committee also scanned the 'international scenario, contrasting it with the Indian scenario, and reviewed models of AI strategy of other countries', clearly suggesting an 'Indian way', rather than aping the global AI process. The fourth committee studying cyber security, safety, legal and ethical issues recommended international discussions to establish standards. It also gave a unique formula to 'give a fair opportunity to the technology',[47] avoiding excessive regulations 'as it may hinder the growth of the technology'.

Though MeitY has various programs like TIDE, MSH and NGIS, it is also aware of significant numbers of start-ups failing within the first few years due to lack of right funding at the right time, market dynamics, limited user perception, inadequate feedback and thus unable to solve the problem at scale – a gap in accessing the growth stage funding to scale up operations. The idea behind this proposal is to provide start-ups which already have brilliant solutions and proof of concept for their product, the more facilities to enhance their product using innovative technologies for the market with a solid business plan and enable them to easily obtain investments from venture capitalists and angel investors. The accelerators in India are focusing more on the longevity of a start-up and seem to be less concerned with how quickly the company grows. MeitY also acknowledged that there was a need for government intervention to support the start-ups to scale, since venture capitalists are shying from investments fearing corona virus-like pandemics and stricter control over FDI. Start-up Accelerators of MeitY for Product Innovation, Development and Growth (SAMRIDH)[48] was created to support existing and upcoming accelerators to select and accelerate potential IT-based start-ups to scale. The scheme will ensure that the chosen accelerators are in the business of incubation for more than the years and supported more than 50 start-ups of which at least 10 have received non-public investment. This will ensure that weak accelerators do not infect the ecosystem. The chosen accelerators shall be tasked and paid separately to developing customised acceleration programs for each start-up and provide customer connect, investor connect and capacity enhancement. To ensure long commitment and legal liability as followed by various accelerators in the world, MSH, acting as an implementing agency for SAMRIDH, will take up equity in start-ups.

AI will also play a crucial role in making digital environments more powerful. AI allows users to interact with the hardware and software in the XR (extended reality) landscape more realistically and paves the way for gesture control to feedback. With special interest in the future of immersive technologies, MSH and Meta also launched the XR start-up program[49] in January 2023, specifically to discover, nurture, and accelerate XR technology start-ups and innovators across India, especially targeting the educated youth bulge of India. At least 20 per cent of the selection in this program has women innovators and start-ups with women founders/co-founders. The first 80 innovators will attend a boot camp, out of which 16 innovators will be provided grants of Rs. 20 lakh each and further support to help them develop minimum viable product (MVP)/prototypes. Such specific programs will improve the development of technologies across the country.

Understanding a need for supercomputing and cloud services, the Centre for Development of Advanced Computing (C-DAC) a R&D organisation of MeitY, released 'Param Siddhi,' a high-performance computing-artificial intelligence (HPC-AI) supercomputer in November 2020.[50] It has Rpeak of 5.267 petaflops and 4.6 petaflops Rmax (sustained) and its software frameworks and cloud platform will help deep learning, visual computing, virtual reality, accelerated computing, as well as graphics virtualisation. On the NITI Aayog's suggestion to set up an AI-based cloud computing infrastructure, C-DAC implanted AIRAWAT[51] (AI research, analytics and knowledge assimilation). AIRAWAT is also supported by the recommendations of the National Strategy for Artificial Intelligence (NSAI), which has identified the 'absence of enabling data ecosystem' as areas that are hampering the growth potential of AI in India. The framework is expected to provide an alternative to Microsoft Azure and AWS, and address concerns related to data privacy. This will also improve Indian supercomputing ability (though there are two supercomputing systems but not designed for AI applications).

The National Artificial Intelligence Resource Platform (NAIRP), envisioned as a national knowledge repository by the expert committee mentioned earlier in this chapter, rechristened as INDIAai and is now a leading portal for knowledge, research, and ecosystem-building initiative for AI in India. Collaboration with the National Association of Software and Service Companies

(NASSCOM), MeitY also launched 'FutureSkills Prime',[52] an innovative online ecosystem to train learners with technical knowledge and skills. These certification programs are approved by National Occupational Standards (NOS) and National Skills Qualification Framework (NSQF), enabling learners to obtain high in-demand skills valued by the market.

Suggestions from NITI Aayog

India's distinctive challenges and aspirations, rapid advancements in AI technology, and its ambition to establish itself as a frontrunner in this emerging field, necessitate a balanced and comprehensive approach to AI strategy that caters to both local requirements and the broader global good. Charting India's path in AI entails a careful consideration of existing strengths and weaknesses, and demands government driven and transformative interventions on a large scale, with unwavering support from the private sector. NITI Aayog identified many hurdles in its lucid analysis.[53] The hurdles that must be surmounted include the absence of conducive data ecosystems, the limited intensity of AI research, adequate availability of AI expertise and skilled workforce, paucity of awareness in business operations, lack of well-defined regulations on privacy, security, and ethical issues, and unfavourable Intellectual Property framework that fails to incentivise research and the integration of AI.

Providing solutions to identified problems, NITI launched the National Data & Analytics Platform (NDAP) in 2022. The NDAP provides open access to public government data by making it interoperable, and interactive on a user-friendly platform for all users such as policymakers, civil servants, university students and researchers, journalists, innovators, and civil society groups. Prior to NDAP, there was a lack of uniformity in data formats and no interoperability and public data in India was difficult to use; users couldn't compare data from different departments or data gathered over time due to differences in format and quality. This lack of interoperability had 'enormous implications for daily operations of the government'.[54] The platform now enables easy access to simplified data linking and use of several types of data at once, merging and cross-sectoral analysis enabling users to create flexible tables and visualisations. All data on the platform has to meet NDAP's in-house 5-star rating framework, thus enabling high quality and verifiable data-driven governance and outcomes.

Support by PSA

In support of the Prime Minister's Science, Technology & Innovation Advisory Council (PM- STIAC)[55] in 2018, the PSA began facilitating the delivery and progress of all nine national missions under PM-STIAC, including those on Natural Language Translation mission and AI mission. The PSA also launches regular challenges for an Indian company registered under the Companies Act 2013 with 51 per cent or more shareholding with Indian citizens or persons of Indian origin. The system so developed or corpus created as part of the challenge will be kept in the public domain for future use by start-ups and researchers. The designated agency is responsible to take necessary steps to protect its intellectual property rights.[56] As part of collating AI development in the country, the office of the PSA notes numerous successful use cases, which may be beneficial for military use, a few instantaneously, and others with suitable changes.

Indian Space Research Organisation (ISRO) Use Cases. ISRO brought out a 'Respond Basket' comprising urgent and most important research areas to prepare detailed solution proposals on a priority basis

- The Indian Institute of Remote Sensing (IIRS),[57] Dehradun, is researching urban spatial growth modelling. With the recent thrust on urban growth modelling using geospatial data and techniques, there is a need to generate Artificial Neural Network (ANN)-based urban growth potential. This is highly usable by the military to understand enemy area growth indicators.
- The North Eastern Space Applications Centre (NESAC),[58] Shillong, is researching mapping elevation using autonomous UAVs in swarms with field acquisition and resolution control for object tracking and recognition. The idea of having personal assistant UAVs is inspired by Google Assistant, Siri, Bixby, and Cortana. The same idea can be extended to UAV hardware for applications in security, surveillance, asset tracking, hands-free photography, and navigation.

AI video surveillance platform by Government of Uttar Pradesh. An AI-enabled video analytics platform monitors inmates in 70 prisons with over 700 cameras.[59] Increasing crime rates make it imperative for the state police department and the CM office to use technology like AI + IoT and video technology for real-time alerts, crime control and any threats to the security of inmates in prisons across states. Using a dataset of approximately one million historic violence videos, the UP Police is now managing crime with AI as a tool. Key UP Police officials access mobile app for real time insights on potential threats across wide parameters like violence, unauthorised intrusions, illegal mobile phones and weapons. The military can seek similar access in target and designated areas to monitor nuisances and threats.

AI-enabled traffic management system by Bengaluru Police. The Bengaluru traffic police have introduced an Intelligent Traffic Management System (ITMS),[60] where AI-enabled cameras detect traffic violations and issue challans through SMSs to mobile phones. A similar measure is possible to control access to strategic roads along the border and manage military traffic through a congested area.

Face Recognition System[61] (FRS) by Government of Tamil Nadu. FRS deployed by the State Government across government schools saw an 85 per cent drop in absentees and stragglers. The education department estimates that FRS saves almost 45 minutes each day otherwise spent on attendance. This system capable of identifying Indian faces even in low light conditions is easily usable in the military for a variety of role, like access control, attendance, and training.

SUMMARY OF MILITARY AI DEVELOPMENT IN INDIA

The end of year review of 2022 by the Ministry of Defence[62] indicated organisational attempts to develop AI tools and the effort resonates in all departments. Over all, the product identification and technology demonstration of AI products has been impressive, be it iDEX framework, the DRDO's Technology Development Fund (TDF) scheme or 'Dare to Dream' innovation contest. iDEX challenges since 2019 have resulted in the signing of 134 contracts (25 AI products) and TDF has helped to sign 64 contracts (seven AI products) in the same period.[63] A detailed analysis of the entire range of AI processes and development in this section will summarise the AI road map for the Indian Army.

Various committees and task forces identified gaps in AI development and recommended milestones. An analysis of the most important from a national security point of view indicates that work is underway and a fair majority of recommendations are either completed or in the penultimate stages:

- The Department of Defence Production issued a government order[64] that was triggered by the Chandrashekhar TF report in 2019. A quick critique indicates that various stages of development high-level committees like Defence AI Council (DAIC) under the chairmanship of Raksha Mantri and Defence AI Project Agency (DAIPA) are fully functional to deliver leadership vision of AI in defence. Defence training centres and institutes are already organising various AI training courses, which will of course undergo midcourse corrections as the military gains more confidence in the AI tools. Frameworks like iDEX and TDF exist to encourage conjoined work with industry to encourage start-ups and develop AI capability for defence. A data management framework, scaling the existing capability of data centres is underway and the defence industry has to undertake a substantial effort to catch up with well-established industry standards.

- The task force on AI for India's economic transformation by the Ministry of Commerce and Industry[65] in 2017 acknowledged few enablers for national security. Negative import lists rechristened as positive indigenisation lists have proved to be useful to create and support a wide-based consortium of MSMEs (subsystems and components), involving multi-disciplinary sectors and fields in emboldening national security. Legislatives like the Digital Personal Data Protection Act, 2023, are in place to provide digital assets and data protection from all cyber threats, though 'genuine data and cyber security tools and methodologies' are still publicly scarce. 'Integrated and unified platform for existing infrastructure' is not yet created and AI-based techniques to provide curated real-time information to various security agencies is under development.

Analysis of DAP 20. DAP is a complex decision-making process to enable expeditious procurements and indigenisation. It contains voluminous aspects of governance and compliance, which are difficult to comprehend, especially to the newly initiated – MSMEs and start-ups of the 'wider defence ecosystem' all fall under this inexperienced category. Clearly, the propensity of delayed acquisition exists and has to be guarded against. The probity and transparency clauses in DAP involve many agencies which inadvertently diffuse responsibility and accountability during the acquisition process. Decision-making is strictly hierarchical and time intensive, which also delays decisions and product deliveries. It is also observed that DAP stresses on probity more than operational factors like logistics, training and time-bound procurement. It would be prudent for all stake holders to find probity in the time domain and take quicker decisions. Additionally, 80 per cent Services Qualitative Requirements (SQRs) entails a change in the SQRs, which is a strict hierarchical process, accumulating time penalty. The Indian Navy is considering a spiral development[66] process (with 80% SQR) that seems to be a good interlude to the delay, with a long-term aim to achieve all SQR subsequently. The Indian Navy experimented with minor tweaking in the existing iDEX procedure to launch 75 SPRINT challenges on 18 July 2022.

Analysis of Technology Perspective and Capability (2018). TPCR offers an insight into the equipment expected to be integrated into the Indian armed forces until the late 2020s. Its primary purpose is to initiate technology development efforts by the industry and emphasises 'Make in India' initiatives. Involvement of private and public sectors, with inclusion focus on MSME will boost indigenous production. The first edition incorporates more informative notes for industry and business decision-making and participation. As an informative document only, it clearly abdicates from any procurement commitments. An Indian military TCPR was released in 2018 prior to the Chandrashekhar TF of MoD and almost in time with AI TF for India's economic transformation and this date-release analysis explains why AI does not find any mention in TCPR 2018. It is assumed that the next edition of the TCPR will contain all the AI-related military expectations.

'AI in Defence' Compendium Analysis. The Ministry of Defence compiled a list of 75 AI at the 'AI in Defence' symposium in 2022.[67] It lists products from various domains and includes new vendors too. The compendium is a good framework to understand the defence AI pathway in India. In-house development of products by the defence services is a typical insight and creates a unique pool of cross-functional teams – a technologist who is a tactician too. The chart below lists standard military vendors, while other products (28) are being developed by yet unknown but capable inventors. AI products for human behaviour analysis, logistics, and supply chain management, process flow automation for large systems and simulators/test equipment are under development by these newer vendors. Various agencies are actively involved in AI product development across a wide range of categories. This reflects the growing interest in AI technology across the Indian defence and security sector. The development of AI tools for C4ISR appears to be a significant focus area, with multiple agencies investing in this category, including BEL, DRDO, and the Indian Army. The fact that multiple agencies are working on the same categories, such as NLP and intelligence monitoring, suggests a coordinated approach to tackle common challenges. Product priorities appear to be set-based on specific needs and capabilities. C4ISR, autonomous systems, and security-related applications seem to be of particular interest, indicating a strong focus on enhancing situational awareness, decision-making, and security measures including the importance of safeguarding digital assets and infrastructure.

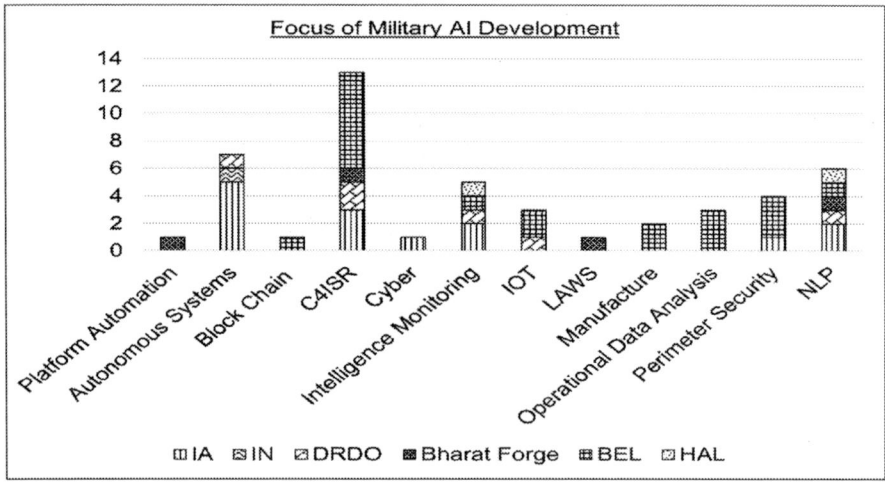

ADB's CPDS 2023 Analysis. The CPDS addresses 30 critical AI problem definitions,[68] aligning with the Army's focus on optimising information dominance and enhancing situational awareness. Eight definitions aim to improve full-spectrum information acquisition from open sources and tactical scenarios, ensuring soldiers have uninterrupted tactical visibility across diverse terrains and environmental conditions. Efforts also include refining tactical picture delivery for legacy observation devices (six problems) and integrating multiple devices for a more informed operational picture. Additionally, an NLP solution is sought for tactical-level communication on the northern borders. Force capability is bolstered by two Manned-Unmanned Teaming (MUT) solutions, combining intelligence, surveillance, and reconnaissance with targeting options in plain and desert sectors. The Indian Army places significant emphasis on AI training tools, spanning from the metaverse to virtual and augmented reality (AR/VR) domains. Four of such statements concentrate on mechanised forces, and one on artillery fire controller training that offer a substantial advantage to Indian soldiers. Three AI-supported forensics tools aim to enhance force protection capabilities, addressing post-injury mobility and complaint management. Logistic operations are anticipated to leverage AI for smart warehousing, with discussions underway for tactical logistic delivery via land and aerial-based drones. The document also gives the work flow of the problem definitions, which will guide and encourage developers to interact for better solutions.

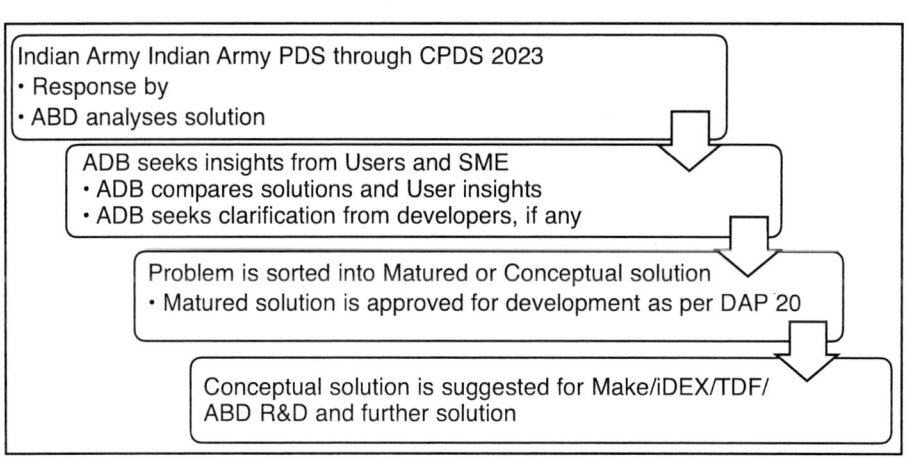

Academia, DRDO, DPSU, Startups, Industry

Analysis of iDEX Challenges. Similar analysis for all 12 DISC challenges iDEX[69] challenges launched till the middle of 2023 including three PRIME indicates that 245 challenges were 33 were related to AI. The chart below indicates 18 per cent are related to logistics, 30 per cent to decision support and 52 per cent to provide force protection.

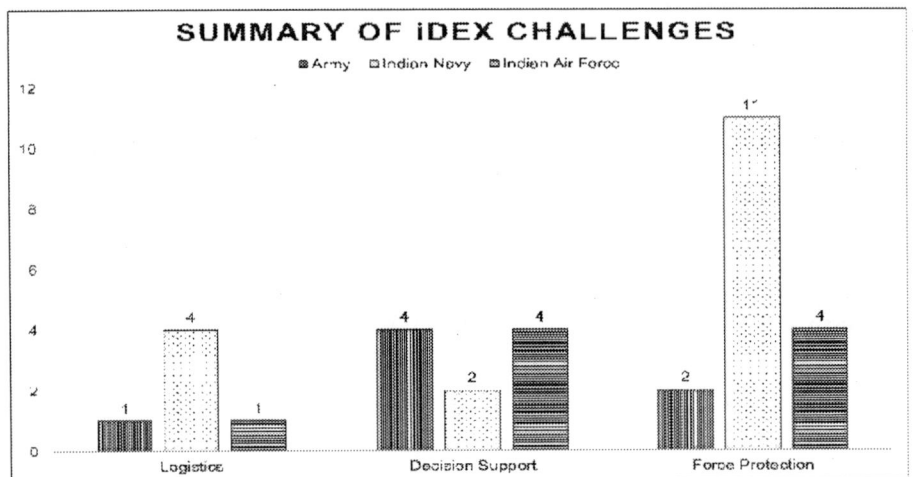

Analysis of CAIR, DRDO. CAIR developed an AI-based COVID detection application software, ATMAN,[70] built with a deep convolution neural network. It is trained on a few hundred sample images and can classify targeted chest X-rays images into normal, COVID-19 or pneumonia classes. This X-ray based diagnostic tool can detect the infection in the early stages in seconds. The software is highly flexible and accessible over Internet and is light enough to be used in a variety of devices like mobiles, tablets, laptops or computers. Its usage is easily relatable to military applications for all types of image analysis, where the users provide training datasets and assist in final validation.

Sharing of High Computing Facilities. Tangible AI products in the previous sections are intrinsically linked to computational power and its availability to developers. Simultaneously, access to secured datasets is also essential. Computational power is the lifeblood of AI, powering its evolution from simple rule-based systems to the intricately woven neural networks of deep learning. The synergy between cutting-edge hardware and sophisticated software algorithms is the driving force behind AI development. Obviously, such secured environment and computational power resource and finance intensive are not commonly available. The following repertoire of resources in India provides the impetus to AI research.

- **AI Research Analytics and Knowledge Dissemination Platform (AIRAWAT) and Param Siddhi AI (PSAI) System.** The exploration to offer shared resources led to the development of the National PARAM[71]

Supercomputing Facility (NPSF), a High-Performance Computing (HPC) capability by the Centre for Development of Advanced Computing (C-DAC), MeitY. Its latest creation is Param Siddhi AI (PSAI)[72] – a high-performance computing artificial intelligence (HPC-AI) supercomputer. It offers a reliable and suitable parallel and distributes processing technologies and resources to AI researchers and developers. Complementing the computing, the AI Research Analytics and Knowledge Dissemination Platform (AIRAWAT) also came into being in May 2023. Combined, the AIRAWAT and PSAI system will provide theoretical peak performance of 410 AI PF/13.17 PF (DP); achieving 8.5 PF (DP) sustained performance. Each of the 82 AIRAWAT-PSAI compute nodes hosts hyper-threaded 128 core AMD CPU with 1TB RAM, 40GB NVIDIA A100 – SXM4 GPU, 14 TB local storage with 1.6 TB/sec networking by Mellanox ConnectX-6 VPI (InfiniBand HDR). Composite system Specification[73] given in the table below are impressive and suitable for many AI development roles.

NVIDIA DGX-A100 Compute Nodes	82 (20992 CPU cores)
Total host (compute node) memory	82 TB (82 nodes * 1 TB per node)
NVIDIA A100-40GB Tensor Core GPUs	656 (82 nodes * 8 GPUs per node)
Total GPU Memory	26.24 TB (82 nodes * 8 GPUs per node * 40 GB per node)
Mellanox 200G HDR InfiniBand switch having 320Tb/saggregate switch throughput (Compute Communication)	800 ports (20 leafs * 40 ports per leaf)
Mellanox 200G HDR InfiniBand switches (Storage Delivery)	400 ports (10 switches * 40 ports per switch)
PFS based storage (network attached) @ 250 GB/Sec, 4M IOPs	10.5 PiB (2-tier storage)

- **Technical Affiliation Scheme.** C-DAC provides cost-effective accessibility to researchers and developers.[74] The utilisation for CPU core is measured and charged on minute basis and GPUs/other accelerators are charged in accelerator hours. High-bandwidth shared storage as a default quota is allocated to each project, while additional storage is charged on a monthly basis. Dedicated allocation and special consultations are also provided at additional cost. Sharing computational power harnesses resources on-demand without the need to invest in expensive hardware and infrastructure. It will boost MSMEs, research projects, and individuals with limited budgets, putting idle or underutilised computing resources to work, maximising the efficiency of hardware. Researchers are provided access to specialised hardware or accelerators, such as GPUs (Graphics Processing Units) or TPUs (Tensor Processing Units), which are crucial for machine learning, deep learning, and scientific simulations. National RARAM Super Computing Facility (NPSF) annual report 2021[75] indicates high success and utilisation of the scheme:

- No users have yet subscribed to more than 252 GPUs,[76] thus indicating availability of a huge computing capability in the system, maximum hours (42,000) were spent using a cluster of 13-28 GPUs.
- The supercomputing is impressive as almost all jobs (19,386) get executed within 0.1 hour and only 1.03 per cent took 7 days for execution.
- Conversational AI has the highest usage of GPU time utilisation (817,710.55 GPU hours) which includes a variety of applications like automatic speech recognition (ASR), text-to-speech (TTS), and natural language processing (NLP).

- **Meghdoot Cloud.** C-DAC also developed a free and open-source Cloud suite based on OpenStack, named Meghdoot,[77] for establishing a Cloud computing environment that offers Infrastructure as a service (IAAS), Virtual desktop infrastructure, platform as a service (PAAS) and software as a service (SAAS) to Indian users. Combining Meghdoot cloud services with AIRAWAT and PSAI systems, creates a sophisticated ecosystem for high computing research across the country.

Analysis of National Artificial Intelligence (AI) Resource Platform (NAIRP). The Ministry of Electronics and Information Technology (MeitY) manages this portal, now known INDIAai (The national AI portal of India) to 'foster inclusion, innovation, and adoption for social impact'. It is expected to become a major content repository for AI in India. It has jumped in global rank[78] (#430,108) by 76,000 places, seeing an increase of traffic by 55.19 per cent. The platform also observed an increase in foreign visitors – USA (+242.51), Canada (+843.92) and Czech Republic (+77.78) while the United Kingdom saw a drop of 21.34 per cent over the last few months.[79] It provides articles, research reports, case studies, AI standards of world-class policy developments and accepted standards in AI development and curated datasets to aid research initiatives.

Analysis of National Data & Analytics Platform (NDAP). It provides 2,001 datasets from across 17 sectors and 49 ministries.[80] The National Data and Analytics Platform (NDAP) improves access to the Indian government's extensive administrative landscape data, allowing users to search, merge, visualise, and download datasets. It has a bounce rate of 44.44 per cent only and accessed by users from the USA and Romania.[81] It provides a comprehensive approach to enhancing data accessibility and analysis. The capacity to merge datasets from various sources enables thorough analysis, uncovering patterns and correlations that might be overlooked in isolated datasets. Visualisation tools enhance data understanding and communication, making intricate information more digestible. Additionally, the option to download datasets supports reproducibility, offline analysis, and integration into diverse research workflows. This platform not only facilitates individual research but also fosters a collaborative community, encouraging engagement, feedback, and educational opportunities for users with varying levels of technical expertise. By leveraging such repositories, the Indian Army can not only enhance individual research capabilities but also foster collaboration, encouraging engagement, feedback, and skill development with varying technical expertise.

Global AI Index. Stanford University publishes an AI Index since 2017. Such indices play a pivotal role serving as essential benchmarks, guiding policymakers in tailoring strategies to fortify the AI ecosystem and foster innovation. They attract vital investments by showcasing India's burgeoning AI capabilities. The global AI indices open doors for international partnerships, to harness shared expertise for accelerated advancements in AI research. In the specific context of the Indian military's AI endeavours, a thorough examination of these indexes is imperative. It unveils nuanced insights into India's global AI standing, providing a roadmap for optimising military applications in alignment with development trends, ensuring technological prowess and strategic advantage. The indicators present a very novel view of the India recognition over time.

- While the first report in 2017 had no mention[82] of Indian capabilities, the 2018 report mentions India's increased growth rate of 297 per cent[83] in robot software download, behind China (1,711 per cent and Japan

581 per cent) since 2017. In the same year, India showed high percolation[84] of AI into various organisations in the field of robotic process automation, machine learning, conversational interfaces, computer vision, NL text understanding, NL speech understanding, NL generation, physical robotics and autonomous vehicles. India showed 197 instances of AI use in various organisations, behind Europe (n=803), USA (n=497) and Asia Pacific (n=263). Developing markets, including China, had a lower penetration of n=189. India increased its annual robot installation from under 2,000 before 2015 to 3,412 by 2017, an impressive sixth global rank.[85]

- 2022 report saw an interesting twist when Indian data was entered due to the country's global importance rather than pure numbers,[86] indicating an undeniable and robust Indian presence in the global technology environment. Indians led the AI skill penetration rates (a relative penetration rate of 2 meant that the average penetration of AI skills in that country is 2 times the global average) at an impressive 3.09 times the global average,[87] increasing to 3.23[88] in 2023 index, followed by the United States of America (2.24) and Germany (1.7). India leads this penetration rate across all the sectors covered in the AI Index report, namely, education, finance, hardware, manufacturing and software.

- Key findings of the 2023 index summarise[89] that the industry now leads academia in developing the most significant machine learning models as a business necessity and AI adoption is pulling industries away from its competitors. AI models are accelerating scientific progress. There are increased demands for AI-related professional skills as also increased number of AI misuse, provoking concerns by policymakers. Indian AI presence has increased substantially; it is now the fifth (5.56 per cent) largest publisher of AI journals[90] and fifth (6.79 per cent) largest AI conference publication[91] printer, a massive jump from only 1.30 per cent in 2010.

- Indians are leading the world in GitHub projects[92] (24.19 per cent) way beyond the nearest contributor – the European Union and United Kingdom (17.30 per cent), although obtaining only 0.46 GitHub Stars.[93] The number of GitHub stars on a repository usually indicates its popularity within the community, representing a user's interest or endorsement of a particular repository. However, the absence or a low number may not necessarily indicate a lack of quality or usefulness. It is possible that newly-created projects may not have gained widespread attention yet, or projects may not have been actively promoted. Niche repositories usually have a smaller user base, leading to fewer stars compared to more general-purpose projects. All in all, this is a major talent pool awaiting exploitation in India.

DPDP Act 2023 and AI. Gazetted in August 2023, the Indian Digital Personal Data Protection Act 2023[94] is unique and seminal. Most importantly, it is not standalone; instead, reasonable amendments to certain other Acts too have been done. The DPDP Appellate Tribunal now finds a mention in Section 14 (clause c) of the Telecom Regulatory Authority of India Act, 1997. The Information Technology (IT) Act 2000, omits its Section 43 A, and Section 87, (sub-section (2), clause (ob)) since the DPDP 23 provides highly clear provisions to recognise both the right of individuals to protect their personal data and the need to process such personal data for lawful purposes and connected or incidental matters. Concurrently, the DPDP also clarifies Section 8 of the Right to Information Act, 2005 (sub-section (1), clause (j)) as personal information. DPDP 23 also finds itself governed under Section 81 of the IT Act 2000 alongside the Copyright Act, 1957 and Patents Act, 1970. AI inferences are found at numerous places; section 2(s)(vii) includes 'every artificial juristic person' (AI agent or a system) within the gambit of 'person', Machine learning or computer vision or natural language processing gets covered under 'automated' (section 2b), Each data processing company (Data Fiduciary – section 2(i)) is expected have a data protection officer (section 2(l)), consent managers (section 2 g) will act as single point authority to provide consent for use of personal information on behalf and concurrence of the data principal (the individual to whom the personal data relates – section 2 (j)). This Act also secures all personal data till the time the individual gives explicit permission to use it in the manner desired by the individual; it can range from free, specific, informed, unconditional and unambiguous with a clear affirmative action, including specific purpose 9, section 6 (1). Companies may not use personal data unless 'accompanied or preceded by a notice' (section 5 (1) to the individual.

The Act provides clear guidelines on the types of data which can be used and the necessity of the consent from individual data which has to be used.

Insofar Indian companies will have to follow legal processes to acquire data to train its AI models, which was almost open-ended till the Act was passed. In 2022, Naor Gilon, Israel's Ambassador to India, mentioned IP problems in India.[95] He advised due caution to respect such legal rights. Tata Constancy Services (TCS) of the Tata Group was hit with a fine of $ 125 million in an intellectual property infringement[96] case filed by healthcare firm Epic Systems. Before its official notification, the Data Protection and Privacy Draft (DPDP) Bill faced significant opposition from privacy advocates and technology companies. Prior to notification, concern centred on the expanded access granted to government agencies and the inclusion of non-personal data under the regulatory purview. Privacy advocates argued that the broader access to data by government agencies could infringe upon individual privacy rights, leading to increased surveillance and potential misuse of sensitive information. The apprehension regarding non-personal data coverage stemmed from the fear of unintended consequences. While the intent might be to safeguard personal information, extending regulations to non-personal data could inadvertently stifle innovation and hinder the free flow of information that has fuelled technological advancements. Tech companies have sought newer models to overcome the restrictions and guidelines of these regulations. The transition from an open data environment to a regulated one hopefully will de-motivate entities with potentially unscrupulous motives that had previously exploited the lack of stringent controls. However, proponents of the DPDP argued that such regulations were necessary to strike a balance between protecting individual privacy and fostering a responsible data ecosystem that encourages innovation while minimizing risks associated with data misuse.

NATIONAL AI MISSION AND INDIAN MILITARY

In January 2018, a 'task force on AI for India's economic transformation recommended concrete steps for AI transformation in India, supporting the launch of the National AI Mission in March 2024 as part of the nine missions under the Prime Minister's Science, Technology & Innovation Advisory Council (PM-STIAC). Set up to advise the Prime Minister on all matters related to science, technology and innovation and to monitor the vision the council is mandated to offer 'action oriented and future-preparedness advice to the government'. The nine PM-STIAC missions include natural language translation, quantum frontier, national biodiversity mission, electric vehicles, bioscience for human health, waste to wealth, deep ocean exploration, AGNIi (Accelerating Growth of New India Innovation) and artificial intelligence.

The National AI Mission is led by MeitY.[97] It focuses on benefits to India 'societal needs in large and specifically on healthcare, education, agriculture, smart cities and infrastructure, smart mobility and transportation' – an extension of 'Making AI in India and Making AI Work for India'. It is intrinsically linked to endure the nine out of 17 UN Sustainable Development Goals: eliminate poverty, establish good health and well-being, provide quality education, create decent work and economic growth, increase industry, innovation and infrastructure, reduce inequality, mobilise sustainable cities and communities, influence responsible consumption and production and build partnerships for the goals.

The AI Mission will receive Rs. 10,300 crore from FY 23-24 till FY 27-32. To harness Indian expertise in management and science, the services of the Digital India Corporation (DIC), a not-for-profit company set up by MeitY is running the AI Mission through INDIAai Independent Business Division (IBD). The DIC seeks a judicious mix of talent and resources from government and the market for successful and timely missions. A structured implementation through a public-private partnership model will boost the AI innovation ecosystem. There is a natural division of the components[98] addressing the large AI ecosystem, collaborating rather than competing within and with each other. Each of these seven collaborating components functions with dedicated initiatives.

- **INDIAai Compute Capacity.** Understanding the need for high-end computing; the pillar will build a high-end scalable AI computing ecosystem. This will cater to the increasing demands from expanding AI start-ups and researchers. This pillar will comprise 10,000 or more Graphics Processing Units (GPUs), built through public-private partnership. An AI marketplace will be also be designed to offer AI as a service and pre-trained models to AI innovators, acting as a one-stop solution for AI innovation. AI is an expensive enterprise and India wants to optimise hardware and software infrastructure to support AI computation. This pillar has specific initiatives for tangible goals.

AIRAWAT, National Supercomputing Mission, and MeitY Quantum Computing Applications Lab offer many development options for start-ups and researchers. As the importance of 'owning compute facility' grows, the technology giants with deep pockets continue to secure vital AI computing assets. Microsoft invested USD 1 billion (2019) and USD 10 billion (2023) in Open AI while IBM has invested USD 6.5 billion (2022) in the research.[99]

- **INDIAai Innovation Centre (IAIC).** As a lead academic institution, IAIC will streamline implementation and retention of top research talent. It will spearhead the development and deployment of foundational models, with a specific emphasis on indigenous large multimodal models (LMMs), domain-specific models, leveraging edge and distributed computing for optimal efficiency. Specific emphasis on edge and distributed computing indicates India-specific architecture and an important guiding beacon for ministries and the military.

- **INDIAai Datasets Platform.** This pillar will utilise a unified data platform to improve access to quality non-personal datasets. Special emphasis to enhance accessibility, quality, and utility of public sector datasets, will assist data-driven governance. It is fair to assume that reliable and available data will fuel the development and capabilities of AI, enabling insights, predictions, and intelligent decision-making. Related initiatives in this component include data management office, India datasets program and India data platform

- **INDIAai Application Development Initiative.** This component will promote AI applications in critical sectors for problem statements sourced from central ministries, state departments, and other institutions. To democratise the solutions and coordinate development initiatives, a common problem statement documentation resource will improve efficiency and delivery mechanism. The initiative will focus on 'developing/scaling/promoting adoption of impactful AI solutions with the potential for catalysing large-scale socio-economic transformation'. Cognisant of intellectual property, and innovative practices as a major confidence-building mechanism, this component intends to prove succour to new developers and permit them the ownership of their respective products. Related initiatives include AI CoEs, National Centre on AI and MeitY Start-up Hub (MSH), that

will assist in AI IP and innovations mechanisms, procedures and control.
- **INDIAai Future Skills.** INDIAai Future Skills will provide the skill and education for AI programmes. By democratising and standardising AI skills, the developers will obtain entry in undergraduate, Masters-level, and Ph.D. programs. Data and AI labs in Tier 2 and Tier 3 cities across India will create wide foundational level skill sets by expanding the reach of AI education. Related initiatives in this component include future skills prime, transformation of ITIs/polytechnics and responsible AI for youth.
- **INDIAai Start-up Financing.** The INDIAai Start-up Financing pillar will support and accelerate deep-tech AI start-ups and provide them streamlined access to funding to enable futuristic AI projects.
- **Safe & Trusted AI.** India understands the need for adequate guardrails to advance the responsible development, deployment, and adoption of AI. This pillar proposes to create adequate and indigenous solutions for standardisation, self-compliance, governances, and audit. As part of ethics and governance, the pillar will also develop and guide responsible and transparent deployment of AI systems to ensure fairness, accountability, and societal benefit. Citizens and government will both share common threads of AI ethics. This pillar supports the initiatives of Digital India Bhashini and India Stack & AI.

The Indian National AI Mission aims for Sovereign AI[100] and wants to create a niche for itself amongst the likes of Google and Meta. Consolidation of effort to acquire financially expensive GPU, appropriate skill management, and application development will assist in dataset and IP management. Public-private partnership shows intent to utilise the best of the two powerhouses for a common goal rather than competing with each other. When examining Indian military AI initiatives alongside the broader AI mission, many parallels emerge with some notable alignments. It would be prudent for military AI to metaphorically 'avian draft' with the National AI Mission, benefiting from consolidated efforts. This collaboration may seem contra-indicatory to military security sense but will have to rework and adapt to such concerns and considerations for improved AI utility.

- The military should utilise the available AIRAWAT, Param Siddhi AI (PSAI) System and C-DAC's Technical Affiliation Scheme (TAS) for cost-effective accessibility. The dedicated allocation and special consultations of these national assets provide an AIAAS as a viable option. Co-sharing computational power harnesses resources on demand without the need to invest in expensive hardware and infrastructure. Conversational AI has the highest utility and includes a variety of applications like Automatic Speech Recognition (ASR), Text-to-Speech (TTS) and NLP, of specific interest to the military. High bandwidth shared storage as a default quota to each project provides exclusive rights to the developers, thus alleviating the military information security concerns. Combining Meghdoot Cloud services with AIRAWAT and PSAI systems creates a sophisticated ecosystem for high computing research across the military region and country.

- Engaging with the INDIAai datasets platform represents another area where the military must reconsider its more reserved approach. The national dataset will be secured with the best minds and systems, over which the military may be able to layer its own secure mechanisms, but India Stack and the associated infrastructure will be highly useful for military needs and may not be forsaken. Universal availability of such datasets will provide larger operating landscape for all military endeavours.

- AI Skilling is yet another facet where the military should hop on to the national resources. Be it the INDIAai Innovation Centre, INDIAai Future Skills or INDIAai Application Development Initiative, the military will never be able to dwell on AI skilling in solitude. As a disciplined force, it will be beneficial for the military to hook on to the largest knowledge gateway of the AI mission in pursuance of its skilling goals. Affiliating the military and technical schools of instructions with these pillars will create human and knowledge bridges for future developers and users. Already mentioned as a challenging enterprise, Indian fears misuse of unreliable LLM[101] and thus the military may also want to stay away and instead leave its development to IAIC. Specific emphasis on edge and distributed computing by this pillar will remain a special interest for the military and may become a major collaboration point with the national resource.

- The military will have a major role to play in the safe & trusted AI

component of the AI Mission. Interestingly, the MoD is represented by DRDO in the national mission, leaving out the Department of Defence (DoD), and Department of Military Affairs (DMA). It would be cavalier to imagine that AI ethics will and should be the sole prerogative of the developer and not the user or the planner. A composite input of the MoD will be highly appropriate, over a purely scientific outlook.
- Placing trust on INDIAai application development initiative will allow the military more flexibility to redefine its critical problem statements in collaboration with other governmental bodies and improve delivery mechanism.

CONCLUSION

The comprehensive examination of AI development in the Indian military underscores the promise inherent in policies and opportunities for young developers. Nevertheless, a labyrinth of challenges emerges, intertwined with the hierarchical structure prevalent within the Army. The potential conflict between the Data Protection and Privacy Act (DPDP Act) of 2023 and the propensity to seek leaner measures through 'Indian jugaad' adds layers of complexity to this landscape.

Nandan Nilekani, often hailed as 'India's CTO' within the tech community, emphasizes a pragmatic approach to artificial intelligence (AI) development. He suggests attention towards efficient and low-cost applications[102] of AI rather than fixating solely on the creation of larger language models, urging a shift in focus from sheer scale to impactful, resource-conscious implementations. As AI models become a commodity, Nilekani cautions against a rat race and suggests 'use of open-source models'. In line with 'AI for All', it is evident Indian developers have to lay emphasis on 'Now AI' aligning with the broader discourse on responsible AI to strike a balance between innovation and practicality and the importance of developing AI applications that are not only scalable but also socially responsible.

An over-zealous reliance on the Defence Acquisition Procedure (DAP) injects additional intricacies into the equation. Positioned as a multifaceted decision-making process, the DAP seeks to expedite procurements and foster indigenisation. However, the formidable governance and compliance aspects embedded within it become formidable obstacles, particularly MSME and start-ups within the broader defence ecosystem. The hierarchical decision-making structure, exacerbated by the probity and transparency clauses, inadvertently diffuses responsibility and accountability, potentially resulting in delays in acquisition and product delivery. A discernible imbalance emerges, with an overemphasis on probity within the DAP, often at the expense of critical operational factors such as logistics, training, and time-bound procurement.

The consideration of a spiral development process, embracing an 80 per cent adherence to Services Qualitative Requirements (SQRs), coupled with the 'formal tweaking' of existing processes, offers a glimmer of hope for addressing these challenges. This strategic flexibility, aligned to address capability gaps, signals a proactive stance toward fostering innovation. It not only aligns with the dynamic nature of AI development but also represents a pragmatic response to the impediments posed by the current acquisition framework. This balanced approach, meeting immediate requirements while allowing the necessary flexibility for future adaptations, is poised to rectify systemic delays.

Understanding the intricacies of the development landscape sets the stage for the subsequent chapter, which will delve into envisaging future conflict scenarios for the military, with specific focus on the army. By marrying existing development capabilities with likely scenarios, the decision to select the most fitting AI tools for future military conflicts, becomes easier.

In anticipation of future conflict scenarios, it becomes imperative to assess how AI can be harnessed to enhance military preparedness and responsiveness. The evolving nature of warfare demands a nuanced understanding of the role AI can play in decision-making, intelligence analysis, logistics optimisation, and strategic planning. Additionally, exploring the potential integration of AI into autonomous systems and unmanned platforms will be crucial in shaping the military's posture for the future.

KEY TAKEAWAYS

- Government initiatives and high-level governance is proving essential for AI development in India. The Ministry of Defence's initiatives like iDEX and TDF are in process of delivering value products.
- The developers, especially start-ups, do not fully understand the military landscape, environment, and geography. This puts them at a disadvantage,

since their proposals dwell in terms of fantastical ideas and low utility. Business development continues to remain a major concern for developers.
- DAP presents hierarchical decision-making challenges; balancing probity and designating operational factors as crucial.
- 'AI in Defence' compendium showcases diverse applications, thus providing for substantial military needs across a wide timeline.
- High-performance computing facilities like AIRAWAT and PSAI offer computing resources for AI algorithm development, though still not fully utilised.
- National AI platforms like NAIRP and NDAP provide valuable datasets for AI research, though still not fully utilised.
- Though India has achieved global recognition in AI development, substantial efforts are still required to build probity and reputation. Laws and legislative measures provide succour and legal robustness to developers.
- The DPDP Act 2023 regulates data use, impacting AI model development and will stretch the efforts of the developers, who are till date 'untethered'.
- India and its military have to stay abreast of evolving data protection regulations around the world to ensure compliance and reorient development in a well-organised manner.
- Explore international collaborations to improve innovation, development, and market for Indian products.

NOTES

1. https://economictimes.indiatimes.com/tech/ites/meet-professor-hn-mahabala-the-man-who-mentored-indias-it-icons/articleshow/53346662.cms?utm_ source=contentofinterest&utm_ medium=text&utm_campaign=cppst, accessed on 11 July 2023.
2. Commerce and Industry Minister Sets up Task Force on Artificial Intelligence for Economic Transformation, Press Information Bureau, Government of India, Ministry of Commerce and Industry, 25 August 2017, https://pib.gov.in/newsite/PrintRelease.aspx?relid=170231, accessed on 11 July 2023.
3. Artificial Intelligence Task Force Constituted by Ministry of Commerce and Industry, Government of India, https://www.aitf.org.in/, accessed on 11 July 2023.
4. Report of Task Force on Artificial Intelligence, https://dipp.nic.in/whats-new/report-task-force-artificial-intelligence, accessed on 11 July 2023.
5. Raksha Mantri Inaugurates Workshop on AI in National Security and Defence, Press Information Bureau, Government of India, Ministry of Defence, 21 May 2018, https://pib.gov.in/newsite/PrintRelease.aspx?relid=179445, accessed on 11 July 2023.
6. Ibid.
7. AI task force hands over Final Report to RM, Press Information Bureau, Government of India, Ministry of Defence, 30 June 2018, https://pib.gov.in/newsite/PrintRelease.aspx?relid=180322,

accessed on 12 July 2023.
8. Task Force for Implementation of AI, Ministry of Defence, 28 March 2022, https://pib.gov.in/PressReleaseIframePage.aspx?PRID=1810442, accessed on 12 July 2023.
9. Updated Version of DAP—2020 (18 April 2023), p. 256, https://mod.gov.in/dod/sites/default/files/wn25423.pdf, accessed on 8 November 2023.
10. Updated Version of DAP 2020, pp. 7-12, https://mod.gov.in/dod/sites/default/files/wn25423.pdf, accessed on 8 November 2023.
11. TPCR 2018—MOD Acquisition Wing, https://www.mod.gov.in/sites/default/files/tpcr.pdf, accessed on 8 November 2023.
12. Updated Version of DAP 2020, p. 29, https://mod.gov.in/dod/sites/default/files/wn25423.pdf, accessed on 8 November 2023.
13. Chapter II, Updated Version of DAP 2020, https://mod.gov.in/dod/sites/default/files/wn25423.pdf, accessed on 8 November 2023.
14. Defence Acquisition Procedure 2020. Government of India, Ministry of Defence, Chapter III, para 7, pp. 323, 324, https://www.mod.gov.in/sites/default/files/DAP2030new.pdf, accessed on 13 July 2023.
15. iDEX Initiative. Press Information Bureau, Government of India, Ministry of Defence, 8 August 2022, https://pib.gov.in/PressReleaseIframePage.aspx?PRID=1849786, accessed on 13 July 2023.
16. Defence Innovation Hubs. Press Information Bureau, Government of India, Ministry of Defence, 11 February 2019, https://pib.gov.in/newsite/PrintRelease.aspx?relid=188372, accessed on 13 July 2023.
17. Defence India Start-up Challenge. Press Information Bureau, Government of India, Ministry of Defence, 24 September 2018, https://pib.gov.in/Pressreleaseshare.aspx?PRID=1547093, accessed on 13 July 2023.
18. Open Challenge. Department of Defence Production, Ministry of Defence, Government of India, https://idex.gov.in/disc-category/5 accessed on 16 July 2023.
19. SPARK—Support for Prototype and Research Kickstart (in Defence), https://idex.gov.in/sites/default/files/2020-09/5d5fc4f2c701def4b72aad9c_SPARK_-Support_for_Prototype_and_Research_Kickstart_in_Defence_framework_under_iDEX.pdf, accessed on 16 July 2023.
20. https://idex.gov.in/sites/default/files/2022-04/i4D%20guidelines%2020-4-22.pdf, accessed on 5 September 2023.
21. https://srijandefence.gov.in/ProductList, accessed on 14 November 2023.
22. https://www.similarweb.com/website/srijandefence.gov.in/#geography, accessed on 14 November 2023.
23. CPDS 2023, https://indianarmy.nic.in/writereaddata/adb-documents/Compendium%20of%20Problem%20Definition%20Statement%202023.pdf, accessed on 4 November 2023.
24. https://mod.gov.in/sites/default/files/DRDOall1412220.pdf, accessed on 25 August 2023.
25. https://ipindia.gov.in/writereaddata/Portal/ev/sections/ps35.html, accessed on 25 August 2023.
26. https://www.drdo.gov.in/adv-tech-center, accessed on 25 August 2023.
27. https://cpdm.iisc.ac.in/ril/research/, accessed on 4 September 2023.
28. https://hitex.co.in/news/iit-hyderabad-and-drdo-collaborate-for-advanced-technologies.html, accessed on 4 September 2023.
29. https://indiaeducationdiary.in/iit-jodhpur-awarded-516-degrees-in-various-academic-programs-in-its-8th-convocation/, accessed on 4 September 2023.
30. http://www.beta.iitkgp.ac.in/files/RD0520235.pdf, accessed on 4 September 2023.
31. https://ai.iith.ac.in/research/ai-research-centre.html, accessed on 4 September 2023.
32. Technology Development Fund, DRDO, https://tdf.drdo.gov.in/scheme, accessed on 17 July 2023.

33 Defence Research and Development Organisation—DRDO, Ministry of Defence, Government of India, Centre for Artificial Intelligence & Robotics (CAIR), https://www.drdo.gov.in/labs-establishment/technologies/centre-artificial-intelligence-robotics-cair, accessed on 17 July 2023.
34 Defence Research Institutes, Ministry of Defence, 5 August 2022, https://pib.gov.in/PressReleaseIframePage.aspx?PRID=1848675, accessed on 17 July 2023.
35 https://www.drdo.gov.in/sites/default/files/inline-files/Guidelines-For-DRDO-Grants-in-Aid-Scheme.pdf, accessed on 25 August 2023.
36 AI/ML Technology, https://drdo.gov.in/aiml-technology, accessed on 17 July 2023.
37 'Dare to Dream' scheme for promoting start-ups, 12 December 2022, https://pib.gov.in/PressReleasePage.aspx?PRID=1882706, accessed on 17 July 2023.
38 https://tdf.drdo.gov.in/daretodream, accessed on 5 September 2023.
39 https://www.drdo.gov.in/transfer-technologies, accessed on 10 September 2023.
40 https://www.meitystartuphub.in/about/, accessed on 26 March 2023.
41 https://www.meitystartuphub.in/tide-1-0/, accessed on 26 March 2023.
42 https://www.meitystartuphub.in/tide-2-0/, accessed on 26 March 2023.
43 https://www.meity.gov.in/writereaddata/files/constitution_of_four_committees_on_artificial_intelligence_0.pdf, accessed on 6 September 2023.
44 https://www.meity.gov.in/writereaddata/files/Committes_A Report_on_Platforms.pdf, accessed on 6 September 2023.
45 https://www.meity.gov.in/writereaddata/files/Committes_B-Report-on-Key-Sector.pdf, accessed on 6 September 2023.
46 https://www.meity.gov.in/writereaddata/files/Committes_C-Report-on_RnD.pdf, accessed on 6 September 2023.
47 https://www.meity.gov.in/writereaddata/files/Committes_D-Cyber-n-Legal-and-Ethical.pdf, accessed on 6 September 2023.
48 https://msh.meity.gov.in/assets/Scheme-Report.pdf, accessed on 6 September 2023.
49 MeitYStartup Hub and Meta shortlists 120 Start-ups and Innovators for the XR Start-up Program, 27 January 2023, PIB Release ID: 1894084, https://pib.gov.in/PressReleasePage.aspx?PRID=1894084, accessed on 7 September 2023.
50 India's AI supercomputer Param Siddhi 63rd among top 500 most powerful non-distributed computer systems in the world, https://dst.gov.in/indias-ai-supercomputer-param-siddhi-63rd-among-top-500-most-powerful-non-distributed-computer, accessed on 9 September 2023.
51 https://indiaai.gov.in/research-reports/airawat-establishing-an-ai-specific-cloud-computing-infrastructure-in-india/, accessed on 9 September 2023.
52 https://futureskillsprime.in/, accessed on 9 September 2023.
53 National Strategy for AI, NITI Aayog, 2018, p. 46, https://www.niti.gov.in/sites/default/files/2023-03/National-Strategy-for-Artificial-Intelligence.pdf, accessed on 9 September 2023.
54 https://www.insightsonindia.com/2023/02/24/giving-data-its-due-national-data-and-analytics-platform-ndap/#:~:text=NDAP%20aims%20to%20democratize%20access,innovators%2C%20and%20civil%20society%20groups, accessed on 9 September 2023.
55 https://www.psa.gov.in/pm-stiac, accessed on 9 September 2023.
56 https://www.investindia.gov.in/innovation-challenge-for-development-of-machine-aided-translation-system, accessed on 9 September 2023.
57 Respond Basket, Respond and AI Capacity Building Office, ISRO HQ, November 2018, pp. 52-53, https://www.nitt.edu/home/Respond-Basket.pdf, accessed on 9 September 2023.
58 Ibid., pp. 77-78.
59 https://indiaai.gov.in/government/government-of-uttar-pradesh, accessed on 9 September 2023.
60 https://indiaai.gov.in/government/government-of-karnataka, accessed on 9 September 2023.

61 https://indiaai.gov.in/government/government-of-tamil-nadu, accessed on 9 September 2023.
62 Ministry of Defence—Year End Review 2022, Press Information Bureau, Government of India, Ministry of Defence, 17 December 2022, https://pib.gov.in/PressReleasePage.aspx?PRID=1884353, accessed on 12 July 2023.
63 Innovation in Defence Production Projects, https://pib.gov.in/PressReleseDetailm.aspx?PRID=1906339, accessed on 17 July 2023.
64 Task Force for Implementation of AI, Ministry of Defence, 28 March 2022, https://pib.gov.in/PressReleaseIframePage.aspx?PRID=1810442, accessed on 12 July 2023.
65 Commerce and Industry Minister Sets up Task Force on Artificial Intelligence for Economic Transformation, Press Information Bureau, Government of India, Ministry of Commerce and Industry, 25 August 2017, https://pib.gov.in/newsite/PrintRelease.aspx?relid=170231, accessed on 11 July 2023.
66 Optimising Defence Acquisition Procedure, 30 July 2023, Vice-Admiral S.N. Ghormade (retd.), https://www.iadb.in/2023/07/30/optimising-defence-acquisition-procedure/, accessed on 9 November 2023.
67 https://www.ddpmod.gov.in/sites/default/files/ai.pdf, accessed 23 March 2023.
68 CPDS 2023, https://indianarmy.nic.in/writereaddata/adb-documents/Compendium%20of%20Problem%20Definition%20Statement%202023.pdf, accessed on 4 November 2023.
69 https://idex.gov.in/challenge-categories, accessed on 23 March 2023.
70 AI-Based Intelligent COVID-19 detector Technology for Medical Assistance (ATMAN), https://drdo.gov.in/ai-based-intelligent-covid-19-detector-technology-medical-assistance-atman, accessed on 17 July 2023.
71 https://www.cdac.in/index.aspx?id=hpc_nsf_npsfidx, accessed on 11 September 2023.
72 https://dst.gov.in/indias-ai-supercomputer-param-siddhi-63rd-among-top-500-most-powerful-non-distributed-computer, accessed on 13 September 2023.
73 https://www.cdac.in/index.aspx?id=hpc_nsf_siddhi-spec, accessed on 11 September 2023.
74 https://www.cdac.in/index.aspx?id=hpc_nsf_siddhi-CP, accessed on 22 September 2023.
75 National RARAM Super Computing Facility (NPSF) Annual Report 2021, https://www.cdac.in/index.aspx?id=pdf_annual_report_npsf_2021, accessed on 24 September 2023.
76 Ibid., p. 37.
77 https://www.cdac.in/index.aspx?id=cloud_ci_cloud_computing, accessed on 24 September 2023.
78 https://www.similarweb.com/website/indiaai.gov.in/#overview, accessed on 13 November 2023.
79 https://pro.similarweb.com/#/digitalsuite/websiteanalysis/audience-geography/*/999/3m?key=indiaai.gov.in&webSource=Total, accessed on 13 November 2023.
80 https://ndap.niti.gov.in/, accessed on 13 November 2023.
81 https://www.similarweb.com/website/ndap.niti.gov.in/#display-ads, accessed on 13 November 2023.
82 https://hai.stanford.edu/sites/default/files/2020-10/AI%20Index%202017%20Annual%20Report.pdf, accessed on 23 November 2023.
83 https://hai.stanford.edu/sites/default/files/2020-10/AI_Index_2018_Annual_Report.pdf, p. 30, accessed 23 November 2023
84 Ibid., pp. 36-37.
85 https://hai.stanford.edu/sites/default/files/2020-10/AI_Index_2018_Annual_Report.pdf, p. 41, accessed 23 November 2023
86 https://aiindex.stanford.edu/wp-content/uploads/2022/03/2022-AI-Index-Report_Master.pdf, p. 143, accessed on 23 November 2023.
87 Ibid., pp. 149-150.

88 https://aiindex.stanford.edu/wp-content/uploads/2023/04/HAI_AI-Index-Report_2023.pdf, p. 182, accessed on 23 November 2023.
89 Ibid,, pp. 3-4.
90 Ibid., p. 34.
91 Ibid., p. 38.
92 Ibid,, p. 67.
93 Ibid., p. 68.
94 https://www.meity.gov.in/writereaddata/files/Digital%20Personal%20Data%20Protection%20Act%202023.pdf, accessed on 28 November 2023.
95 https://www.businesstoday.in/latest/trends/story/israel-warns-india-on-ip-related-issues-while-pledging-to-share-tech-on-make-in-india-339830-2022-06-30, accessed on 28 November 2023.
96 https://timesofindia.indiatimes.com/city/bengaluru/tcs-to-face-125-million-loss-in-ip-infringement-case/articleshow/105400072.cms, accessed on 28 November 2023.
97 https://www.investindia.gov.in/pm-stiac#:~:text=The%20Artificial%20Intelligence%20(AI)%20Mission,which%20will%20include%20international%20collaborations, accessed on 19 April 2024.
98 Cabinet Approves Ambitious IndiaAI Mission to Strengthen the AI Innovation Ecosystem. https://pib.gov.in/PressReleaseIframePage.aspx?PRID=2012355, and ID 2012375, both accessed 19 April 2024.
99 https://www.indiatoday.in/business/story/cabinets-nod-to-india-ai-mission-with-outlay-of-rs-10372-crore-2512045-2024-03-07, accessed on 20 April 2024.
100 Jyoti Panday and Mila T. Samdub. Promises and Pitfalls of India's AI Industrial Policy, 1 March 2024, https://ainowinstitute.org/wp-content/uploads/2024/03/AI-Nationalisms-Chapter-4.pdf, accessed on 20 April 2024.
101 Anirban Ghoshal. India's advisory on LLM usage causes consternation, 5 March 2024, https://www.cio.com/article/1311757/indias-advisory-on-llm-usage-causes-consternation.html, accessed on 20 April 2024.
102 https://www.moneycontrol.com/news/technology/indias-cto-nilekani-unveils-ai-strategy-focus-on-use-cases-not-my-model-bigger-than-yours-11853961.html, accessed on 5 December 2023.

Chapter Four

Future Indian Military Conflict Environment

INTRODUCTION

In an era defined by rapid technological advancements, intricate geopolitical complexities, and ever-evolving threat landscapes, the anticipation and management of future conflicts pose a paramount challenge for nations worldwide. This chapter embarks on the crucial task of identifying conflict environments, abstracting the multi-dimensional nature of impending conflicts, and elucidating the pivotal role that AI tools play in ensuring effective military responses. Understanding the intricacies of future conflicts is not merely a strategic imperative; it is a proactive approach to safeguarding national interests. The significance of identifying conflict environments lies at the core of strategic foresight and preparedness.

As India navigates the competitive terrain of global politics, the ability to recognize potential conflict hotspots becomes instrumental in implementing pre-emptive measures, diplomatic interventions, and, when necessary, decisive actions. The consequences of unanticipated conflicts extend beyond national security, affecting economic stability and geopolitical influence. Through meticulous analysis, incorporating content studies, expert opinions, and survey data, this chapter aims to illuminate areas of heightened risk, providing policymakers with insights to craft nuanced and effective strategies.

The landscape of future conflicts is inherently multi-dimensional, transcending traditional notions of warfare. While conventional military engagements remain a definite possibility, the spectrum has expanded to include hybrid warfare, cyber threats, asymmetric tactics, and the involvement of proxy actors. Understanding this complexity is imperative for military planners, necessitating versatile approaches that span not only the physical battlefield but also the digital realm and influence operations. The evolving nature of conflict demands a paradigm shift in military preparedness, acknowledging the interconnectedness of various domains. Notably, the immediate future may not witness direct military involvement in every conflict. The contemporary geopolitical landscape is characterised by a myriad of non-military disputes, economic rivalries, and cyber confrontations. However, as nations progress, so does their attack surface area. The likelihood of military involvement increases commensurately with the expansion of national interests, making it imperative to adopt a comprehensive approach to conflict anticipation and resolution. The ability to discern when military intervention becomes unavoidable is a delicate yet vital task for national security.

In the event of military involvement, the swift resolution of battles becomes a critical factor. Here, the integration of AI tools can emerge as a force multiplier, given their ability to process vast amounts of data in real-time, identify patterns, and provide actionable insights. This capability is invaluable in dynamic and unpredictable conflict scenarios, enabling quicker decision-making, enhancing situational awareness, and optimizing resource utilization. The Environment, Values, and Resources (EVR) congruence, on the other hand, emerges as a crucial determinant in debating military involvement and the immediacy of AI tools for such conflicts. Understanding the specific requirements of each conflict scenario, considering factors such as terrain, adversary capabilities, and geopolitical nuances, enables the judicious selection and application of AI tools. This encompasses not only military assets but also diplomatic and humanitarian endeavours. A comprehensive understanding of potential conflicts facilitates the prioritisation of resource allocation, ensuring maximum impact and effectiveness.

METHODOLOGY

Within this intricate puzzle, there exists a mosaic of sources and mediums that offer valuable insights into a nation's potential path. This chapter details the content survey of the publications, survey forms and interviews with Subject Matter Experts (SME).

Content Survey of the Publications. The content survey concentrates on the period from 2016 to 2022. The year 2016 has been a watershed for India. In 2016, India witnessed several significant and unexpected developments that had far-reaching implications for its economy and governance. These events signified a mix of economic reforms, policy changes, and transformative decisions that aimed to reshape India's economic landscape and governance practices. While they brought both challenges and opportunities, they demonstrated the government's commitment to address pressing issues and drive India's progress on the global stage.

- In November, high-denomination currency notes (Rs. 500 and Rs. 1,000) were withdrawn from circulation. This bold move was aimed at curbing black money and promoting digital transactions.
- In August, the Rajya Sabha (Upper House) unanimously approved the Goods and Services Tax (GST) Bill replacing a complex system of indirect taxes with a unified tax structure.
- The Indian government initiated a significant change in budget presentation. The presenting of the Union Budget was shifted to the first working day of February and the 92-year-old colonial practice of presenting a separate railway budget was discontinued, integrating it into the main budget.
- India achieved the distinction of being the fastest growing major economy globally, surpassing China in February.
- The series of terrorist attacks in India during 2016, underscored the persistent security challenges faced by the nation. These events reaffirmed the nation's ongoing efforts to combat terrorism and maintain its sovereignty, The surgical strike in September is considered an assertive Indian military action. Terrorist initiated actions include attacks on the Pathankot Air Force Station (January), Uri military camp (September) and Nagrota military camp (November). These attacks by terrorists were grave infringements and shook up the national security thinkers. Multiple terrorist attacks at Pampore (February and June), Baramulla, Handwara, Shopian, and Zakura resulted in casualties, triggering a series of reorganised security opinions and processes.

Justification for Content Survey. The content analysis of publications by autonomous bodies, namely, the United Service Institution (USI) of India and the Manohar Parrikar Institute for Defence Studies and Analyses (MP-IDSA) is valuable since they have a robust foundation of high-quality outputs. Both think tanks are repositories of scholarly research, strategic analysis, and expert opinion. Their outputs offer a degree of reliability and credibility that is invaluable for discerning future trends. Key justifications for this approach are as follows.

- **Peer Review and Credibility.** The cornerstone of this approach is the high-quality peer review of think tank publications. In the realm of scholarly research, peer review is the gold standard for quality assurance. A thorough review process ensures that the publications meet the rigorous standards of accuracy, relevance, and methodological rigour. Only content that has withstood the scrutiny of experts is considered for analysis. This ensures that predictions are based on credible, well-researched information.
- **Vision and Policy Implications.** These think tanks are architects of visionary ideas and policy recommendations. Their publications often provide a roadmap for the nation's future, addressing key aspects such as political ideology, diplomatic alliances, economic strategies, and military doctrines. The articulation of a nation's aspirations, goals, and strategies can be found within the pages of these publications.
- **Political, Diplomatic, Economic, and Military Will.** The strategic posture of a nation is deeply intertwined with its political, diplomatic, economic, and military will. Think tank publications, reviewed and selected for their quality, serve as a treasure trove of information that sheds light on these critical aspects. They encapsulate not only what the nation aspires to be but also the practical means by which these aspirations are to be achieved. Consequently, think tank publications offer invaluable insights into the nation's will to engage in conflict, should it become necessary.

Survey Form. This form intends to contrast and compare the concurrent content analysis with a survey to validate the results. Strategic consolidation focused the content analysis to manageable environments; Territorial threats, collusion, internal security, hybrid threats, cyber, economic, and maritime challenges, managing foreign policy issues, out-of-area contingencies (OAAC), synergy challenges and Strategic Alliances. The survey sought inputs on the

LIKELIHOOD of occurrence and IMPACT of each, separately, to create a risk matrix. While Likelihood was arranged on a scale from 1 to 7 (Very Likely, Likely, Somewhat Likely, Neutral, Somewhat Unlikely, Unlikely and Very Unlikely) the Impact was ranged on a scale of 1 to 7 (Very High Impact, High Impact, Significant Impact, Moderate Impact, Low Impact, Least impact and No Impact). Additional points and suggestions were sough as long responses at the end of the survey. The volunteers were also asked to provide their experience in national security. Created on Google Forms[1] it was shared from 14 October to 31 December 2023 and received 142 inputs.

Interviews with SMEs. SMEs offer intrinsic insights on future pathways. They are professionals with in-depth knowledge who provide valuable insights into the future pathways and likely conflict areas of India. They offer a nuanced understanding of the challenges faced by India and recommendations to address these challenges. By analysing the responses of multiple SMEs, common themes and patterns emerge to develop a more comprehensive understanding of the future and likely conflict areas of India.

CONTENT ANALYSIS

USI Published Text. The USI was founded in 1870 by a soldier-scholar, Colonel (later Major General) Sir Charles MacGregor for 'furtherance of interest and knowledge in the art, science, and literature of the Defence Services.' It was initially housed in the Old Town Hall at Shimla, in the foot hills of the Himalayas. It is now located at New Delhi. It has emerged as India's pre-eminent think tank on matters of national security. Post-independence, the USI transformed into a typical track 1.5 institution that has rendered 'yeoman' service in developing strategic culture amongst the policy-makers of modern India. Its activities, range from historical research to publications of diverse literature, career progression of military officers, and a niche in net assessment, scenario building and strategic gaming. The Vision of 2030 notifies; 'Transform USI as a tri-service military institution with a niche for a multi-disciplinary progressive policy research and narrative building in comprehensive national security with military focus on a wider global geopolitical context, while preserving its rich heritage and unique character as India's oldest think tank.' Though the USI has numerous publications, the research however focussed on keystone publications, namely, Strategic Yearbook (ISBN 978-93-95675-87-1), USI Journal (ISSN 0041-

77OX) and Monographs (ISBN 978-93-88161-12). Selected special studies, national security papers and seminars are published as monographs as original academic research resource, generally dealing with one, highly focused, research question. The USI Journal is a peer reviewed research periodical, a credible resource for knowledge in the art, science and literature of national security in general and of the defence services in particular. The Strategic Year Book enables writers and readers to reflect, articulate and debate on contemporary and futuristic national security issues against a global backdrop.

MP-IDSA Published Text. The MP-IDSA is a non-partisan, autonomous body dedicated to objective research and policy relevant studies on all aspects of defence and security. To achieve its goals it undertakes scholarly, policy-oriented research, dissemination of research findings, training and capacity building and public education. Its vision encompasses 'To promote national and international security through the generation and dissemination of knowledge on defence and security-related issues.' The Institute strives to generate policy options for national and international security and provides inputs for management of the country's security apparatus. Strategic Analysis (ISSN 0970-0161) is the flagship bimonthly journal that provides a forum for independent research, analyses, and commentaries on national, regional and international security issues that have policy relevance. The Journal of Defence Studies (ISSN 0976-1004) is a quarterly that encourages further research on the core issues of defence, provides the latest perspectives, and serves as a platform for sharing research findings and opinions of scholars working on defence-related issues. It contains a mix of research articles, essays, topical commentaries, opinion pieces and book reviews.

Data Consolidation. The USI Library provided printed copies of all publications between 2016 and 2022. Monographs (21), USI Journals (28) and Strategic Year Books (07) published by the USI and Strategic Analysis (42), and Journal of Defence Studies (26) published by the MP-IDSA were analysed. The analyses initially sought to identify direct reference to future 'conflict scenarios'; however, the publications did not provide sufficient data points. The author then widened the search to include 'security concerns', 'challenges', 'competitions', 'conflations' and 'competitions.' The endeavour was not a word search or 'word cloud' creation, but rather to understand the relevance of the article and the writer's intent. All articles were read carefully to identify the reference points and then converted to numerical values. No automation technique was involved in this initial part of the search. A large number of data points surfaced during the search to include: Threats from China along the boundary, competition from the growing Chinese economy, threats from Pakistan along the boundary, threats from Pakistan-sponsored proxy war, collusion between China and Pakistan, threats to island territories, Indian Ocean Region – challenges, competition, conflicts and opportunities, Indo-Pacific Region – challenges, competition, conflicts and opportunities, threats from hybrid wars, civil and military liaison, cyber war and crime, challenges in inter service jointness and synergy, necessity of a national security strategy, partnered cooperation with strategic allies, challenges to Indian economy, internal security situation, radicalism, communalism, Out of Area Contingencies (OOAC) in realms of lethal and human assistance and disaster relief. A stabilised international order and challenges and opportunities in Afghanistan, the Central Asian Republics, West Asia, and South-East Asia also surfaced as points of interest.

It was observed that there were too many data points to make a concentrated focus on the research problem. Correlated data points were combined in a data consolidation strategy to enhance the precision and focus of analysis. Specifically, the amalgamation of certain data heads within the dataset streamlined and sharpened the analytical process. This strategic data condensation extracted more meaningful insights from the dataset by reducing redundancy and highlighting key relationships. The consolidated dataset now concentrates on the following.

- Territorial threats consolidate all data points relevant to threats from China and Pakistan (singularly and collusively) along the border. It also includes threats to island territories.
- Internal security combined the threats from terrorism, insurgency, LWE, radicalism, internal strife and communalism.
- Hybrid threats cover a wide spectrum including cognitive domain.
- Cyber threats covered all aspect of cyber warfare including cybercrimes to induce perception modification.
- Economic threats consolidate attempts against economic policies, economic houses and proxy companies undermining the national economy
- Synergy challenge covers inter-service, inter-organisational coordination and intra-governmental cooperation. It also covers civil-military synergy towards common understanding of national security issues.

- Maritime challenge covers the wide area of the Indo-Pacific and Indian Ocean regions which are fast becoming a competition arena for energy, security, environment, human trafficking, and other illegal activities.
- Indian challenges in managing foreign policy issues towards areas of interest and concern cover Afghanistan, South-East Asia, West Asia and Central Asian Republics individually. All these regions have proved to be major points of conflation and result in conflicts, if not managed well.
- Strategic-partner relationships have also emerged in the publications, indicating attempts to organise wide ranging international partnerships, not necessarily military.
- OAAC contingencies are also mentioned in the publications, signifying interest and capability development attempts.

Summary of Content Analysis

Comprehensive content analysis presents few divergent yet complementary viewpoints on critical issues.

USI Content Survey. Monographs are published to document the discussions on national security issues. Though the publication is centred around one research question, two publications (01/2021 and 01/2018) documented more than one topic, and thus find reference in multiple locations. The research analysed 305 USI Journal articles, discounting book reviews and military history. One USI Journal (2021) was purely dedicated to Hybrid/Grey Zone analyses and another (2020) to 150 years of USI publications. The former has been included in the content analyses and the latter has been excluded. About 305 articles were analysed for this research. The Strategic Year Book is a yearly publication consisting of 30 articles and 230 articles are referred in this research. Similar to monographs, articles in the USI Journal and Strategic Year Book also provide multiple references to likely conflict and competition situations.

The scatter chart below summarises the analyses of 556 articles published by USI. The challenges and threats emerging from the analysis pertain to territorial threats (n=50), hybrid threats (n=35), challenges in the Indo-Pacific Region (n=34), challenge of synergy (n=29), internal security (n=27), and economy challenges (n=22). The lowest references mention competition in

South-East Asia (n=2), cyber threats (n=7), competition in CAR (n=8) and OOAC (n=8).

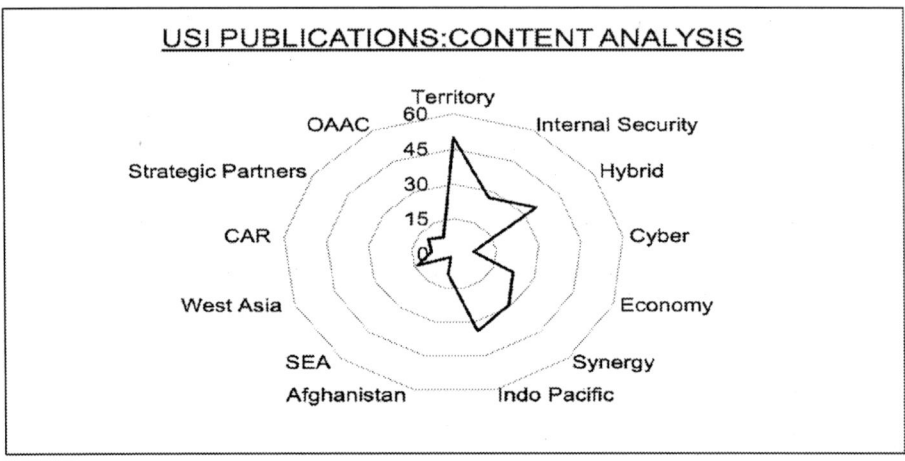

MP-IDSA Content Survey. Nearly 42 Strategic Analysis provided research inputs with a few exceptions. The 2018/3 issue celebrating centenary of Pokhran nuclear tests and is not considered for this survey; the 2019/4 issue is dedicated to BRICS and thus finds reference in the Strategic Partnership field. Similarly, the 2020/5 issue is dedicated to the UN and four articles mention the usefulness of Indian deployment as an energiser to international partnership; the 2021/6 and 2022/4 issues focus on the Bangladesh War and 'Asianism', respectively, thus they do not provide any input for the research. The Strategic Analysis issue of 2016/6 pertains to Russia only and is left out of the research. 28 Journal of Defence Studies provided detailed inputs, barring a few, which did not provide reference data for the research. Articles from these two journals also refer to multiple aspects and thus find mention in the respective subheads.

The scatter chart below summarises the analyses of 329 articles published by MP-IDSA. The challenges and threats emerging from the analysis pertain to Strategic Partners (n=26), Indo-Pacific Challenge (n=18), Economy (n=16), Internal Security (n=13), Synergy Challenge (n=10) and Territorial Threat (n=8). The lowest references mention OOAC (n=0), Cyber Threats, with South East Asian and Central Asian Republic Challenges sharing (n=2).

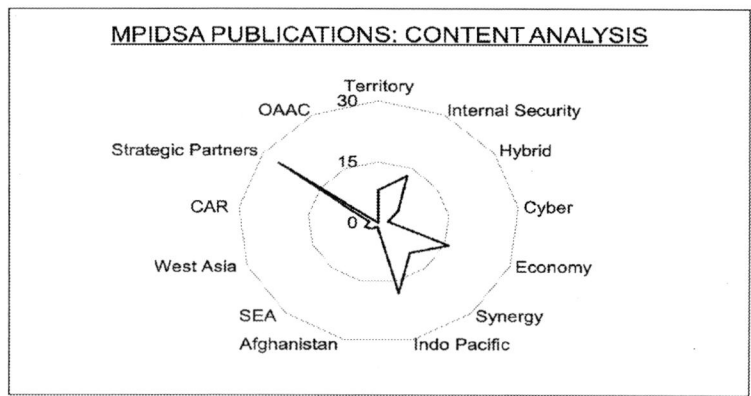

The combined scatter chart obtained from the publication history of the two premier think tanks realigns the challenge areas as Territorial Threats (n=58), Challenges in the Indo-Pacific Region (n=52), Hybrid Threats and Internal Security (n=40), Synergy (n=39), Economy (n=38) and Strategic Partners (n=37).

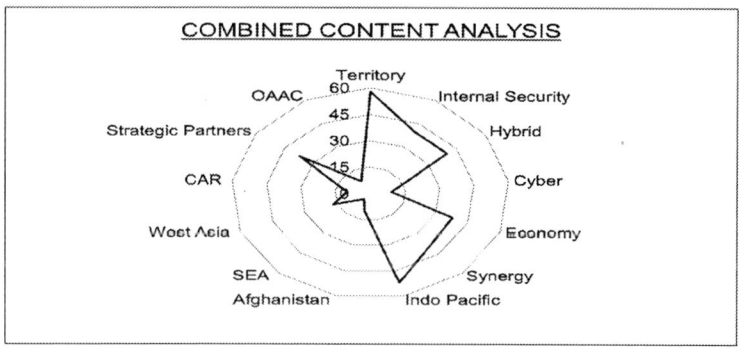

Analysis of Commercial Text. Although not initially part of the research, commercially published military magazines also serve as critical tools for envisaging future threats and bolstering national security. They offer a wealth of benefits, including expert insights from seasoned military professionals and defence analysts, coverage of current events in the defence and security sectors, and a window into the latest advancements in military technology. However, it is essential to recognize the potential pitfalls, which include the likely absence of peer review. In many cases the articles may be pure opinion pieces. These may serve to persuade and advocate a particular viewpoint, often incorporating emotional appeals and rhetorical techniques. Academic work aims to contribute to the field's body of knowledge, whereas opinion pieces target a broader audience to provoke thought and stimulate discussion.

In spite of inherent research shortfall, their increased and unrestricted circulation promotes and generates debates, thus offer novel evidence for this chapter. Analysis of 166 issue of prominent military magazines (Force, Indian Military Review, Salute, Indian Defence Review, India Strategic, Geopolitics and South Asia Strategic Review) from the USI library highlights threats as given in the table below. The frequency of conflict (from highest to lowest) is territorial (n=44), foreign policy (n=30), maritime (n=20), synergy (n=19) ending with hybrid and strategic partnership (n=11) and internal security (n=10).

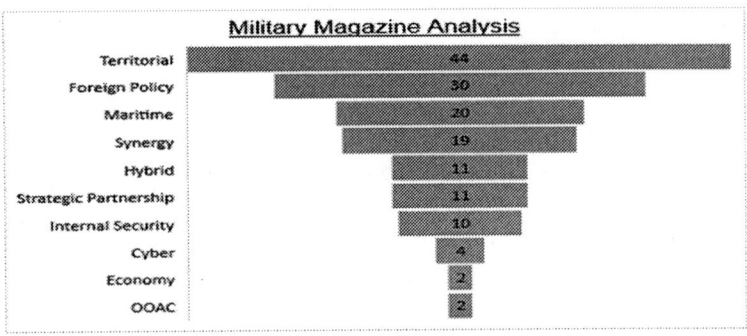

SURVEY ANALYSIS

The survey form made on Google (https://docs.google.com/forms/d/e/1FAIpQLScg2ooU1ErIV6ZYgNq5JQHeJ3gmOsh1lQSm613w9Dr2TkmO2A/viewform?usp=sf_link) was shared from 14 October to 31 December 2023 and received 142 inputs. It provided a large dataset indicating interesting data points.

Likelihood Analysis. Nearly 84.51 per cent of the respondents graded cyber as most probable to occur as a conflict environment followed by hybrid (76.76), internal security (75.35), economic threat (67.61), strategic partnerships (65.49), territorial threat (64.79), maritime (58.45), synergy needs (56.34), foreign policy (54.93) and OOAC (35.21). It is not alarming that the higher probability of occurrence is allocated to conflicts (internal security, economic threats) which may affect all citizens in real time, and in fact is a serious concern already. The large and diverse Indian population has a vast attack surface and is easily affected by a hybrid nature of warfare and cross-domain functions of cyber operations, thus naturally putting it highest on the stack. Internal security is also read as an overarching domain which harbours within itself the functions of terrorism, insurgency, left wing extremism (LWE), radicalism, internal strife and communalism, often working across the geographical landscape, sometimes concurrently. Conflicts like strategic partnerships, territorial threats, maritime, synergy needs, foreign policy and OOAC are highly specialised threat sectors and not fully understood by non-practitioners, probably leading to their lower 'ratings'. The low rating of the OOAC indicates the benign thoughts of the respondents, desiring India to largely steer away from such contingencies, a fact almost corresponding to present EVR congruence matrix. The data points are especially essential for security architects to understand that special communication efforts will be required to convey strategic messages with regard to the lower end of the stack.

Impact Analysis. The impact assessment of the threats corresponds to probability assessment with minor deviation. Cyber impact rides high at 80.99 per cent, followed by hybrid (79.58), territorial threats (72.54), internal security and economic threats (each at 69.72), synergy needs (68.31), strategic partnerships (65.49), maritime (57.04), foreign policy (50) and OOAC (38.73). The impact of territorial threats shows up higher as all respondents consider it a major symbol of sovereignty. Internal security and economy share a common high concern since it affects the lifestyle, freedom and security of the citizens. Foreign policy and OOAC remain low due their specialised nature and are probably less understood. The responses clustered tightly around the mean showing the least variability for hybrid threats, internal security and cyber threats in descending order. This indicates a common threat concern indicating easier strategic messaging. Consolidated bar chart below presents the likelihood and impact definition of the survey.

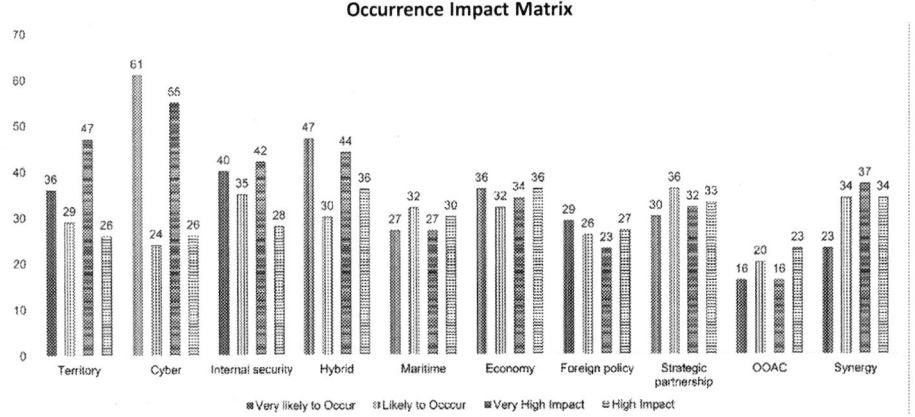

Summary of the Survey. The survey gave options to the respondents to provide additional suggestions. Only 15 respondents provided additional comments to the survey, but the sentimental analysis of the suggestions indicate a mix of concern, criticism, and a call for proactive measures in addressing various threats to India's security.

- Positive sentiments include the importance of security and the importance of risk mitigation efforts. Notable are a whole-of-government approach as essential to thwart all future conflicts. CCS and higher defence organisations should have meaningful dialogues, organise perspective plans and allocate sufficient funds. Indian foreign relations could be bolstered by strong economic, trade, cultural and spiritual exchanges, supported by durable sovereignty protection capability
- Concern sentiments are indicated by lack of integration between various organisations and institutions leading to adverse effect on national security. Suggestions include creation of an overarching organisation to collaborate between 'common goals – different path'. Synergy within defence services, PMF, intelligence agencies, and foreign policy experts should be robust and dynamic. Common training methodologies and documentation is found essential for complete synergy.
- Negative sentiment iterates corruption, caste and minority orientation of the political elite, lack of skill development and unemployment. One respondent also suggested short-sighted and reactive governance.
- Suggestions also include reduced dependence on conventional kinetic conflicts and increased non-kinetic capability including cyber and information warfare, non-lethal and non-contact warfare, quoting the hybrid nature of Hamas and the Israel conflict. Then spread of conflict across the entire spectrum indicates multiple domains of economic, cyber, maritime and internal unrest. Suggestions about the geopolitical relevance of diplomacy and strategic communication are also pertinent to the survey result.

SUBJECT MATTER EXPERTS (SME)

Key insights obtained through interviews with subject matter experts are included in the subsequent paragraphs. These experts provided practical perspectives and valuable insights into the future conflict environments and role of AI in the Indian military. The effort obtained a straightforward and clear overview on strategic considerations, operational implications, and the changing nature of conflicts. The opinions are summarised in the succeeding paragraphs.

- **Geographical Dynamics.** Distinct regional theatres play a crucial role in shaping the frequency and intensity of conflicts. The northern and western theatres, for instance, witness more frequent clashes compared to the peninsular regions. Understanding the geographical nuances is imperative for strategic planning and resource allocation, ensuring a matured approach to conflict management in diverse terrains.
- **Maritime Challenges and Diplomacy.** The protection of exclusive economic zones (EEZ) emerges as a formidable challenge, both diplomatically and militarily. The increasing presence of China's maritime militia in the Indian Ocean Region (IOR) demands a proactive stance from India. Undertaking dissuasive actions becomes crucial to protect sovereign interests and maintaining stability in the maritime domain.
- **Strategic Reach and Global Power Status.** India must extend its reach to strategic locations, including the South China Sea, the Eastern African seaboard, and as far as Australia to safeguard its sovereignty and assert its influence globally. While such aspirations may face scepticism, they are justified. It is imperative for India to position itself and participate in the global dynamics. Strategic partnerships and issue-based cooperation will define the future contours.
- **Non-State Actors and the Geopolitical Balance.** The democratisation of AI technology empowers non-state actors and private entities to develop and deploy sophisticated systems, potentially upsetting the geopolitical balance. Geopolitical analysts have to remain vigilant to unforeseen developments, since AI introduces an element of strategic surprises in confrontations.
- **Stages of Conflict.** Conflict in the contemporary era transcends traditional boundaries along the entire spectrum of operations, enmeshing kinetic and pre-kinetic stages. An often overlooked aspect is the targeting of non-military infrastructure, which significantly impacts military operations as supply chains intertwine with predominantly

civilian structures. This interdependence necessitates a comprehensive approach to conflict resolution to protect interconnectedness of military and non-military domains.
- **Grey Zone Campaigns and Decision Dilemmas.** The emergence of newer domains, such as space and cognitive, intensifies grey zone campaigns. Decision-makers grapple with the delicate balance between excessive retaliation and no retaliation, weighing the potential loss in the cognitive battle against the risk of diminishing power. Striking this balance becomes crucial for effective conflict resolution.
- **Tactical Battles.** India's operational philosophy, with a focus on immediate missions, necessitates advancements in intelligence, surveillance, and reconnaissance (ISR). AI emerges as a pivotal enabler, enhancing the focus and success of tactical battles. Specialised AI applications for line of control (LC), line of actual control (LAC), and Counter-Insurgency (CI) operations become essential for managing isolated tactical situations (Galwan, 2020) with global implications.
- **Changing Nature of Battle with AI Decision Support.** The integration of AI in military operations introduces a transformative shift in the nature of battles. AI empowers commanders with enhanced transparency, enabling them to judiciously choose their battles. The AI transparency assists in strategic decision-making, allowing commanders to assess situations with unprecedented clarity. The concept of a fused decision-making (DM) environment, where AI augments human decision-makers, emerges as a pivotal development. This fusion supports commanders in overcoming operational surprises, providing a critical advantage in dynamic and unpredictable conflict scenarios.
- **Unforgiving Nature of Military Conflicts.** Amidst the transformative potential of AI, a cautionary note underscores the inherent brutality and unforgiving nature of military conflicts. While AI offers invaluable decision support and operational advantages, an over-reliance on technology, including AI, should not be misconstrued as the sole determinant of victory. In the crucible of war, where human lives are at stake, the role of human judgment, adaptability, and resilience remains irreplaceable. The human element retains its significance at the forefront. Regardless of technological advancements, including AI, it is crucial to acknowledge that humans are, and should, remain at the pointy edge

of the battle. The intricacies of warfare, influenced by human experience, demand a balance between technological innovation and the inherent qualities that defines a human agent on the battlefield.

- **CI Operations and the Battle of Hearts and Minds.** AI assumes a critical role in counter-insurgency (CI) operations, where it contributes immensely to understand internal dynamics, external drivers, and cognitive effects. The battle for hearts and minds, central to CI, highlights the unique use case for AI in shaping narratives and countering subversion efforts.
- **Cognitive Health.** AI becomes a robust tool in managing the cognitive health of soldiers, their families, and citizens. Recognising the psychological aspects of conflict underscores the need for AI applications that contribute to the overall well-being of individuals involved in or affected by conflicts.
- **AI Governance.** Drawing parallels with nuclear regulations, the proactive governance of AI is imperative due to its potential to impact organisations, countries, and citizens worldwide. The intangible nature of AI ingredients like algorithms, compute power, and training datasets pose challenges akin to controlling physical components to deny a subterfuge nuclear device. There should be a collective and international effort to regulate its use.
- **Societal Impact and International Cooperation.** The effects of AI will swiftly reverberate through societies, compelling governments to collaborate on common governance frameworks and compliances. As citizens witness the rapid changes brought about by AI, international cooperation becomes vital to address the ethical, legal, and societal implications of this transformative technology.
- **Collusion.** All experts were of the opinion that China will drive the degree and type of collusive cooperation with Pakistan. China will ensure that collusion does not harm their international image of a strong country and they cannot be seen as seeking help from a failed state. Pakistan's inability to safeguard the CPEC and its resident engineers is a major concern for this trepidation. Unruly and uncontrolled Pakistan establishment, governance, political personalities and routing terror organisations are shaking China's faith. Alliance between with highly controlled state and Chinese society is thus a matter of discussion in

both the strategic circles.
- **Core Capabilities.** The Indian military should restrict its deployment within its specialised kinetic abilities. A different decision can only be taken once the military is directed to reorganise, re-equip, and train itself for other non-kinetic capabilities.

Special Comments. There was a common understanding amongst experts that the present India may not want to choose conflicts to support economic claims; however, it is also understood, that a time will appear on the horizon when India will have to cast away its qualms and choose the unsavoury decision. Necessity to fiercely protect the EEZs is akin to protecting the frontiers of the sovereign territory. While internal and territorial seas (12 NM beyond internal waters) have full sovereignty, the EEZs, (including 12NM of continuous zone) up to 200 NM from the baseline have sovereign rights but not sovereignty. The EEZ was created for the sole purpose of granting coastal states greater control over the living and non-living resources adjacent to their coasts.[2] The right to explore, exploit, conserve and manage living and non-living natural resources, and jurisdiction over off-shore installations and structures, marine scientific research, and protection and preservation of the marine environment is enjoyed by the coastal state. EEZs concurrently permits 'internationally lawful uses of the seas' by any other state and which may be undertaken without coastal state's notice or consent and include a broad range of military activities. This argument is considered offensive by many countries including India and expects to be consulted before any such enterprise. This EEZ excursion is a likely trigger conflict in failed negotiations when a belligerent state takes an adverse position.

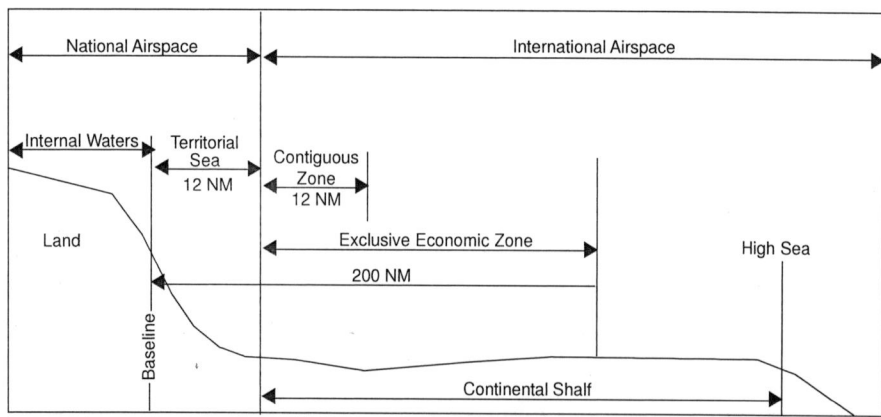

SUMMARISING CONFLICT ENVIRONMENTS

These interviews, publications and survey provide trends, patterns, and indicators that can aid in the prediction of future conflict. This combined approach relies on both qualitative and quantitative analysis to extract meaningful insights to bridge the gap between academic rigour and real-world application to shed light on the potential conflict that will shape the destiny in the years to come. The conflict milieus will be used as building blocks to identify and design the AI tools required by the military in the following chapters. It is important to note that neither think tank publications nor interviews or surveys can be viewed in isolation. They complement each other and provide a holistic view of the future path. This combination allows for a more robust prediction of future conflict.

A consolidation summary assumes the role of weaving together insights derived from multiple primary researches. It highlights diverse facets of various conflict environments, surpassing the boundaries of singular study. By strategically juxtaposing one environment against another, this approach has deliberately broadened the scope of analysis and presents a cohesive narrative of future conflicts.

Territorial Threat

India has a land border of 15,106.7 km and 7,516.6 km coastline (including island territories). A full topography spectrum exists along Indian borders ranging from deserts to highly populated areas and fertile plains, riverine, hills, dense forests, mountains and glaciers. The shared land border length with neighbouring countries is given in the MHA document.[3]

Country	Length of the Border (km)
Myanmar	1,643
Bangladesh	4,096.7
China	3,488
Bhutan	699
Nepal	1,751
Afghanistan	106 (illegally occupied by Pakistan as Occupied Jammu and Kashmir)
Pakistan	3,323

Along the national periphery in Jammu and Kashmir, Ladakh and Arunachal Pradesh land borders of India face threats from the *Pakistan-China strategic nexus* that seeks to change the status quo at the contested borders and undermine India's security. China is developing CPEC in Pakistan (through Indian territory in Pakistan Occupied Kashmir), building the China-Nepal Economic Corridor, China-Myanmar Economic Corridor and other dual-use infrastructure in the littorals of the Indian Ocean. China is attempting to alter the balance of power in its favour in India's strategic neighbourhood.

The Chinese strategic community is divided in the way China must approach India. One faction advocate inducing cooperation, while the opposing faction leans towards a strategy involving coercion or the use of force, drawing inspiration from the prolonged effectiveness they attribute to the 1962 border war. This dichotomy results in China's engagement with India oscillating between periods of cooperation and conflict. The fear of an openly hostile India, actively obstructing China's access to the Indian Ocean via Pakistan or Myanmar, adding complexity to this dynamic. Despite these divergent perspectives, there appears to be a prevailing consensus within Chinese strategic circles emphasising the importance of preventing a complete rupture in China-India relations. China's aggressive actions, including Galwan in May 2020, to change the status quo at the LAC, have severely damaged Sino-Indian relations and it has got the military experts in China to rethink future military conflicts with India. In so much, China will be forced to accept greater US-India cooperation under the Indo-Pacific framework as reactionary. It is possible that China will intensify its engagements in South Asia to isolate India in the region and build pressure. China will also make border confrontation or conflict a new normal, desensitising the Indian polity, strategic community and media. China may thus be expected to undertake hybrid options against India to selectively apply subtle pressure on the weak points while indicating cooperation concurrently.

Maritime Challenge

Indian concerns cover the wide area of the Indo-Pacific and Indian Ocean which is fast becoming the competition arena for energy, security, environment, human trafficking and other illegal activities, undermining the cohesion of the nation. The region has 38 countries in Africa, Asia and Oceania, it is home to 64 per cent of the global population and contributes 62 per cent of the Global GDP.

Nearly 50 per cent of the global trade and 40 per cent of oil passes through this region and in the Indian context it is 90 per cent of trade and 80 per cent of critical freight. Over 23 lakh sq km of EEZ have both potential and challenges. China, though not a maritime neighbour, has a near continuous presence of 7-8 warships or research vessels in the IOR. Development of naval bases indicates China's intent to create permanency in the region.

Combined with territorial threats, China would continue to be a source of concern and possible conflict, for which military readiness would be essential. Illegal military development and expansion of reefs and small islands in the disputed territories of the South China Sea may embolden China for similar action in the IOR, an eventuality which will affect India directly. Prone to piracy, trans-national crimes related to human and drug trafficking and natural disasters, the region acts as a zone of hybrid threats too. Jagdeep Dhankhar, Vice-President of India, phrased concerns of the region succinctly during the Indo-Pacific Regional Dialogue (2023)[4] 'India has serious stakes in the peace, prosperity and stability of the Indo-Pacific region to ensure maritime trade security'.

Internal Security and Hybrid Threats

Internal security is a combination of threats from terrorism, insurgency, LWE, radicalism, internal strife and communalism; in fact, defining many contours of hybrid threats. A hybrid threat is a diverse and dynamic combination of conventional and unconventional methods employed by state or non-state actors to achieve strategic objectives. Adversaries blend military, political, economic, informational, and other non-military means to create a complex and multifaceted threat. The hybrid threat model recognises that modern conflicts often transcend traditional boundaries and involve a unique mix of any or all capabilities. Improved technology has increased the transformational pace from the traditional physical battle space to informational and cognitive domains. Adversaries exploit this ambiguity to achieve their objectives through a combination of methods. Beyond traditional military means, hybrid threats encompass a wide range of non-conventional tools, such as economic coercion, cyber-attacks, disinformation campaigns, proxy warfare, and subversion.

The preference for hybrid options stems from several strategic advantages which are high level of deniability, lower escalation risks, and high precision in targeting with new technologies like AI allow better manipulation of information and narratives to achieve objectives. Understanding hybrid threats is crucial for contemporary security analysis, as it requires a comprehensive approach that goes beyond traditional military planning. National integration of various elements is essential to effectively counter and respond to hybrid threats. Within hybrid spans, grey zone warfare encompasses all the means adopted by states and non-state actors in the operational space between peace and all-out conflict to achieve national objectives. It keeps the intensity of warfare below the threshold of all-out conflict and does not cross the threshold and use a different mix of methods, staying clearly short of bloodshed. The crime and socially disruptive behaviour also offer a fusion of novel capabilities with the fluidity of grey zone warfare.

Synergy

Synergy covers inter-service and inter-organisational coordination and intra governmental cooperation. It also covers civil-military synergy and common understanding of national security issues. Rajnath Singh, the Defence Minister of India, has highlighted the imperative for civilian-military synergy to enhance national security in the face of evolving global challenges. This recognition comes against the backdrop of an expanded understanding of national security that includes non-military dimensions. Conflicts such as Nagorno-Karabakh, Russia-Ukraine, Israel-Hamas, and the Red Sea escalations by Houthis demonstrate the shift towards unconventional warfare. The Raksha Mantri, as late as June 2022[5] recognised the importance of civil-military jointness, and stressed the significance of breaking silos between civil administration and the armed forces to address hybrid threats effectively. He also recognised the necessity of cooperation without compromising autonomy, describing it as working together while respecting individual identities, akin to the 'colours in a rainbow'.

India's post-independence governance initially led to the creation of various institutions to address social, economic, and political aspects. However, over time, these institutions became isolated and operated independently, hindering effective collaboration. Traditionally, there has been a perceived separation between politicians, bureaucrats, and military professionals, each considering their domains as distinct. In the current scenario, a strategy without connected efforts between military and political spheres is deemed unlikely to succeed.

The East Ladakh standoff of 2020 underlines the need for concurrent military and political efforts, emphasising the importance of coordination across various domains.

There is a growing recognition of overlapping, particularly in non-kinetic, cognitive security challenges. A cohesive, multi-domain response is deemed essential for strategically resolving these challenges. To address these diverse security threats, the Indian government has taken steps like appointing a Chief of Defence Staff (CDS) and establishing the Department of Military Affairs (DMA). These measures aim to foster a 'whole-of-the-nation' and 'whole-of-the-government' approach. Many democracies appoint military officers to high governmental and political positions with a good reason; to bring forth a military opinion about nation security issues. While some argue that India has attempted to integrate military officers into top posts in administration and government, the effect is different. Unlike other democracies where military officers are directly appointed to political positions,[6] the Indians are retired officers who entered politics after their service.

Achieving effective governance and national security demands a profound understanding and synergy between military, political, and bureaucratic domains. The military, comprehending political imperatives, must align strategies with national goals, ensuring a proactive contribution to security while respecting civilian authority. Simultaneously, politicians should grasp the intricacies of military operations, appreciating the strategic implications of their decisions. Bureaucratic engagement with military perspectives involves recognising the specialised knowledge and operational insights that the military brings to national security. Stepping down from the high table will signify a willingness of bureaucrats to acknowledge the unique expertise of the military in matters of defence and strategy.

This collaborative approach, an intuitively difficult one, will enhance decision-making by incorporating diverse viewpoints, fostering a more comprehensive understanding of security challenges. Respecting military views will stretch bureaucratic decisions to align with operational realities and optimizing the effectiveness of policies. It is expected to promote a culture of cooperation where bureaucratic processes accommodate military expertise, ultimately contributing to well-informed and holistic governance in the realm of national security. A mutual understanding among these pillars of governance is paramount, fostering an environment where collective insights lead to comprehensive, well-informed policies and actions.

India is also planning to release its national security strategy (NSS),[7] ending a long debate of why a formal document is not required. It is India's assertion of arrival on the stage and confidence in its strategic culture to articulate in a published document. The NSS provides guidance to all agencies to act in a 'coordinated manner'. It is an effect of 'whole-of-government approach' to face all future challenges. It will inform citizens in India or abroad, about their welfare, their territorial integrity, and sovereignty, ensuring economic growth and approaches to achieve national objectives. It conveys messages to allies and redlines to adversaries – it will strengthen deterrence. Articulation of NSS is acknowledging the complex relationship and yet synergistic realisation of all the elements of Comprehensive National Power.[8]

Indian Economy

India will emerge as the world's third largest economy by 2027 with GDP crossing US$ 5 trillion dollars and a developed economy by 2047. Navigating geopolitical uncertainties and the slowdown in the global economy, will, undoubtedly, not be easy. The MSME (likely to grow to have 75 million units by 2024) sector has contributed 30 er cent to the GDP, 43.6 per cent of merchandise exports, and close to 123 million jobs. It has a strong presence in rural areas and provides the much-needed income generation at the grassroots. The governmental push during the COVID pandemic compelled many to adopt digital services to sustain themselves. Digitisation with embedded finance start-ups and aggregators now provide faster credit access. The emergency credit line guarantee (ECLG) scheme and production-linked incentive (PLI) schemes provide good financial assurance. Special thrust on sunrise sectors such as defence, space, semiconductors and e-mobility are opening newer opportunities.

All 1.4 billion Indians are driving consumption and investments. This large population also invites global companies, with a promise of skill and talent, and technology. With the largest pool of English-speaking Science, Technology, Engineering and Mathematics (STEM) graduates, India had an impressive annual increase of 2.14 million each year. The digital public infrastructure (DPI) has made digital payments and accessing essential documents a mere routine for Indians. The model is being studied by other countries; after all, India's payments are more than the digital payments made by the next four leading countries combined. The world is looking to invest in India and 25 of S&P 500 companies' Indian-born CEOs are helping build global connectivity to Indian businesses.[9]

India will do well to nurture its MSME sector along the growth highway and guard it from the challenges and conflicts.[10] The sector is threatened by structural issues, such as limited access to formal credit, skill and technology gaps and insufficient infrastructure and taxation. There are other barriers to this growth too; unequal distribution of wealth between regions and citizens, exacerbated by linguistic fault-lines, divisive politics, minority appeasement by political leaders that 'routinely exhaust political capital, which could otherwise be leveraged for social and economic transformation'.[11] Ease of doing business continues to irk investors; there are attempts to improve but foreign investors are seeking relief from bureaucratic land acquisitions, contract enforcement, intellectual proprietary rights and local labour issues. All such pillars of a vibrant economy are susceptible to malicious intent to derail the growth. The large surface area of Indian economy is susceptible. The minds of domestic consumers, and foreign investors are vulnerable to cognitive attacks; the digital architecture is prone to technology attack. The security of the Indian diaspora and their businesses are also part of the economic surface of the country and need protection. Economic threats also include malicious attempts against economic policies, economic houses and proxy companies, cognitive of the human capital; individually and collectively.

Strategic Partnership

As an alternative to military alliances, strategic partnerships are agreements to interact and cooperate with each other on a long-term basis in specific fields, without prejudice to other partnerships. It may exist exclusively or collectively over mutual interests covering economic, political, defence and security. India has signed strategic partnerships with more than 30 countries including the USA, UK, Japan, France, China, etc. India is zealously guarding its 'strategic autonomy' in a web of 'complex interdependence'.

In the context of the challenges posed by the dysfunctionality of international institutions, especially the UNSC, strategic partnerships become crucial for several reasons. These partnerships offer a more agile and responsive mechanism for conflict prevention and management, unburdened by the bureaucratic complexities often associated with international bodies. By strategically aligning with like-minded partners, states create a network capable of addressing emerging security challenges that may fall outside the purview of traditional international bodies. These alliances allow nations to form flexible coalitions tailored to specific issues without much global repercussions. By pooling diverse resources, expertise, and perspectives, nations can develop multifaceted strategies that address the root causes of conflicts and promote long-term stability and conflict prevention. In situations where historical animosities and deep-seated mistrust hinder effective cooperation within international institutions, strategic partnerships allow nations to build trust incrementally through focused collaboration on specific issues.

In the face of dangerous military posturing and volatile flashpoints, strategic partnerships facilitate coordinated responses that deter potential aggressors. Shared intelligence, joint military exercises, and synchronised diplomatic efforts contribute to stabilising regions prone to conflict. Military-to-military training alongside such partnerships facilitates direct communication channels between armed forces of nations, reducing the likelihood of misunderstandings, miscalculations, and unintended escalations. This training also provides additional communication channels with other states, which, in turn, prevent conflicts in their nascent stages.

Indian Foreign Policy

India is now scoping the global stage consisting major points of conflation which may result in conflicts, if not managed well. A growing trend of multi-domain and hybrid warfare blur the lines between a classic state of war and peace. There is contestation over ideology between liberal democratic countries vis-a-vis other models of governance. Contestation over technology will create rifts between the leaders and challengers. Expectations to harness newer energy resources and minerals, may create conflicts. Strategic shocks like COVID and the Ukraine-Russia conflict have exerted and disturbed the global supply chains forcing nations to become self-reliant and create redundancies.

The Indian leadership is singularly instrumental in shaping India's contemporary international presence. This dynamic and charismatic approach has garnered attention and admiration on the global stage. India's robust diplomatic outreach, its proactive engagement in forums like BRICS, QUAD, and G20 highlight its commitment to foster regional stability and international cooperation. These relationships bolster India's strategic capabilities, open avenues for defence cooperation, and promote economic ties. As a voice for the Global South, India acknowledges its responsibilities and actively participates in international forums. India had pointed out that Europe cannot expect support and solidarity while remaining indifferent to the challenges of the Global South, which is now resonating across many governments. Membership of African Union to G20 stems for this very logic; if the European Union with a population of less than 450 million is a member, why can't the African Union with more than 1.3 billion?

In an era where global dynamics often demand dramatic, highly visible actions, the international community appears to be awaiting India's strategic moves.[12] Historically, the Western world's international relations calculus has often been punctuated by events of profound visibility and destructiveness, from Pearl Harbour to the bombing of Japan, and the invasions of Iraq and Afghanistan. Such events are etched in the collective memory of humanity, often epitomising a nation's 'arrival on stage.'

How will India, a nation now notably present on the global stage, define its role? The question, however, is not why India has arrived; it already has, stoically and resolutely. Rather, the question is whether the world should view this arrival with apprehension or applause. It is a question of assessing whether India's rise embodies a hegemonic ascent or a transformational growth. To 'explain' India's geopolitical evolution and its impact on the global stage, its foreign policy has matured, is robust and stands strong. India's foreign policy has consistently emphasised diplomacy, cooperation, and non-aggression. War is never a desired course of action, and India's historical experiences, have instilled a deep commitment to resolving conflicts through peaceful means. Prime Minister Modi addressed 'all the people of the world' on 23 August 2023, after the successful landing of Chandrayan-3 on the moon.[13] 'India's successful moon mission is not India's alone', he said and hoped it would 'help moon missions by other countries in the future'. India's guiding principle of *Vasudhaiva Kutumbakam* is deeply rooted in its cultural heritage and conveys the necessity to freely applaud such a rise and participate in Indian transformational growth. Foreign policy and strategic communications have never been more important for India.

Out of Area Contingency (OOAC)

Indian growth has its appreciators and distracters, both of whom are concerned about Indian OOAC capabilities. It will be the duty of the Indian government to ensure the physical safety of its diaspora in the face of conflicts or natural disasters and in the military context, it refers to the employment of the armed forces beyond borders. India has to possess the capabilities for humanitarian assistance and disaster relief (HADR) and lethal OOAC options for the future. India's economic growth and prosperity have beyond border manifestations too, be it trade, access to energy, critical components, or the diaspora, and their businesses; all are sources of significant remittances, and may need protection or evacuation. There are costs associated with economic growth and India has to rise to this responsibility too.

The MP-IDSA Report in 2012[14] defined the issue considerably well. India has done well to evacuate its citizens from conflict zones, be it Ukraine or Gaza and it stems from a universal understanding of India being a responsible nation without prejudice. The report indicates lack of regional specialisation and insufficient analysis of field reports from participating units, causing underutilisation of the vast experience gained. There seems to be a lack of politico-military analysis capability of OOAC in terms of lethal operations. The report also underscores the importance of training commanders and troops to adapt swiftly to changing geopolitical dynamics.

Cyber Threats

It is cross-domain function, which affects all sectors, infrastructure and people and covers all aspects of cyber warfare including cybercrimes. It is targeted towards national cyberspace; a collection of all computer systems and intranets and the information residing on them that are owned by a nation. In the context of this research, cyber threats exist to the infrastructure, the information on the infrastructure and also the people of the nation. While the people are affected by cyber information operations, cyber attacks may be classified into cyber-crime, cyber-hactivism, cyber-espionage, cyber-terrorism, and cyber-war. Cyber war refers to cyber attacks carried out by a state or sponsored through non-state actors, both with the intention of achieving strategic objectives (Estonia–2007). Cyber-espionage and terrorism is defined by cyber-attacks against a business or government entity (GhostNet Network–2009).

The beginning of 2023 saw 480 DDoS attacks by hacktivists[15] making India the most targeted country. India lost Rs. 63.40 crore in FY 2020-21 and Rs. 58.61 crore in FY 2019-20, to cybercrimes, making it a grave concern.[16] Cyber information operations, also referred to as cognitive operations, exploits cyber domain to manipulate information reaching the target population. Cognitive operations include distinct methodologies; psychological operations convey selected information to influence foreign audiences, deception operations that mislead an adversary and public information that aims to inform domestic and foreign audiences. It is easy to fathom that cyber warfare in the hybrid context has a huge potential for employment in future conflicts.

Spectrum of Conflict

The contemporary security landscape is marked by a notable shift towards sub-conventional conflicts, characterised by irregular warfare, asymmetric, hybrid tactics, while blending military and non-military elements. Newer conflicts thrive on ambiguity and non-attributable tactics, where the actions are designed to remain below the threshold that would trigger a conventional military response. Within this form, hybrid warfare fuses conventional and unconventional methods, combining military, economic, diplomatic, and informational tools to achieve strategic objectives.

There is a fairly wide and common understanding about threats to India's integrity[17] by accentuating casteism and artificial clubbing of grievance-seeking groups. There are attempts to denigrate Indian claims to civilisational history and heritage. The Indian government, educational institutions, industry, and society are also targeted. Agitations over the Citizenship Amendment Act and the National Register of Citizens forced the Ministry of External Affairs to reach out to countries and justify them as internal matters.[18] Farmers' agitations during 2021 found support from entertainer Rihanna, environmental activist Greta Thunberg and adult movie actor Mia Khalifa, to name a few. Serious commentators and former Indian envoys hinted at a vast international conspiracy to defame India.[19] The term 'toolkit' gained notoriety when it became widely known among the general public during the farmers' agitation. It was associated with efforts to influence the narrative and amplify the protests. It was supporting and coordinating a campaign against India.

The sub-conventional spectrum of conflict is experiencing an increased likelihood of occurrence coupled with a lower intensity of violence. This dynamic is shaped by a combination of global interconnectedness, political fragmentation, and economic inequality. The trend towards non-kinetic methods and a global aversion to high-intensity violence allows conflicts to persist without overwhelming oversight. The global community has become increasingly averse to high-intensity violence, particularly when directed against civilian populations. This lower intensity conflicts aligns with evolving norms that emphasise the protection of civilian lives.

By incorporating both kinetic and non-kinetic with military and non-military aspects, into contour of the sub-conventional spectrum of conflict,[20] the multifaceted nature of modern conflicts emerges. Of special relevance for future conflicts are the non-military and non-kinetic methodologies.

- Non-military conflict methodologies extend beyond traditional warfare, encompassing diplomatic, economic, and informational strategies. Diplomacy plays a crucial role, utilising negotiation and dialogue to resolve disputes. Economic tools, such as sanctions or trade policies, are employed to influence the adversary's behaviour. Informational methods involve propaganda, cyber influence, and public relations campaigns to shape perceptions. Non-military approaches highlight a diverse array of tools available in addressing conflicts without resorting to armed force.
- Non-kinetic conflict emphasises non-physical means to achieve strategic objectives. This includes diplomatic negotiations, political interference, economic sanctions, economic coercion and psychological warfare. The focus is on influencing the opponent's decision-making without resorting to direct physical force. Cyber attacks, information warfare, and economic sanctions are techniques of non-kinetic conflict methods.

The strategic communication in modern conflicts has evolved into a dynamic and essential component of statecraft and military operations. From shaping narratives to building alliances and countering disinformation, strategic communication is the linchpin in the information-centric landscape of conflicts. Nations must master the art and science of strategic communication for achieving strategic objectives and maintaining global stability. It stands out as a primary tool of choice against both non-military and non-kinetic methodologies. The shaping of narratives and the management of perception is essential to guard against proliferation through social media and instant global connectivity. Governments and military entities have to leverage strategic communication to control the narrative, presenting their version of events and influencing public opinion, domestically and internationally.

KEY TAKEAWAYS: INDIAN MILITARY IN FUTURE CONFLICTS

In line with the main theme of the research, prophesising conflict environment is necessary to identify the likelihood of future military confrontation and, subsequently, the AI tools required for successful battles. In an effort to quantify the intangible contours, the table below consolidates and ranks the conflict environments from all the methodologies utilised in the chapter. The final column 'Means' the table and positions the final probability.

	USI	IDSA	Survey	SME	Mean Rank	Priority
Territory	1	6	5	1	3.25	1
Hybrid	2	8	2	2	3.50	2
Internal Security	5	4	4	2	3.69	3
Maritime	3	2	8	4	4.19	4
Strategic Partnership	9	1	6	6	5.50	5
Economy	6	3	4	10	5.75	6
Synergy	4	5	7	9	6.13	7
Cyber	10	9	1	6	6.50	8
Foreign Policy	8	9	9	4	7.50	9
OOAC	9	10	10	8	9.25	10

With the existing military capability and Indian EVR, three levels of military involvement emerge.

- **Known Employment.** The Indian military has been and will be employed in the well understood conflict environments of territorial conflicts, internal security situations, maritime conflicts, cyber threats, synergy efforts and out-of-area contingencies. Synergy is defined as a challenge which will affect future capabilities, claims and conflicts; hence, it requires procedural methodology to incorporate all the instruments of national power including the military. Synergy is acknowledgement of collaborative efforts across various national instruments to address challenges, even if not inherently conflictual.

- **Partially Known Employment.** The military has the capacity to be employed for hybrid threats and strategic partnership environments. Indian hybrid capability is in a state of development as also the strategic partnerships. Strategic partnerships are analogous with diplomatic and economic efforts and define a methodology to prepare collaborative participation in conflicts. This partnership sponsors a chosen common concern for a specified period of time, despite other differences. Hybrid capability is a special competence development to prosecute non-kinetic and non-lethal effect and thus may not dwell completely in the military domain.
- **Unknown Employment.** The present India has yet to quantify employment of the military in foreign policy development, other than military diplomacy and strategic partnerships proposed earlier. Similarly, military employment to safeguard economic interests are also an unknown factor in Indian strategic thought. In the distant future, there may be a probability where India is committed to protect its diplomatic and economic interests, military employment may be envisaged, but as the research shows it is going to happen soon.

Defining Realms for Military AI Tools

Based on the researched data and analysis of future conflicts, priority AI developmental efforts emerge. High impact and frequently occurring environments prioritise development of AI now. It is essential to align the development of AI tools with the specific needs and challenges in each operational context, ensuring practical and effective integration.

- **Territorial and Maritime Conflicts.** AI tools to enhance situational awareness and decision-making and war fighting in territorial conflict scenarios.
- **Internal Security Situations.** Prioritise AI applications for surveillance, intelligence analysis, and threat detection in internal security operations.
- **Cyber Threats.** Develop AI-based cyber security solutions to detect, prevent, and respond to cyber threats targeting military infrastructure and communication networks.
- **Hybrid Threats.** Invest in AI technologies for analysing and countering hybrid threats, with a focus on non-kinetic and non-lethal approaches.

- **Synergy.** Though not military AI, exploring applications that facilitate coordination and information sharing among various national instruments will improve in synergic efforts.

A whole-of-nation approach will drive AI tool development for strategic partnerships, foreign policy development and safeguard economic interests. It is not a pure military arena, but a call from the 'far future' may involve the military; thus, it is essential to participate in developing AI tools in coordination with lead agencies. In such environments, the military will have to create capacities to function in a supporting role. AI tools in such cases will support a collaborative effort, data sharing, and joint decision-making processes to focus on risk assessment and strategic planning.

CONCLUSION

It is imperative analyse complexities of future military engagements and connect with appropriate AI solutions. This chapter, while drawing insights from diverse methodologies and expert perspectives, yields a nuanced understanding of known, partially known, and unknown military employment scenarios within the Indian context. This analysis provides a solid foundation for prioritising the development of AI tools to meet the evolving military challenges on the horizon. Recognising the urgency associated with certain high-risk scenarios, immediate action becomes imperative. Accelerating the development and deployment of AI solutions in these critical areas, such as territorial conflicts, internal security, and cyber threats, is not merely a strategic option but a pressing necessity. This strategic approach to prioritise AI development for the Indian military aligns seamlessly with the identified conflict environments and associated risks. By adopting this approach, AI tools are not only conceptualised but swiftly operationalised to fortify national security, enhance preparedness, and foster collaborative endeavours amid the dynamic future military landscape.

The next chapter embarks on a crucial journey, delving into the identification and design of cutting-edge AI tools tailored for high-priority conflict areas elucidated in the preceding analysis. It will focus on practicality and effectiveness with the aim to conceptualize AI tools that not only align with the unique demands of each conflict scenario but also redefine the paradigm of military operations in the next century.

The next chapter will also draw from previous chapters (technology, military case studies, inherent AI expertise and conflict prediction) to envisioned AI tools that are not theoretical constructs but tangible solutions. Most crucially, it seeks to bridge the theoretical foundations laid in the current discourse with action that can be translated into real-world applications. It aims to provide a roadmap to integrate AI into military operations.

NOTES

1. https://docs.google.com/forms/d/e/1FAIpQLScg2ooU1ErIV6ZYgNq5JQHeJ3gmOsh1lQSm613w9Dr2TkmO2A/viewform?usp=sf_link, created on 14 October 2023.
2. https://marineregions.org/gazetteer.php?p=details&id=8480, accessed on 21 January 2024.
3. https://www.mha.gov.in/sites/default/files/BMIntro-1011.pdf, accessed on 22 November 2023.
4. PIB Delhi. Vice-President's address at the 2023 edition of the Indo-Pacific Regional Dialogue, 15 November 2023, https://pib.gov.in/PressReleasePage.aspx?PRID=1977077, accessed on 1 January 2024.
5. Greater Civil-Military Jointness must to Further Strengthen National Security, https://pib.gov.in/PressReleaseIframePage.aspx?PRID=1833513, 13 June 2022, PIB, Ministry of Defence, accessed on 15 January 2024.
6. For India's National Security, Time for Civil and Military Synergy, Lt. Gen. P. R. Kumar, 7 February 2021, https://www.indiandefencereview.com/spotlights/for-indias-national-security-time-for-civil-and-military-synergy/, accessed on 15 January 2024.
7. Work starts on shaping first national security strategy, long wait ends, 4 November 2023, https://indianexpress.com/article/india/work-starts-on-shaping-first-national-security-strategy-long-wait-ends-9012566/, accessed on 15 January 2024.
8. https://timesofindia.indiatimes.com/blogs/ChanakyaCode/a-national-security-strategy-for-india-documenting-strategic-vision-prudently/, accessed on 15 January 2024.
9. https://www.economist.com/international/2023/06/12/indias-diaspora-is-bigger-and-more-influential-than-any-in-history, accessed on 31 December 2023.
10. Rumki Majumdar. India economic outlook, October 2023, https://www2.deloitte.com/xe/en/insights/economy/asia-pacific/india-economic-outlook.html, accessed on 31 December 2023.
11. https://hbr.org/2023/09/is-india-the-worlds-next-great-economic-power, accessed on 31 December 2023.
12. Pawan Bhardwaj. India's Geopolitical Evolution: A Rise to Transformational Growth, 26 September 2023, https://www.usiofindia.org/strategic-perspective/India-Geopolitical-Evolution-A-Rise-to-Transformational-Growth.html
13. English rendering of Prime Minister's address after the successful landing of Chandrayan-3 on Moon via video conferencing. https://pib.gov.in/PressReleasePage.aspx?PRID=1951491#:~:text=I%20am%20 confident%20that%20all,journey%20beyond%20the%20Moon's%20orbit, accessed on 25 September 2023.
14. https://www.idsa.in/system/files/book/book_NetSecurityProvider.pdf, accessed on 15 January 2024.
15. https://www.radware.com/blog/security/2023/05/india-one-of-the-most-targeted-countries-for-hacktivist-groups/, accessed on 26 December 2023.
16. https://timesofindia.indiatimes.com/business/india-business/cyber-crime-losses-rise-to-rs-63-crore-in-fy20-21-govt/articleshow/90532711.cms?utm_source=contentofinte&from=mdr, accessed on 26 December 2023.

17. Malhotra, Rajiv and Viswanathan Vijaya. *Snakes in the Ganga*, Occam (An imprint of BluOne Ink), Garuda Prakashan Pvt., Ltd., Central Delhi, Delhi, India, 26 September 2022.
18. https://www.business-standard.com/article/pti-stories/india-has-reached-out-to-countries-across-the-world-on-caa-nrc-mea-120010201069_1.html, accessed on 16 January 2024.
19. https://www.newindianexpress.com/galleries/nation/2021/Feb/04/twitter-war-who-are-the-international-celebrities-supporting-the-farmers-protest-in-delhi-meet-ri-103073.html, accessed on 16 January 2024.
20. Hoffman Frank. The Contemporary Spectrum of Conflict: Protracted, Gray Zone, Ambiguous, and Hybrid Modes of War, https://www.heritage.org/sites/default/files/2019-10/2016_IndexOfUS MilitaryStrength_The%20Contemporary%20Spectrum%20of%20Conflict_Protracted %20 Gray%20Zone%20Am

APPENDIX A

SURVEY FORM: FUTURE CONFLICT ENVIRONMENTS FOR INDIA

Introduction

Though this research is predominantly to identify AI tools for future battles, it emerges out of Indian future aspirations and concerns. The defence of territorial integrity and internal security situations will always need military participation, so does the employment of AI tools. India does not harbour expansionist interests, but her global prospects may get affected by competitions and conflicts. In many such future conflicts, the Indian military will also get involved in collaboration with other instruments of power. There are some obvious questions to this problem; What are the conflict zones that India intends to enter? Does India have global ambitions or do regional aspirations satisfy? Will a conventional war ever happen on the borders? Is low-level conflict intrinsic to conventional wars now? The survey intends to answer these within 'India's realm'. India's prospective conflict situations carry profound implications for global stability, peacekeeping efforts, and national security.

The analysis initially sought to identify direct mention of future conflict scenarios; however, the publications did not provide sufficient data points. The research then widened to include 'security concerns', 'challenges', 'competitions', 'conflations' and 'competitions.' A large number of data points surfaced during the search which diverted focus. Strategic data condensation was done to extract more meaningful insights from the dataset by reducing redundancy and highlighting key relationships.

Survey Form

This form intends to contrast and compare previously-mentioned content analysis with a survey to validate the results. Strategic consolidation focused the content analysis to manageable environments; territorial threats, collusions, internal security, hybrid threats, cyber, economic, maritime challenge, managing foreign policy issues, Out of Area Contingency (OAAC), synergy challenge and strategic alliances.

The survey sought to seek inputs on LIKELIHOOD of occurrence and IMPACT of each separately to create a risk matrix. While Likelihood was arranged on a scale from 1 to 7 (Very Likely, Likely, Somewhat Likely, Neutral, Somewhat Unlikely, Unlikely and Very Unlikely) the Impact was ranged on a scale of 1 to 7 (Very High Impact, High Impact, Significant Impact, Moderate Impact, Low

Impact, Least impact and No Impact). Additional points and suggestions were sough as long responses at the end of the survey. The volunteers were also asked to provide their experience in national security. The following expansions are drafted for the survey in the first section.

- Territorial threats include those from China and Pakistan (singularly and collusively) along the border and threats to island territories.
- A special question sought opinion on LIKELIHOOD of collusion between Pakistan and China against India?
- Internal security combines the effects from terrorism, insurgency, LWE, radicalism, internal strife and communalism.
- Hybrid threats cover all the options including cognitive domain actions which may be utilised against the country.
- Cyber threats cover all aspect of cyber warfare including cybercrimes and hacktivism.
- Economic challenge consolidates targeting attempts against economic policies, economic houses and proxy companies undermining the national economy.
- Maritime challenge covers the wide area of the Indo-Pacific and Indian Ocean regions which are fast becoming the competition arena for energy, security, environment, human trafficking and other illegal activities, undermining the cohesion of the nation.
- Foreign policy issues in areas of interest cover Afghanistan, South East Asia, West Asia and the Central Asian Republics which have proved to be points of conflation and result in conflicts, if not managed well.
- Out of Area Contingencies (OOAC) – both HADR and lethal versions.

The second section of the form sought to understand the role of enablers to achieve desired results in future conflict environments.

- Synergy challenge covers inter-service, inter-organisational coordination and intra-governmental cooperation. It also covers civil-military synergy, national security concerns, documentation, decision-making and whole-of-government effort.
- Strategic alliances with other countries to organise international, multi-national or regional partnerships in all spheres, not necessarily military.

Chapter Five

Tailored Tactical AI Solutions for Indian Army

TACTICAL BATTLES

Integration of AI into tactical military operations is the prime aim of this research. Understanding the military context of AI is a major perquisite for all technology developers. This chapter will attempt to undertake a complex and simultaneous effort to demystify the military context and identify the AI integration points in this landscape to provide insight into its practical applications. It is a well-known fact that military operations are intrinsically dynamic. Human-to-human interactions in life/death situations make it a complex environment with a split second decision making requirement.

- Militaries have developed strict codes of conduct (called battle drills) to speed up reactions during combat. These battle drills define concurrent and automatic actions by each member of the team, without waiting for orders or distribution of responsibilities and obviously there are many numbers of such battle drills that soldiers have to learn, perfect and execute during combat. Tactical battle is fought by a military unit with its integral and organic equipment in a defined area and tactics is this lowest form of military decision-making and operations put together. A series of such tactical decisions and engagements when sequenced and synchronised in a given time and space matrix to achieve a stated objective are known as operations.
- The military commander is also handed over a well-explained and demarcated space over which full control is expected. The space known as area of responsibility (AOR), is defined by boundaries and depth

(the space also extend well into own area which remains the responsibility of the commander). The commander is also allocated military units (with organic weapons and equipment) for this AOR. Geographical space beyond this, till where the commander and exert influence with organic capability, is termed the area of influence. It is thus limited by the reach of the weapons and equipment allocated to the commander. Beyond this, extends the area of interest, where any military activities (of the enemy) will endanger or jeopardise the mission accomplishment of the commander. It is thus normal for military commanders to request additional inputs about such areas from their superior commanders.
- All military manoeuvring done during execution of battle drills is called battle field mobility, while tactical mobility enables commanders to move additional assets (also organically available) into a battle space to win a particular operation.

This landscape of battle drills and operations with organic assets to achieve a well define objective are considered tactical-level operations by military experts. In the realm of military operations, tactical level engagements represent the ultimate execution of battle directions. This level of operation is characterised by its focus on achieving well-defined objectives through the coordinated use of organic assets where the intricacies of warfare unfold in real-time. Military experts define tactical level operations as the orchestrated deployment of forces and resources to accomplish specific goals within the complex and dynamic environment of the battlefield. Despite the inherent challenges and dangers, tactical-level operations remain the cornerstone of military strategy, representing the frontline where victory is forged.

The tactical-level operations define the concept of 'pointy edge'—the tip of the spear that thrusts furthest into battle. This metaphorical point represents the vanguard of military forces, where soldiers and units engage the enemy directly, facing the brutal realities of combat. It is here, amidst the chaos and uncertainty of the battlefield, that the true nature of warfare reveals itself, driven by the unrelenting forces of human conflict and the unpredictable dynamics of the physical and electromagnetic environment. In this unforgiving environment, soldiers have to rely on their training, instincts, and camaraderie to overcome adversity.

In this melee of battle drills and operations, military commanders must navigate various challenges like terrain obstacles and enemy defences through the fog of war (unknowns that exist in intelligence, communication, and decision making) complicated by constraints of time and resources. The relentless nature of tactical-level engagements demands a level of adaptability and resilience that is unparalleled, as soldiers contend with the ever-changing dynamics of the battlefield and the inherent risks of combat.

Characteristics of Tactical Battlefields

Military tactical operations remain dynamic and complex that involve planning, execution, and coordination of actions aimed at achieving specific military objectives within a defined operational area. These characteristics collectively define the nature and challenges of tactical military operations, highlighting the need for agile, adaptive, and integrated approaches to achieve success on the modern battlefield.

- **Terrain and Weather.** Tactical operations are supported and influenced by terrain and weather conditions, which impact mobility, visibility, and the effectiveness of weapons systems. Commanders take special measures to exploit suitable conditions to achieve quick success.

- **Objective Orientation.** Tactical operations are typically driven by specific objectives, such as seizing a key terrain feature or neutralizing enemy forces and serve as the focal point for planning and executing associated military actions.

- **Integration of Manoeuvre and Firepower.** While manoeuvring elements seek to gain positional advantage and exploit enemy weaknesses, the firepower assets provide the means to suppress enemy forces, control terrain, and achieve desired affects.

- **Tempo.** Tactical success depends on maintenance of the tempo to seize and maintain the initiative while disrupting the enemy's cohesion and ability to react.

- **Lethality.** Tactical operations are lethal and involve inherent risks due to enemy fire, human casualties and damage to equipment.

- **Uncertainty.** Tactical environments are often unpredictable and subject to rapid changes due to enemy and battlefield developments.

- **Technology Characteristics.** Tactical forces operate in battlefields where large captive power capability and assured Cloud connectivity (to AI central servers) is routinely interrupted.

INDIAN TACTICAL BATTLES

The previous chapter identified the conflict environment and the likelihood of military confrontations and this combination indicates the prioritises for military AI tools development. Among all the possibilities established earlier, this chapter will confine to tactical operations during territorial conflicts, internal security situations and maritime conflicts, hybrid being integral to all possibilities. 'AI Now' is aligned to high impact and frequently occurring conflict environments to enhance situational awareness, decision-making and war fighting in territorial and maritime conflicts, emphasising surveillance, intelligence analysis, and threat detection in internal security operations and investing in AI technologies for analysing and countering hybrid threats, with special attention on non-kinetic and non-lethal approaches. OOAC encompass tactical engagements beyond borders and thus are intrinsic to the analyses. Cyber is an independent field and a matter of a separate research while synergy between organisations and governmental departments requires procedural efforts supported by all the instruments of national power including the military and beyond the scope of this chapter. The table below summarises the operational aspects, tactical landscape, and tactical conditions of Indian battles. Each of the tactical landscapes is deliberated in the following paragraphs.

Tactical Landscape	*Operational Aspects*	*Tactical Conditions*
- Riverine	- Surveillance	- Temperatures –45° to +45°
- Deserts	- Intelligence	- Sandstorms
- Semi Deserts	- Reconnaissance	- Snowstorms
- Build-up	- Decision Making	- Lightning
- Hills	- Engagement	- Humidity 20% to 85%
- Mountains	- Operational Movement	- Captive Power capability
- High Altitude	- Logistics Planning	- Cloud Connectivity
- Glaciers	- Logistics Movement	- Training Data
- Forests	- Administration	- Form factor
	- Health Care and Treatment	
	- HR Management	

Counter Insurgency (CI) Operations

CI operations involve deployment of the military once the state has evoked the Disturbed Areas Act.[1] The Indian Army deploys with strict control under the legal ambit of the Armed Forces Special Powers Act.[2] The area nominated as 'Disturbed' is populated by Indian citizens and thus is not equated with an interstate conflict. The Indian Army is restrained from using full combat potential and the concept of operations dictate priority on intelligence collection, analysis, and highly surgical tactical actions. Ab initio decisions to go lethal are an exception rather than the norm. The population of this area is considered the centre of gravity and all emphasis is focused to address the grievances of the people which are used as fuel by inimical elements and proxy players. It is thus absolutely essential for the Army to focus its attention on intelligence collection and analysis.

- Under the unified command structure, operations in such zones, the Army performs its own functions and has a parallel need to coordinate and cooperate with other state instruments like the governance, state police, education system, intelligentsia, and legal authorities. During such population-centric activities, the Army always feels the need to understand the local language and communicate effectively. This translation role is fulfilled by local police representatives and volunteers. Over time, the Army has developed in-house communication expertise but AI support will alleviate much of the miscommunication issue.
- The Indian Army is the largest military volunteer force and prides itself of being 'forever in battle' in active combat. Typically, an infantry battalion is allocated an area of responsibility for an average period of two years. It is relieved by another battalion at the end of the tenure. All essential aspects are then formally passed over to the incoming unit. Previous tenure experience of the newly-inducted battalion, in many cases, is a bonus. And so is the individual experience of officers and men from their personal tenures in specialised units like Rashtriya Rifles (RR), Assam Rifles (AR) battalions and headquarters. While the units may rotate, the formation headquarters (brigade, sector, division, force, area, corps, and command) are static and retain a large volume of operational, geographical and background information. These large volumes of data and operational diversity is ideally served by AI

information retrieval systems. This information retrieval system may have a large area footprint but its special ability lies in providing tactical inputs. An AI system can be enabled to generate predictions and will be highly valuable in CI operations.

- The Army undertakes lethal operations during CI operations, which strictly follow the concept of minimum force. This is to ensure minimum collateral damage to the infrastructure and no harm to innocent bystanders. Operations in built-up areas result in room interventions against holed up terrorists and possibly with hostages. Traditional Army responses may cause collateral damage to the building and injuries to civilian hostages. AI solutions will improve the better success rate of tactical operations in such a situation.

Hybrid Operations

In all its manifestations, the Army is typically burdened to fight against the cognitive effect of hybrid wars. Identification of non-lethal and non-contact events is also a challenging aspect. A typical example of cognitive operation will manifest through social media platforms in the form of propaganda, disinformation, and psychological warfare. While the belligerents aim to target the faith in leadership and the organisation the zero trust concepts of the cyber realm are cannot be considered a mitigating solution here.

- All soldiers and their families are highly dependent on social media and mobile phones and complete removal of personal devices is not possible. Instead, the Army lays restrictions on carriage to prohibited zones or operational areas. During non-office periods or rest times, soldiers flip through their devices and may get targeted by hybrid effort. AI solutions is needed to debunk fake news, propaganda, and digital crime will secure the military cognitive.
- Hybrid operators have worked against India for a long time. Their directions have resulted in numerous and unpalatable instances along the LC. Efforts to call them out at international fora is a routine job for the Ministry of External Affairs. The Indian Army has also observed the handlers of terrorists and their henchmen close to the LC, coordinate events during the final execution. A case in point thus exists to monitor such persons of interest (POI) for pre-emptive or decisive action when deemed necessary.

Hills, Mountains and Glaciers

Mountains are characterised by rugged terrain, high elevations, and steep slopes. They often present obstacles to military operations due to the challenging topography, which can impede movement and limit communication. Harsh weather conditions, extreme temperatures and frequent storms, further complicate operations. Glaciers are massive sheets of ice characterised by freezing temperatures, crevasses, and unstable ice formations. Military operations in glacier environments are exceptionally challenging due to the extreme cold and dangerous terrain. Movement is slow due to lack of oxygen, with the risk of avalanches and crevasses. Logistics become complicated due to limited accessibility and therefore the need for specialised equipment.

Occupation of high features, security of passes, denial of the valleys and long logistic chains through terrestrial routes are typical characteristics of such areas. Mountains provide natural defensive positions and strategic vantage points. Troops stationed in mountainous regions must be trained in mountain warfare tactics and equipped with specialised gear to navigate the terrain effectively. High vulnerability to extremely cold climate, wind, and snow conditions reduces the equipment capability and pushes the soldiers to their physical limits. Dependence on air mobility becomes a major prerequisite in such areas. Non-availability of civilian infrastructure like electricity and communication in far-flung areas forces the Army to arrange for captive power. Restive LC and LAC have led to permanent army deployments, where additional utilities are required.

- Logistics is affected by challenging terrain and road conditions. Load switching from one type of transport to a variety of smaller forms in combination with aerial supply becomes a norm. Logistic supply to troops engaged in active combat is a typical situation of concern, where tactical operations have to be supported by uninterrupted supply. Traditional methods may fail to evince complete confidence. Tactical commanders are routinely forced to divide their attention between operations and supply. Logistic supply during active contact in tactical battle areas can provide relief to this important aspect.
- Dispersed deployment of troops and restricted movement ability is overcome by quick tactical mobility, in which the army moves soldiers, weapons, and ammunition from one place to another, where the quantum of movement is a secondary factor to the quality and the

frequency of mobility. A tactical commander would be highly satisfied if his AI system delivers one medium machine gun quickly across a river gauge, rather than moving ten soldiers over a day. Small delivery systems can move faster, require less power, and have flexible utility.

- A tactical commander is always eager to look behind the hill to observe the enemy vulnerabilities. A standard reconnaissance patrol, which is the epitome of infantry skills, will do better if backed with AI solutions. The infantry patrol is restricted to its hide (a camouflaged hideout behind enemy lines) and expected to tactically move around to gain information. A stealth AI platform will increase the observation and decision-making skills of this patrol.
- Radio (wireless) communication in battle is affected by screening and weather conditions. Traditional underground communication lines, may get damaged by rockslides, landslides, or avalanches. AI solutions can provide autonomous communication relays, without affecting the tactical speed of operations.
- LC and LAC create a highly active tactical landscape where small skirmishes like Galwan can usher uncontrolled escalation and strategic ramifications. Tactical battle space analysis of the area of interest is a well accepted military process to predict impending situations. Tactical commanders employ a variety of sources to obtain information and stitch a complete picture. Information is obtained through human intelligence (HUMINT), signal intelligence (SIGINT), open source intelligence (OSINT), social media intelligence (SOCMINT), and routine military analyses. Sifting this multi-spectral information is highly manpower intensive. In spite of dedicated efforts over a long period of time, it can leave out vital indicators of impending conflict. Tactical commanders are concerned about loss of surprise and initiative since army manuals suggest 'tactical battles are best fought at own place and time of choosing'. AI solutions to integrate all information and provide predictive analysis will assist tactical commanders to pre-empt, or at the worst, prepare their troops to execute a successful battle.
- In spite of the Army's best efforts at tactical mobility, rugged terrain will always require additional troops and equipment to monitor and thwart enemy designs. Enemy violations of border agreements or

ceasefire obligations result in incidents like Galwan (2020) and Kargil (1999). The enemy was able to occupy unheld heights since there is no early warning of such nefarious aspirations. In addition, these sectors are devoid of human habitation and long-range patrols are regularly launched to keep such threatened areas under observation. Sourav Kalia was leading this type of a patrol before the Kargil war and it is well known that his infantry patrol was lynched by enemy soldiers.[3] AI solutions can provide autonomous monitoring of such vulnerable areas and tactical decisions can be taken immediately. Monitoring can be accompanied with lethal action too.

- During a tactical operation, soldiers vie to secure advantageous positions to outwit the enemy, much like chess. This is primary infantry skill, especially in mountains. At times, a soldier has to make a fatal decision, which then guides the others in the group to reroute themselves. AI Bot capable of geographical survey in real time will provide added situational awareness to soldiers during contact battle, improve their firing position, and attack the enemy through his weakened position. AI Bot can also provide covering fire to soldiers to abandon a compromise location.

- Surveillance of valleys and gorges is yet another infantry skill of pride. Called long-range patrols, infantrymen trek across vast distances and altitudes to survey large swathes of land for military intelligence and enemy attempts to enter own territory. This patrol is self-contained (all the load is carried on one's back) for the entire duration. AI solutions can increase the reach of these patrols and improve observation across deep valleys and gorges, cracks and caves where traditional UAV and aerial platforms fail to offer much assistance. Standalone, autonomous systems can provide focused attention to a specific piece of ground too.

Deserts

Deserts are characterised by low precipitation levels, extreme temperatures, and sparse vegetation. They often feature vast expanses of sand or rocky terrain with little or no water sources. The lack of cover and limited resources pose significant challenges for military operations. Troop movement can be hindered by the difficult terrain, while the harsh climate puts personnel at risk of dehydration and heat-related illnesses. The absence of natural barriers makes it easier for adversaries to spot and engage military forces. Deserts also offer strategic advantages, such as wide open spaces for manoeuvrability and surprise attacks using sandstorms as cover.

Indian desert battles are fought across wide expanses and the Army has to undertake surveillance of these areas with land-based and aerial systems. AI solutions co-opting and stitching information through various sources could provide adequate inputs for special focus on most threatened sectors. Wide area surveillance by AI systems combined with focused manual reconnaissance will improve capacity within resources.

- Extended defence of the deserts also forces the Army to prioritise kill system deployment like missiles, tanks, guns and attack helicopters. Autonomous systems slaved to humans will increase the kill range while keeping the humans safe. Specialist vehicles like tanks and anti-tank guided missiles will benefit with grouped AI vehicles, allowing continuous engagement without the fear of getting outflanked by enemy tanks.

- Combat teams operate across large expanses, aiming to outmatch enemy teams. Many times, this manoeuvring takes the combat team beyond the reach of traditional supply echelons. An autonomous AI system will enable low-silhouette vehicles to provide ammunition, fuel and rations to these widely dispersed combat teams.

- Enemy combat vehicles can be engaged by an autonomous system guiding and controlling artillery fire well beyond the tradition reach of traditional army aviation airborne observation. Swarm-controlled multiple AI systems can provide more accurate enemy coordinates. Minimum or low landmark battle space of deserts is at best suited for such tactical operations.

- Wide open desert expanses favour long-range infiltration for a variety of kill tasks. The raiding groups are however constrained by fuel capacity and thus strike only a few targets before returning to base. Teaming with autonomous systems will reduce the need for physical verification of the target where the AI systems undertake wide area reconnaissance and human kill teams choose appropriate targets for destruction. This saves fuel and improves the stay and the kill capability of the mission.

Built-up Area

Build-up areas are characterised by urban or suburban environments with dense populations, infrastructure, and buildings. Military operations in built-up areas present unique challenges due to the complex urban terrain, which includes narrow streets, tall buildings, and underground infrastructure. Troops must navigate through crowded streets while minimising collateral damage to civilian structures and population. Close quarter combat becomes common, requiring specialised training and tactics. The presence of civilians complicates targeting and intelligence gathering. Built-up areas offer opportunities for cover and concealment, as well as access to resources and infrastructure for military operations.

Built-up areas are most unforgiving regions for all armies to operate and are generally avoided unless absolutely necessary. Fighting in built-up areas is a weary effort with street-to-street fighting, and house-to-house clearance. It is as brutal and fearsome as shown in military movies. The soldiers in the traditional sense are always overwhelmed when attacking such an area. The Ukraine and Gaza experiences will remain vivid long after the battles are over.

- The soldiers desire that AI solutions and a swarm system provide 360-degree view of streets, buildings, and houses to observe threats, while human teams are fighting and clearing houses. Stealth monitoring by AI-driven physical systems can provide latest update of tactical situations prior to an operation.
- Solutions are also required to provide nonstop communication, despite terrain masking by tall buildings.

Maritime

Maritime islands are characterised by their isolation, surrounded by water bodies. They vary in size and terrain, ranging from small coral atolls to larger land masses with diverse ecosystems and human population. Their coastal location offers strategic advantages for naval and amphibious operations. Control of maritime islands can provide access to crucial sea lanes and maritime resources, making them valuable assets for power projection and defence.

- Security of island territories is a serious responsibility, far away from the main forces; small units are expected to defeat ingress into the islands which are sovereign territory of India. Numerous islands spread wide

across seas and rivers have to be kept under surveillance and defended. Military operations in maritime islands are influenced by their geographic isolation, which can make resupply and reinforcements difficult. The Army will never find enough troops to physically man all such islands and applying logistics to such far-flung detachments is a difficult proposition.
- AI solutions are necessary to fill the gaps of traditional tactics. Continuous monitoring with combination of aerial, terrestrial, surface, and subsurface assets will provide advance warning to concentrate troops on threatened islands. The AI system can also be prepared to monitor and direct kill resources on the advancing enemy in case the traditional human-centric approach fails to deliver due to vast distances and unpredictable weather and sea conditions. Lethal autonomous system may also be required to be deployed in certain sensitive territories where human reaction could be relatively slow.
- Troops will require robust and flexible fail-safe AI communications systems during all times, especially when other types of communications fail.

Forest

Forests are characterised by dense vegetation, high rainfall, rivulets and rivers, and diverse ecosystems. Military operations in forested areas are complicated by the dense foliage, which restricts visibility and mobility. Troops must contend with uneven terrain, obstacles such as fallen trees, and limited lines of sight. Communications can be hindered by the thick vegetation, making coordination challenging. However, forests provide natural concealment and cover, allowing for stealthy movement and ambush tactics.

- Reduced mobility and communications are hallmark of tactical operations in forests. Thick foliage provides excellent ambush possibilities. Considered yet another important infantry skill, jungle and forest fighting will benefit from small, highly mobile auto stabilised terrestrial AI systems. These small bots will provide a moving screen of surveillance and lethal support to soldiers on tactical operations in the jungle. These bots will scout ahead for suspected areas and if required on instructions of infantry soldiers, convert into a kamikaze system.

- Tactical operations would also benefit from specialist bots, guiding the soldiers in and out of operations. Reducing the burden of navigation on soldiers, allowing them to perform higher order tasks for mission accomplishment. Communications relays across highly forested battle areas in the internal aspect area system will provide value.

TACTICAL BATTLE PRIORITY MATRIX

Nuanced resource allocation and prioritisation is required in the conflating nature of modern warfare. Complex interactions between the tactical landscape and operational dynamics is served by exhaustive critical analysis of various factors such as terrain characteristics, operational imperatives, and strategic priorities to obtain the insights into the interconnectedness.

Based on the environment analyses and effect of army tactical operations, the following priorities emerge.

- Surveillance and intelligence play pivotal roles throughout the entire spectrum of military operations, permeating every stage with their indispensable contributions, regurgitating essential insights into enemy capabilities, intentions, and vulnerabilities. This early awareness empowers military commanders to formulate informed strategies and allocate resources effectively. As conflict transits to the tactical level, real-time surveillance becomes paramount, offering commanders crucial situational awareness to adapt and respond swiftly to dynamic battlefield conditions. Intelligence analysis enables the identification of high-value targets and potential threats, guiding precision strikes and facilitating the neutralisation of enemy assets with minimal collateral damage. The seamless integration of surveillance and intelligence capabilities across all stages of military operations is fundamental to achieving mission success and ensuring the safety and security of personnel.
- In mountainous, high-altitude, glacier, and forest environments, reconnaissance operations face distinct challenges necessitating the development of specialised AI tools. The rugged terrain and adverse weather conditions often impede conventional reconnaissance methods, leaving gaps in situational awareness. Standard tools may falter in providing accurate intelligence, compromising operational effectiveness, hence, military strategists and technologists have to prioritise AI solutions

tailored to these environments. AI-driven capabilities offer autonomous analysis of terrain features, anomaly detection, and real-time threat identification. AI-powered sensors and drones navigate difficult terrain more effectively, minimising risks to personnel.

- The development of AI for reconnaissance missions in riverine, desert, urban built-up areas, and hilly terrain is undoubtedly an essential priority for military strategists. However, it may not always be accorded the highest priority due to several reasons. Firstly, immediate threats or pressing operational needs in other areas could take precedence over reconnaissance in these terrains. For instance, if there is an imminent threat in an urban area, resources might be allocated to developing AI solutions for urban warfare instead. Secondly, the complexity and variability of these terrains demand thorough research and development efforts to ensure the effectiveness and reliability of AI systems. Therefore, while reconnaissance in riverine, desert, urban, and hilly areas is crucial, it may require more time and resources for comprehensive development compared to other priorities. Advancements in AI technology for reconnaissance may also benefit from initial groundwork in more conventional terrains before being adapted to these challenging environments. Consequently, while reconnaissance in these terrains remains a critical priority, it may not always be assigned the highest priority in the immediate term, ensuring a balanced allocation of resources across various operational needs.
- Engagement in diverse terrains such as deserts, mountains, high altitudes, glaciers, and forests is a critical aspect of military operations. The unique configurations of these terrains demand flexibility and precision in combat tactics to ensure mission success and personnel safety. In such environments, where visibility may be limited and manoeuvrability constrained, achieving a high single-shot kill probability is paramount. This necessitates the development and deployment of specialised weaponry and tactics tailored to the specific challenges posed by each terrain type.
- Logistics planning, healthcare provision and treatment in mountainous terrain require meticulous consideration, especially given the challenges posed by limited transportation infrastructure. Unlike more accessible

environments, mountainous regions often lack efficient means of transportation, hindering the movement of personnel, medical supplies, and essential equipment. This restricted connectivity severely affects the flexibility and responsiveness of logistical operations, potentially endangering the success of tactical engagements and healthcare delivery. Consequently, military planners have to leverage advanced technologies to overcome these obstacles. Innovative solutions such as rugged communication systems, autonomous supply drones, and AI-driven predictive analytics algorithms can enhance logistical efficiency and healthcare resilience in mountainous environments.

- Human Resources management within military operations stands distinct from other facets in its AI requirements, often necessitating less intensive implementation efforts. Unlike other critical areas such as logistics, intelligence, and combat support, where AI applications are integral for enhancing efficiency and effectiveness, HR management predominantly relies on human judgment and interpersonal skills. While AI technologies can certainly augment certain HR functions such as training and leadership development, it has low priority in aspects of HR management. Moreover, the ethical considerations surrounding personnel management and the need for empathetic understanding further emphasise the limitations of AI in this domain. Therefore, while AI may offer incremental improvements in HR management, its role remains supplementary rather than transformative, highlighting the importance of human expertise and leadership in fostering a cohesive and resilient military workforce.

The preceding analysis forms the foundation of a tactical battle priority matrix, which serves as a simple framework for allocating priorities to tactical battles. Crafted to leverage AI technologies, the priority matrix aims to enhance tactical engagements and improve operational outcomes. This tool is designed to streamline decision-making processes, ensuring that resources are allocated effectively to optimise military effectiveness on the battlefield by nominating each relationship as a function of V – Vital, E – Essential, and D – Desirable.

	Riverine	Maritime	Deserts	Semi Deserts	Built-up	Hills	Mountains	High Altitude	Glaciers	Forests
Surveillance	V	V	V	V	V	V	V	V	V	V
Intelligence	V	V	V	V	V	V	V	V	V	V
Reconnaissance	E	V	E	E	E	E	V	V	V	V
Decision Making	D	E	E	E	D	D	E	E	D	D
Engagement	D	E	V	V	D	D	V	V	V	D
Operational Movement	D	E	E	E	D	D	E	E	E	D
Logistics Planning	D	E	D	D	D	D	E	E	E	D
Logistics Movement	D	V	E	E	D	D	V	V	D	D
Administration	D	D	D	D	D	D	D	D	D	D
Health Care and Treatment	D	E	D	D	D	D	V	V	V	D
HR Management	D	D	D	D	D	D	D	D	D	D

INDIAN ARMY RAPID AI DESIGN ARCHITECTURE (IRADA)

Lack of a common designing framework affects successful delivery of military AI tools. Such a demanding and designing framework will play a crucial role in fostering mutual understanding and collaboration between military users and developers, particularly in overcoming inherent knowledge gaps. A special framework is required to provide standardised protocols and interfaces, facilitating clearer communication and alignment between stakeholder expectations and developers ability. It will ensure that developers gain a comprehensive understanding of operational needs and constraints, enabling them to tailor AI solutions effectively to meet military requirements.

A common framework offers significant benefits by promoting efficient resource utilisation and fosters a shared knowledge base. Developers can leverage established guidelines and best practices within the framework to bridge knowledge gaps and accelerate innovation. This will create AI tools that are better aligned with operational requirements, are more adaptable to changing contexts, and ultimately more effective in supporting military missions

To improve the effectiveness of AI deployment in harsh field conditions and address challenges such as mistrust, over-expectations, and hallucinations, a comprehensive framework is therefore proposed in this chapter. Named as the Indian Army Rapid AI Design Architecture (IRADA), it will give a boost to *irada* (intent) of the Indian Army. The framework should be able to assist tacticians to define AI tools and developers to understand the intrinsic environments in which such tools are expected to work.

This framework provides a structured approach to understand the problem, identify key components, and develop AI solutions. Detailed discussion, validation, and utilisation of this framework is expected to improve it further. Each of the following aspects of the framework is discussed in detail in the subsequent paragraphs.

- **Requirement Specification.** This framework aids in specifying the functional and non-functional requirements of AI tools. This involves determining the capabilities, performance criteria, and interoperability requirements.
- **Problem Definition.** The framework guides military planners to 'over-define' the problem for a clearer understanding of the developers, who may not be aware of the military environment and its characteristics. AI problems are highly focussed (till AGI levels are achieved) and thus cannot reflect the generic needs of the military. This also urges military users to prioritise AI tools.
- **Design and Development.** The framework provides a structured approach to developers in selecting an architecture design, algorithms, data acquisition, model training, validation, and deployment.
- **Technology Assessment.** This framework assists developers to evaluate various AI technologies and methodologies for their suitability to address the problem at hand. This also involves identifying relevant algorithms and techniques.

Is AI Required, if Yes, Where?

Relevance of AI in military stems out of the very nature of AI's output. Known to be best utilised for all that is dull, dangerous and drudgery, AI has a wide work scope in the military. AI technologies offer transformative capabilities that enhance operational effectiveness and precision, reduce collateral damage, improve decision-making, and mitigate risks to military personnel.

Firstly, and naturally, AI plays a critical role in tackling dangerous tasks, often involving high levels of risk to human life, such as reconnaissance in hostile environments, bomb disposal, or surveillance in contested airspace. Militaries can enhance operational safety and maintain a tactical advantage while minimising casualties. Secondly, AI addresses the drudgery associated with repetitive tasks that are vital but time-consuming in military contexts. ML algorithms can analyse vast amounts of data to extract actionable insights to support all types of decisions, reducing inefficiencies and freeing humans for more important roles. AI technologies alleviate the burden of dull and monotonous tasks like base protection and border surveillance that can lead to fatigue and reduced cognitive performance among military personnel. Military training is repetitive in nature to build in the muscle memory and can tend to become uninspiring, dull, and boring drills. AI-based simulations and training systems provide engaging and realistic skill development and readiness. By harnessing AI, the military can achieve operational superiority, improve mission success rates, and reduce human casualties in diverse and dynamic operational environments.

Having understood the benefits of useful AI and, as in case of any other tool, the military identifies its 'honest' needs, which stem from national objectives, aims, and threats. A simple layout below establishes a broad but essential decision point. The military is expected to defend national territory and act as a deterrent. The army is already sufficiently equipped in matters of strategy and weapons to accomplish such tasks. Advent of AI should therefore enable the army to accomplish these tasks efficiently and economically. AI should inter alia assist the army to overcome previous deficiencies (though they should not exist), prevent threats to territory or failure of deterrence. This reasoning guides towards higher utility and priority of ready–to-use technology (AI models). It is also certain that AI will deliver results in a small arenas, where ready–to-use technology overlaps with essential military necessities. Developers have to be truthful in revealing their immediate abilities and the army has to be realist enough to correctly define its requirements. These specific demands will focus developers' ability to consolidate efforts and deliver in time.

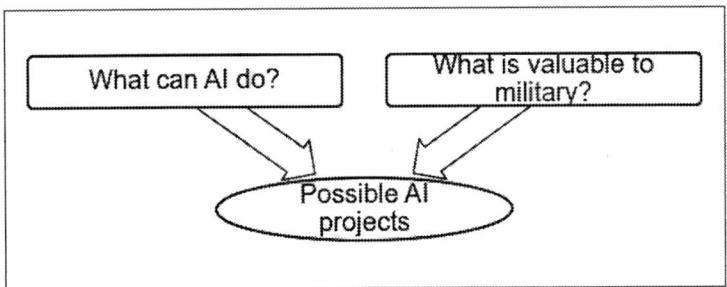

Militaries, including the Indian Army are constantly struggling between 'AI

Now' and 'AI for the Future' constantly. Released in 2023, 'Oppenheimer', introduces Edward Teller as an early member of the 'Manhattan Project', which developed the first atomic bomb; a fission weapon. During the development, he made a serious push to develop a fusion weapon, but was politely sidelined by the main protagonist and allowed to continue with parallel research, without delaying the development of the fission atomic bomb. The leader knew the difference between 'Now' and 'Future'. Fusion bombs came up much later after World War II. In the context of AI, answering two fundamental questions will somewhat reduce the dilemma. Firstly, which AI tool will improve delivery of what is valuable to the military? And, secondly, how does the PEAS framework (initially explained in Chapter 1) recommend a workable AI tool to create 'Bang for Buck'? The following two highly valued fundamentals are instructive in this case.

- Sensitive borders including coastlines require constant monitoring and firm physical control so border management systems are primary requirements. There has to be a good mix of surveillance and retaliation systems. Rough terrain supports infiltration, and small team operations and salami slicing. All these can be easily 'controlled' by autonomous systems (AS). Operational logistics and administration in such areas also affects combat efficiency. This uncertainty and adversity thus require well planned prediction and delivery mechanism, served best by AS.
- Grey zone and cognitive operations are a constant threat to India. It is vital to identify and monitor future perpetrators and this is the truest interpretation of 'AI Now'. Monitoring persons of interest (POI) with adversarial interests will be a major value addition to the military arsenal.

Defining AI Problems

Similar to any other tool, the military identifies its requirements based on strategic goals, objectives, and threats. An AI tool, due to its advanced analytical capabilities, and efficiency in automating tasks, should be capable of effectively addressing the most critical military challenges. Creating a succulent problem statement is the first and essential step. The military has to choose wisely and allocate AI development resources to improve the output. AI tools in all cases should provide substantial edge to 'what is valuable to the military'. At this initial stage of 'proving the concept' military verticals are likely to compete rather than cooperate amongst themselves to 'embed AI'. It will rest on the highest decision-making leadership to prioritise between the desirable and the do-able. Two definitions of a problem statement for predicting an avalanche in a snow-bound road stretch are exemplified below to highlight the improvement from good to excellent.

- **Good Problem Statement.** "Develop an AI-powered predictive model for avalanche forecasting in a snow-bound roadway, leveraging real-time weather, snowpack, and terrain data to enhance safety measures and minimise travel disruptions." This problem statement outlines the scope, goals, and data sources for the project, providing a clear direction.
- **Better Problem Statement.** "Develop an AI-powered avalanche prediction and mitigation system tailored to snow-covered roads, integrating meteorological, snowpack, and terrain data for real-time risk assessment. This system should provide timely alerts and also recommend proactive safety measures, such as road closures or controlled avalanches, to safeguard lives and ensure uninterrupted transportation services." This problem statement is highly specific, outlines the objectives, and emphasises the importance of proactive safety measures, setting a clear and ambitious goal for the project.

Logical abstraction of the problem to its contextual relevance[4] as an AI tool is the first major step, which military decision-makers have to make. This abstraction focuses on understanding the environment in which the AI system will operate and the context in which it is planned and executed. This deliberation has to answer why an AI tool is required, what is the scope of work in its development, what are the key drivers. It should always align to the key principles of the business, in this case, the military. The theory of architecture development quantifies 'principles as rules and guidelines, enduring and seldom amended, support the organisation fulfilment of its mission'. The military decision makers thus have to clearly indicate the principles of AI use before commencing development. It is a top-down responsibility, which will trigger the bottom layer.

Tactical diligence of military AI systems refers to deployment and utilisation discussions and deliberations. Mission relevance and aligning AI capabilities with specific operational requirements are the most unsavoury responsibilities of military commanders. Is there a necessity to enhance decision-making or situational awareness or mission effectiveness? Can all the requirements run concurrently or will they have to be prioritised? This will require customising AI algorithms and applications to address the unique challenges and constraints of military operations.

Design and Development through PEAS Analysis

An AI system is a complex integration of various attributes. Developers aim to keep the complexities to a manageable level by undertaking PEAS (Performance Measure, Environment, Actuator and Sensor) analysis.[5] It helps a rational agent (colloquially AI tool) to consider all possibilities and select the most efficient action it chooses, and the shortest path with least cost for high productivity. Unsurprisingly, this framework also offers alternate options to create a simpler tool. A modified framework is attached as Appendix B to this chapter. Case studies of use case with regard to PEAS analysis serve as practical examples for the development of AI solutions, ensuring a more informed and successful integration of AS into military practices.

In the context of military, AI can play a pivotal role in operations by streamlining essential tasks and separating aspirational features from critical necessities. AI can reduce clutter in complex systems, enabling forces to focus on vital objectives efficiently, allocating resources to the most valuable tasks, and ultimately strengthening military effectiveness. There is a necessity to de-clutter 'needs' and simplify the AI tasks, even breaking them down into small forms and structures. A theoretical discussion attempts to prove this argument.

- **Composite Border Surveillance System.** It is obviously the most important AI tool for border management along the entire stretch. A quick PEAS analysis provides insights into the mega structure required for such an enterprise. In itself it is a highly complex, multi-agent system, dependent on a huge dataset, affecting communications and power management. It is clearly not an optimal tool. The following details prove that this type of a system is a complex interplay of multiple intelligent agents.
    - **Performance Measure.** Identify movements of vehicles and personnel towards the border without cluttering normal civilian movement. Identify armoured vehicles, artillery guns, military load carriers, specialised vehicles and communication vehicles separately; group the movements into enemy formations; identify lateral

movements of all vehicles and personnel, which can later join up to a common area/location near the border; identify exodus of civilians, if any; and co-relate all EM spectrum inputs to generate a composite picture across the border.
  - **Environment.** Physical environment will include sandy areas, developed, riverine, low and high hills, forests and shrubs, valleys and high ridges, finally snow-bound and glaciated areas.
  - **Actuator.** 'AI at Rest' actuators include algorithms to indicate an exodus, conterminous lateral movements, and concentrations of armoured and artillery vehicles finally co-opting inputs into a composite picture. AI will also create 'Intelligence Void' request for query.
  - **Sensor.** All EO sensors of military formations, troops in contact reports (SITREPS), intelligence analysis reports from neighbouring formations and higher HQ, Drones, aircraft reports, LEO, aerostats, EW assets and open-source reports.
- **Decluttered Border Surveillance System.** It is also possible to model a light and equally quicker AI system by redefining the optimal performance measure to finite objectives and outputs. The end user has to analyse and identify this 'optimal measure' with a motto: 'deliver what is valuable.' The abstract model could be developed as follows:
  - **Performance Measure.** Identify movements of all military vehicles and personnel towards the desert border between Points A and B. Identify lateral movements of all vehicles and personnel, which can later join up at a common area/ location near the border.
  - **Environment.** Physical environment will include desert border between Points A and B and 50 km (BFSR-XR) into enemy territory
  - **Actuator.** 'AI at Rest' actuators include algorithms to indicate exodus, conterminous lateral movements, and concentrations of armoured and artillery vehicles finally co-opting inputs into a composite picture. This AI will autonomously create an 'Intelligence Void' report for query.
  - **Sensor.** Only EO sensors of military formations and troops in contact reports (SITREPS).
- **Missing the Bigger Picture.** Often quipped for missing the bigger

picture, these decluttered systems/tiny AI systems are in fact otherwise. They will continue to provide a similar bigger picture as envisaged in the 'traditionally' complex system. Decluttering simplifies the AI architecture without decreasing its utility. An AI designer retains the desired value by creating another AI to read the insights of all other 'mini-AI' systems mentioned in the preceding paragraph. Known as 'Federated Learning AI',[6] it trains on multiple local datasets contained in local nodes without explicitly exchanging data samples. It is the perfect AI solution for problems of large datasets, multiple-agents, dynamic environments, and communication-power management. AI is decluttering its own responsibilities, and yet providing a common operating picture to decision makers at the end of the cycle.

AI Play Book. As a universally recommended approach, emphasis on the role of a playbook is essential. A playbook serves as a central tool for establishing a clear direction and standardised reference point within the organisation. It empowers all members of the organisation to align their efforts towards shared objectives, a perspective highlighted by Andrew Ng,[7] who underscores the importance of an AI playbook.

- The launch of a pilot project should be considered an initial step. This project, with a gestation period ranging from 6 to 12 months, can be outsourced if the in-house capacity is insufficient. To achieve this goal, it is advisable to assemble a robust, cross-functional team consisting of AI/ML practitioners and domain experts. This team's responsibility will be to thoroughly evaluate the relationship between the organisation's aspirations and its capabilities. It is worth noting that the pilot project should primarily focus on demonstrating the feasibility of the AI product rather than striving for full usability at this initial stage with an aim to enhance traction and user engagement.
- Following the successful completion of the pilot project, the next suggested step is to provide training for in-house cross-functional teams (CFT), subsequently expanding AI training initiatives to encompass all levels and roles within the organisation. This broader approach ensures the widespread dissemination of AI knowledge, making it an integral part of the organisational culture.
- The playbook finally emphasises the importance of clear and effective

communication in the success of AI initiatives. It is advisable to communicate the AI strategy through robust two-way internal and external channels to ensure alignment and engagement throughout the environment.

Over Engineering. At the stage of design diligence and conceptual abstraction, answers to the question; 'what is required to solve the problem', roots the team to the development cause. This abstraction has to decompose the requirements to understand the problem, and what is needed to address the problem. In more specific terms—what is just enough to resolve the problem, all the while staying away from the 'How' of the problem. In the AI realm, a problem can be solved with a statistical, ML or a Deep Learning (DL) model. In an exuberance to fulfil functionality, unnecessary complexity and over-engineering will occur. Such convoluted code structures may also create technical debt in future maintainability. Predictive maintenance problems of military workshops can be studied to understand the aspect of over engineering. For five vehicle repairing workshops, each with eight types of vehicles and each vehicle having 50 types of repair issues, theoretically both statistical and machine learning models could provide predictive maintenance solutions.

- The choice depends on relationships between vehicle characteristics, repair issues, and maintenance needs, which are relatively simple and well-understood, for which a statistical model may be sufficient.
- With a high-quality structured data and clear relationships between variables, a statistical model might suffice; however, for a large volume of diverse data, including text data or sensor readings, a machine learning model may be better suited to handle it.
- If interpretability is a priority, a statistical model may be preferred to understand the factors driving maintenance needs, while machine learning models are less interpretable.
- Machine learning models can scale to handle large datasets and complex relationships like pan-country scales and pushing forward of spares.

Technology Assessment

Indian AI developers do not possess military knowhow like their Western counterparts, partly due to segregated military and civilian technology development and partly due to the volunteer nature of the Indian Army, where dual-use knowledge is generally not shared. Recent civilian military cooperation decisions will improve this deficit, but till such time it becomes successful, the army will have to share the burden of 'designing' the AI tools. The under mentioned decision matrix points will assist in this designing and answer the questions of 'How' and 'With What'.

- **AI Output.** A clear expression of the desired outcome will bring focus on the design of the AI system. This will also define if the AI system will consist of single or multiple agents, where the latter type of stacking is likely to create more intrasystem conflicts.
- **Environment.** Will the AI system operate in a digital or physical mode, or both?
- **Cloud.**
  - Does the military environment allow access to a central server?
  - If there is no access to a central server, can an Edge or Fog AI System deliver the desired output?
  - Does an Edge device exist, or is there a requirement to develop one?
  - Will the Edge device development (an academic pursuit) address the immediate military needs?
- **Military Robustness.**
  - Will the AI system reside in safe environments?
  - Will the AI system reside in rough military combat conditions?
  - Is it a mobile or a static system?
  - If mobile, will it be a manipulator, wheeled or tracked system?
  - Is there a need to ruggedise it? If yes, can it possess a small form factor to reduce power management and utility?
  - Will any new development (an academic pursuit) address the immediate military needs?
- **Model and Algorithm.** This is by far the deal sealer. Existing and usable algorithms will deliver quicker as online directories provide a bouquet of choices, provided the AI and ML engineers are convinced of its utility and military security concerns.
  - Does the relevant algorithm and model exist?
  - Will there be a need to try a combination of readymade algorithms to create a model?

- Can a tailor-made algorithm or model be created and will this development (an academic pursuit) address the immediate military needs?
- Can the solution deliver the right output without over-engineering?

Dataset. A well curated dataset provides multiple options to create AI models. The military designer will have to undertake a detailed review of this ability. Is the data existing, if yes, then is it curated or noisy; if not, will it be internal curation or outsourced; if outsourced, how are the security concerns alleviated? If there is no dataset, can novel techniques be applied to develop the AI system and will the novel techniques delay the development cycle? Will this novel technique (an academic pursuit) address the immediate military needs? Military data centres and network operation centres catering for data storage[8] are expected to improve data utility across the entire spectrum. All military units (data generators) will store their data at the centres and data creators, creating large data lakes to enhance operational preparedness. By themselves, the data centres may not be quipped for AI applications, but they can store curated data for the applications. The Indian Army is already seeking a data dictionary policy,[9] in which data creators are provided guidance and data generation aligns to the usable dataset.

DESIGNING AI TOOLS FOR INDIAN ARMY

The highlight of this chapter is to design AI solutions based on the IRADA mentioned in the previous section. A structured and systematic methodology is essential for developing effective solutions. Tool designing will commence with tactical diligence, which implies a careful and thorough examination of the specific requirements and challenges faced by the Indian military. This step is crucial for ensuring that AI tools address the unique needs and constraints of the context. This will be followed by a PEAS analysis to strengthen the design process. This framework will evaluate the AI system's performance and its interaction with the environment. This analysis helps in identifying the key components and functionalities required for the AI solution to operate effectively in real-world scenarios. The designing will also include a logic map to facilitate a clear understanding of the development. This visual representation will help stakeholders to observe the relationships between different stages of development, identify key components, and plan strategies for implementation.

In the context of AI, the previous section introduces IRADA to somewhat reduce the dilemma struggle between 'AI Now' and 'AI for Future'. The process of design creation is a deliberate step supported by a continual and iterative study of factors mentioned in the framework steps: Requirement specifications (specifying the functional and non-functional aspects of AI tools), over-defining the problem definition, structured approach in the design and development stage to select an architecture design, algorithms, data acquisition, model training, validation, and deployment and finally an honest technology assessment.

Based on the analyses of military necessity mentioned earlier, some tactical problems can be resolved by AI solutions. The designs mentioned below are initial sets of deliberations, which may not be fully acceptable. The list is a vision of what is possible and is not a completed work in any manner. Continued contemplation of the few fundamental questions mentioned earlier will present several more options, a few hopefully better. The army can utilise this novel assembly methodology of creating small manageable AI systems to work independently and then utilise their outputs to feed a federated AI for wider scoped and more complicated goals.

Language Translator

A translator for the Union Territory (UT) of Jammu and Kashmir will be tailored for five major spoken languages. The model will be specific to language pairs (Hindi-Kashmiri, Hindi-Pahari, Hindi-Dogri, Hindi-Gojri, Hindi-Punjabi), where Hindi is the universal spoken language of army troops deployed in the UT. Energy-efficient and compact components, such as a low-power processor and a small form-factor battery with high energy density will conserve energy when not in use. The PEAS framework for a language translator is attached as Appendix C. A similar model is also required for the contested borders of Arunachal Pradesh in Northeast India. In this case, the AI model should obviously address new language pairs (Hindi-Nishi, Hindi-Adi, Hindi-Apatani, Hindi-Tagin) and Hindi-Nagamese since it is a widely spoken connect language.

Information Retrieval (IR) System

The desired IR system is expected to sift through a large corpus of data from military systems and provide requisite information. The data may be a combination of images, video, text and audio formats, but it is also possible that the data is available in one format only. As suggested by the IRADA framework, the problem can be abstracted to smaller AI solutions, with one federated system working as an IR system and yet another as a prediction system. The PEAS analysis would consequently have seven concurrent analyses to define the AI output for each of the AI systems.

Dealing with a combination of images, videos, text, and audio formats poses significant challenges in data curation, which will in effect delay the projects. The heterogeneity of formats requires careful consideration of integration, as each data type demands unique processing techniques and tools. The Indian Army is seeking a data dictionary that is essential for effective data management and analysis. An Army data dictionary policy is already underway to reduce irrelevant artefacts, but till such time as the generated data is consistent with the dictionary, the solution will lie in abstraction to lower levels of capability and simplicity.

Utilising federated AI in this project will improve the accuracy of the system, since multiple agents (AI systems) are providing independent analyses and not creating multi-agent adversarial environments. Developers get lost in resolving situations where multiple autonomous agents interact with each other, potentially with conflicting objectives, strategies, or interests, leading to competitive or adversarial interactions. Accuracy is a primary requirement of a prediction system and hence such multi-agent adversity has to be avoided. A rough schematic layout could be described as the image below and the PEAS framework for language translator is attached as Appendix D.

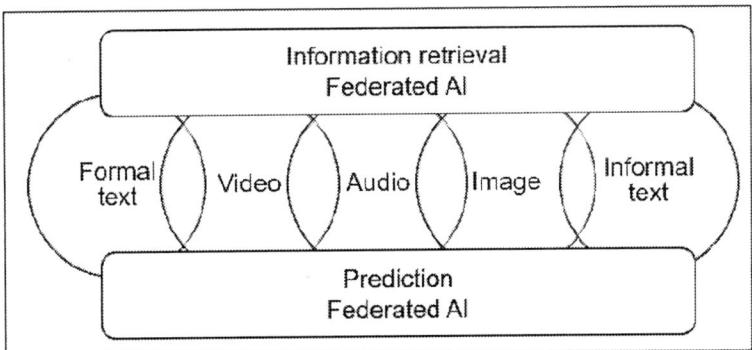

Urban Combat Buddy

Urban combat, an essential facet of modern warfare, unfolds amidst the densely populated landscapes of urban environments, necessitating precision and restraint amid chaos and violence. The presence of civilians adds enormous complexity. An unseen enemy on other side of the wall or street or the building adds uncertainty and heightens stress, reducing mission success. A soldier will seek to peer through such man-made obstacles, spot the enemy, and engage in a more favourable battle. Extant solutions such as Explorer Ball Cameras[10] exist. The ball consists of six cameras and provide an instant stitched image to the tactical commander outside the room. However, in a military combat scenario, devices with multiple cameras have a better survivability.

Based on two broad modules, computer vision-based image stitching is a simple first stage solution for the model. A tailored programme can pick up relevant portions from image generators to create a live stitched image. A second module will isolate and highlight human forms (background segmentation), masking the background. A schematic layout of such a system is given below and the PEAS analysis is attached as Appendix E.

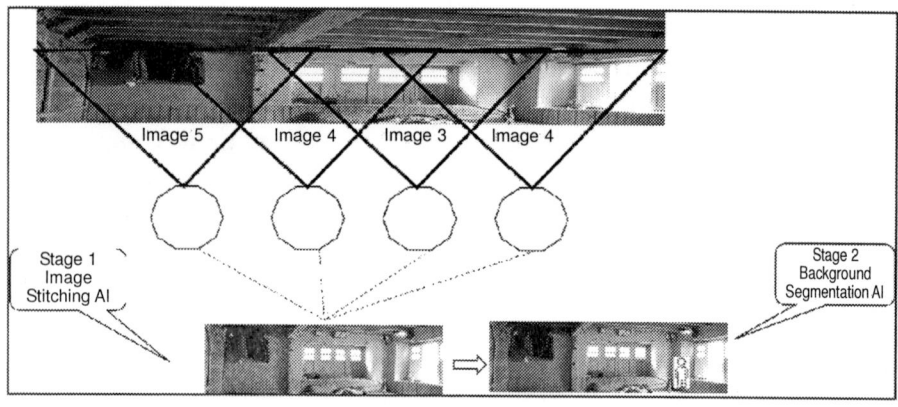

Source: https://in.pinterest.com/pin/319474167297560948/

Autonomous Tracking of Persons of Interest (POI)

Given the persistent actions of adversarial neighbours engaging in grey zone operations against India and its military, exemplified by incidents like Pathankot, Pulwama, and Uri, it becomes imperative to maintain continuous surveillance over persons of interest (POI) across both the physical and digital realms. The military must adopt a mindset wherein virtually any human activity could potentially serve as a tool for military objectives. This expanded perspective includes non-military entities like social media, financial institutions, and critical infrastructure, as they can also be exploited. The increasing prevalence of grey zone operations, strategically positioned between peace and war, underscores the urgency of identifying these POI and their proponents. To effectively counter such threats, the military should proactively monitor and track POI, thereby enhancing its non-contact and non-lethal capabilities. The Indian Army has struggled to predict the unpalatable instance on the LC. This AI tool will provide a recipe to remove this nuisance all together. This strategic augmentation will also render the military a more versatile instrument of power, at the disposal of the political leadership, in confronting these emerging challenges.

The scope of the work expected out of the system indicates it to be a multi-agent system, which is a complex task. The system has to collect data from various social media platforms. The collected data has to be formatted into a unified dataset for further use. Or, as mentioned earlier, Federated AI is utilised to deconstruct the requirement and a workable AI is created. The most critical part will be the user interface in which the output provides an interactive interface to search, browse, and visualise the tracked person's social media activity and information. It should importantly provide timeline visualisation and location mapping. The PEAS analysis is attached as Appendix F.

Tactical Logistic Bot
Bulk logistics is broken down into smaller palettes for further supply to the troops in contact. The tactical commander is responsible to send this small load to his troops while controlling the actual tactical battle. Most of the time this load is carried by fighting porters and sometimes the troops themselves get involved in it. This arrangement divides the attention between combat and logistic resupply. Modern technology has to provide a solution so that the troops' attention remains decidedly on the conduct of battle. A typical solution should consider carriage of small loads, since it directly affects the form factor and power train of the delivery machine. The contested tactical battle space also demands narrow width systems (0.6 m is considered as a lane wide enough for soldiers to walk). The existent load carriers (radio controlled) systems vary in track width of 1.8 m to 3.2 m, which has little tactical utility at the pointy edge of battle. Such wide body systems may suit logistic supply at the rear of the battle but not into it.

Recommended systems could be articulate vehicles for better stability, with low and power management. Such designs will also carry more loads in general terms. AI navigation solution is a necessary feature for such tactical load carriers. In case the bot identifies movement restrictions between Points 1 and 2, it will route directly to Point 3 and deliver the load to its destination. This navigation ability of such a bot can be applied easily since the Army is deployed on known borders and the terrain is well known. A schematic layout is given below and the PEAS analysis is attached as Appendix G.

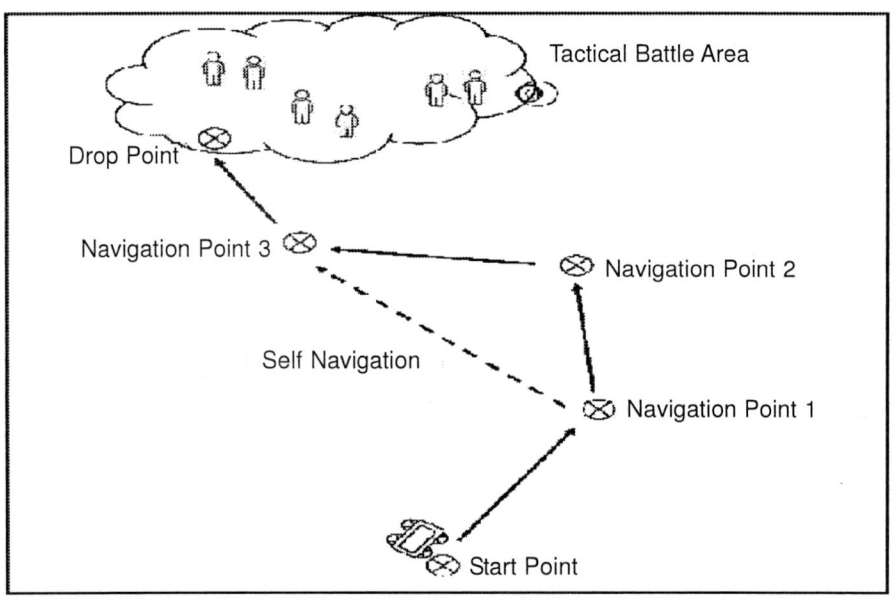

Tactical Support Bot

Tactical operations are the mainstay of securing most advantageous positions to eliminate the enemy. This is done by one group firing on the enemy, forcing him to take cover from fire, and thus losing control of the battle space. At this time, the other group quickly moves to a new position from where the enemy can be fired upon. This is called fire and move battle drill; fire to keep the enemy's head down and move to a new position without his knowledge, engage and eliminate him. Simple to explain but most difficult to execute since the enemy is also attempting a similar advantage. A tactical support bot, slave to the soldiers, moves around observing the enemy, engages him, giving time to own soldiers to move to a new position. The bot and own troops can be secured with a RFID against any friendly fire. A schematic layout is given below and the PEAS analysis is attached as Appendix H.

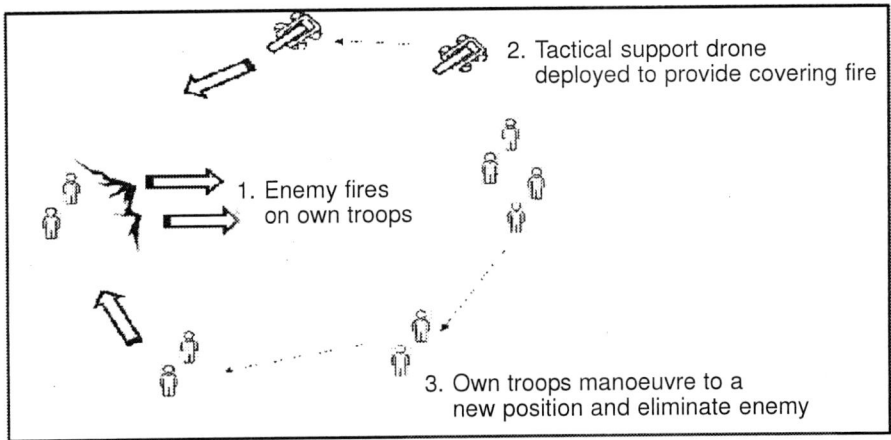

Communication Relay Bot

Tactical battles are fought with battle drills mentioned in the beginning of the chapter. The communication is however essential for sharing tactical pictures and change of decisions to secure advantageous positions between battle drills. The tactical commander may also want to redeploy his forces to exploit enemy weaknesses reported by troops. Tactical communication is line of sight (LOS) based on VHF. VHF communication can be transmitted and received using lighter, portable equipment, thus advantageous for small, separate units that need to communicate with one another across open space within an LOS range. But this LOS communication is prone to screening due to natural or man-made obstacles. Dense foliage, trees, hills and mountain ranges, buildings and basements can cause screening. At such times, an AI communication relay system will provide continuous and autonomous relay of communications and keep the tactical commander connected to his troops.

As soon as one tactical group gets out of communication from its commander, AI drones, which are monitoring the connectivity, get airborne and manoeuvre to find a position to re-establish communications. Standby drones note the battery power of the deployed system and autonomously take off to relieve in time. The drones could be either aerial or land-based bots. A schematic layout is given below and the PEAS analysis is attached as Appendix I.

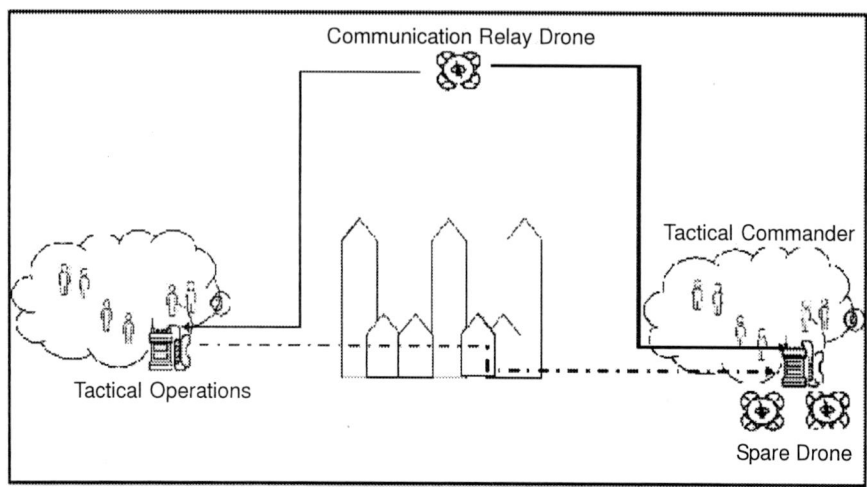

Autonomous Weapon System

Theoretically speaking, this is the simplest and most necessary AS in the Indian context. Contested borders, rugged terrain and out-of-proportions ramification of local conflicts make a strong case to produce autonomous weapon systems. Unlike other countries fighting for regime change and GWOT, Indian military decisions are remarkably sublime and always oriented to 'Defence of India'. In that context, any violation of the borders is violation of the 'Defence' and any action there off is fully constitutional and legal. Appendix J measures three versions of autonomous weapon systems.

Prediction of Landslides and Rock Falls

Though large-scale warning systems are well established, they, however, cannot be easily adapted or scaled down to smaller regions and tracts due to the inherent complexities of data aggregation, localised environmental factors, and resource limitations. A few small-scale but flexible systems provide alerts ahead of time via hooters and blinkers installed on the road and remotely via SMS on the phone.[11] Similar systems are desirable for the loose shale mountainside on Indian borders, where the roads are prone to landslides which reduce tactical mobility. An AI solution is a good option to predict landslides, rather than sending out an advance team to physically scout the area for roadblocks. The system will require a database of images or videos of roads in different conditions, including some that show signs of landslides or rock falls. Simple and light algorithms, namely, logistic regression, decision trees, random forests, and gradient boosting, are powerful and widely used in various machine learning tasks. These algorithms can build simple and light prediction models, even with limited data.

By using simple and low technology sensors like QR codes, the user can identify exact spots on the ground as a feasible alternative to GPS for location identification. QR codes can be placed at regular intervals, such as every 250 metres, along mountain roads to serve as markers for precise location data. It is a cost-effective and accurate method for collecting location data without relying on GPS technology. However, the QR should be securely placed, remain visible and scannable, even in adverse weather conditions or challenging terrain.

Insights are highly dependent on the data received from the sensors and the AI systems' access to its database. Deploying edge-computing devices will reduce the need for extensive data transmission and can provide real-time responses. It is also possible to train an AI model to identify initial indicators of a landslide or rock fall along a road and then deploy it offline to generate alerts, reducing the need for communication. IA can combine multiple communication protocols for redundancy and robustness, like IOT SIM cellular communications for critical real-time data and mesh networking for backup data transmission. Appendix K is a PEAS analysis attempt of such a system.

KEY TAKEAWAY: TACTICAL AI TOOL DESIGNING FRAMEWORK

The Indian Army stands as a bastion of resilience amidst the complex geopolitical landscape of South Asia, where contested borders and revisionist neighbours have attempted to degrade the national security. Coupled with the presence of proxy players instigating internal security challenges, the Indian Army finds itself in a perpetual state of battle, an experience unfamiliar to many other military organisations around the world. This continuous engagement in conflict, characterised by tactical surprises and asymmetric warfare, has endowed the Indian Army with unparalleled combat experience, earning it the reputation of being one of the most battle-hardened forces globally.

The LoC with Pakistan and the LAC with China serve as volatile flashpoints where even minor skirmishes have the potential to escalate into full-fledged conflicts. The presence of proxy actors, funded and supported by hostile entities, adds another layer of complexity to India's security calculus, requiring constant vigilance and robust responses. This constant state of tactical conflicts commonly referred to as the 'battle of a thousand cuts' imposes significant strain on the army. In this context, the deployment of AI tools emerges to mitigate the challenges posed by historical deceit, contested borders, tactical surprises, and proxy threats. AI offers a range of tactical capabilities that can enhance the army's operational effectiveness, resilience, and situational awareness in dynamic and hostile environments.

The army can leverage AI is its ability to process vast amounts of data to generate actionable intelligence in real-time and counter the threat before it manifests. This capability is particularly valuable in border surveillance and counter-terrorism operations, and internal security situations where timely information can mean the difference between success and failure.

A parallel discussion also ensues to identify AI tools for the army. This chapter has created a framework which identifies the critical tactical situations that will benefit from technology solutions. The known environs of the borders provide a physical reference to design the tools. Most plausible and re-occurring tactical situations describe the performance measure of such tools, a few of which have been documented in the chapter. Most of all, the chapter provides a logic pathway to identify and design specific AI solutions for Indian tactical situations. It is a unique endeavour in line with the hypotheses of this research. Are the tools defined in the chapter ideal? No, they are not. But they follow a logical step and possibly support a development theory.

There is more to tool designing. The tools are irrelevant till they are supported by a plan. The induction and deployment plan include various other vital factors such as decision points for successful induction and utilisation of AI in the army. The next chapter will dwell into such supporting factors – the factors of DOTMLPFP.[12] Doctrine, Organisation, Training, Material, Leadership and Education, Personnel, Facilities, and Policy for decision-making to determine all the non-material or material approaches required to fill a capability gap. As culmination of the two years of research, the next chapter will also attempt to summarise the answers to the research questions.

NOTES

1. The Disturbed Areas (Special Courts) Act, 1976 Act No. 77 of 1976, https://lddashboard.legislative.gov.in/sites/default/files/A1976-77.pdf, accessed on 8 March 2024.

2. The Armed Forces (Special Powers) Act, 1958, https://www.mha.gov.in/sites/default/files/armed_forces_special_powers_act1958.pdf, accessed on 8 March 2024.
3. Lt. Saurabh Kalia: The Patrol Leader, https://www.indiandefencereview.com/spotlights/the-patrol-leader/
4. TOGAF® Fundamental Content, 3.7.1 Contextual Abstraction Level, 'https://pubs.opengroup.org/togaf-standard/introduction/chap03.html' accessed on 7 March 2024.
5. Stuart J. Russel and Peter Norvig. *AI A Modern Approach*, 3rd edition, Pearson India Education Ltd., Noida, India, 2010, pp. 40-43.
6. What is Federated Learning? https://research.ibm.com/blog/what-is-federated-learning, accessed on 23 February 2023.
7. Dr. Andrew Ng is a globally recognised leader in AI. He is the Founder of DeepLearning.AI, Founder & CEO of Landing AI, General Partner at AI Fund, Chairman and Co-Founder of Coursera and an Adjunct Professor at Stanford University's Computer Science Department. Many students have benefited from his Coursera programmes—Machine Learning, AI for Everyone, Neural Networks and Deep Learning.
8. https://www.lntsmartworld.com/defence#:~:text=We%20are%20executing%20a%20state,Centres%20which%20are%20under%20implementation, accessed on 7 March 2024.
9. Ministry of Defence Government of India. Request for Information by Directorate General of Information System Army Integrated Decision Support System, https://mod.gov.in/sites/default/files/Final-RFI-of-Proj-AIDSS-on-27-Oct-22.pdf, accessed on 7 March 2024.
10. Michael Zhang. The Explorer is a Tactical 360 Ball Camera That Can Be Thrown Into Danger, 2 July 2015, https://petapixel.com/2015/07/02/the-explorer-is-a-tactical-360-ball-camera-that-can-be-thrown-into-danger/
11. Landslide Monitoring and Early Warning System, https://www.indiascienceandtechnology.gov.in/node/167820, accessed on 25 April 2024.
12. https://acqnotes.com/acqnote/acquisitions/dotmlpf-analysis#google_vignette, accessed on 3 February 2024.

APPENDIX B

Peas Analysis Framework

Task	Name of the AS or an Agent or a Tool			
Performance Measure	What is expected to be done? What is the desired output or action? List out the desired tangible outcomes.			
		Options		
		1	2	3
Environment — Fully & Partially Observable	Do agent's sensors give it access to the complete state of the environment at each point in time?			
Episodic & Sequential	The agent's experience is divided into atomic "episodes." Does gent receive precepts and perform single actions while next action is not dependent on present percept or action			
Static & Dynamic	Does the environment remain unchanged while an agent is deliberating?			
Discrete & Continuous	Is there a limited number of distinct, clearly defined precepts and actions?			
Deterministic & Stochastic	Is the next state of the environment completely determined by the current state and the action executed by the agent?			
Single & Multiple Agent	Is the agent operating by itself in the environment?			
Know & Unknown	Does the designer of the agent have knowledge about the environment makeup?			
Sensors	A device which detects the changes in the environment (physical or synthetic) and sends the information to the AI system.			
Actuators	Converts electronic current (AI decision) into physical action within a system. A physical action affecting the environment is also termed Effectors.			
Overall priority				

Note

- The actual framework assists in designing an intelligent agent, however in the present instant the author has included 'Options' columns much like 'Options in Military Appreciation' to identify which 'Course of Action' would provide optimal benefit. In the context of AI; this modified framework will suggest architecture of 'quick-build' AI tool.
- This framework also defines the tool as one entity, even if there is no requirement of 'analysing multiple options to identify which 'Course of Action' would provide optimal benefit

APPENDIX C

Language Translator

Performance Measure		Live translation of languages in the binary pairs of Hindi-Kashmiri, Hindi-Pahari, Hindi-Dogri, Hindi-Gojri, Hindi-Punjabi and Hindi-Nishi, Hindi-Adi, Hindi-Apatani, Hindi-Tagin, Hindi-Nagamese. Autonomous selection to the primary language in case of noisy environment. Continuous translation and recording of 6 hours and sleep mode power retention of 10 hours.
Environment	Fully & Partially Observable	Fully observable
	Episodic & Sequential	Sequential
	Static & Dynamic	Dynamic
	Discrete & Continuous	Continuous
	Deterministic & Stochastic	Stochastic, because it is a live language translator
	Single & Multiple Agent	Single
	Known & Unknown	Yes, well known
Sensors		Receiver mike, speaker.
Actuators		Loudest language pair selector

APPENDIX D

Information Retrieval (IR) System

The IR system will source information in various formats. It is possible to create a multi-capability system which scours through images, video, text and audio data, cleans the entire corpus and extracts information. Another method could consist of independent AI systems extracting information from a restrict data, be it image or video or text or audio. Each of such AI will produce an output, which is consolidate by Federated AI and provide a holistic output. Yet another AI will learn from this level of AI and provide predictions. The PEAS investigation would consequently split into seven concurrent analyses to define the AI output for each of the AI systems. Image Data PEAS is given as a sample in this appendix and the other six analyses will follow the same process.

Performance Measure		Provide images for a textually query. Provide content based image retrieval.
Environment	Fully & Partially Observable	Fully observable
	Episodic & Sequential	Episodic, since the output is related to a fixed query which will subsequently be utilised by Federated AI
	Static & Dynamic	Static
	Discrete & Continuous	Discrete
	Deterministic & Stochastic	Deterministic
	Single & Multiple Agent	Single
	Known & Unknown	Yes well known
Sensors		Keyboard, Camera for QBE (Query By Example) query
Actuators		Nil

APPENDIX E

Urban Combat Buddy

The images will be captured by multiple by ball cameras, thrown into a room, which communicate with each other and create a panoramic image of the target room. All the ball cameras send back the images to a central controller with the tactical commander and the AI stitches the images together instantaneously. Another AI module highlights the human targets in the room.

Performance Measure		Module 1 – Stitch images to provide panoramic view. Provide images despite low-latency processing. Module 2 – Highlight the humans being hiding in the background.
Environment	Fully & Partially Observable	Fully observable
	Episodic & Sequential	Sequential
	Static & Dynamic	Dynamic
	Discrete & Continuous	Continuous, with emphasis on hardware acceleration (e.g., GPUs or specialized AI chips) and efficient algorithms to achieve low-latency processing.
	Deterministic & Stochastic	Stochastic, since it is live image processing
	Single & Multiple Agent	Double agents in each of the modules
	Know & Unknown	Yes, well known
Sensors	Multiple camera balls with gyro stabilisation	
Actuators	Object detection algorithm to highlight the humans in the room	

APPENDIX F

Autonomous Tracking of Person of Interest (POI)

	Open Source AI	*Social Media AI*	*HUMINT AI*	*Special Operations AI*	*Control AI/ Federated AI*
PerformanceMeasure	• Collate, highlight common time and place stamp. • Generate linkages with other POI.	• Collate, highlight common time and place stamp. • Generate linkages with other POI.	• Collate, highlight common time and place stamp. • Generate linkages with other POI.	• Collate, highlight common time and place stamp. • Generate linkages with other POI.	• Collate, highlight common time & place stamp from other AIs. • Generate linkages with other POI. • Generate waning. • Generate query over intelligence gap
Environment	• Print and Digital media, vernacular media.	• Social Media Accounts • Hashtags • Photographs • AV and Reels	• Physical realm • Most authentic and confirmatory in nature	• Physical realm • Highly contested, and surreptitious • To acquire confirmed inputs	• Controlled environment of the insight generated by other four AI systems
Actuator	• Generate insights • Generate variations.	• Generate insights • Generate variations.	• Generate insights • Generate variations.	• Generate insights • Generate variations.	• Generate common insights • Generate common variations. • Prioritise insights
Sensor	• Web scrapper • NLP for radio transmissions • CV and NLP for print media	• Web scrapper • NLP • CV	• Human agents	• Camera • Digital bugs • Trojan Horse	• Insights from previous four AI systems

APPENDIX G

Tactical Logistic Bot

The bot will navigate through pre-fed turn points and able to rework best route; if the designed route is unusable then the bot will be able to chart a new route towards the drop point. Well-defined Indian borders can be artificially mapped by physical Navigation Points (NP). The Bot communicates with NP, and maps the best route to drop point.

Performance Measure		Move along the designated route and navigation point (NP) Observe the route to the navigation point and re-route, if the route is unclear. Be able to correlate all navigation points in vicinity and chart best route. Navigation points act as POP and relay their vicinity details to the bot. Safety anchor, in case the bot is getting unbalanced
Environment	Fully & Partially Observable	Fully observable
	Episodic & Sequential	Sequential
	Static & Dynamic	Dynamic
	Discrete & Continuous	Continuous
	Deterministic & Stochastic	Stochastic, since it is live observation of the battle space.
	Single & Multiple Agent	Navigation points and Bot act as multiple agents and communicate with each other to identify best route for the Bot.
	Know & Unknown	Yes, well known
Sensors		Camera on NP and Bot to observe the route. Transmitter and receiver on NP to observe the immediate vicinity and communicate with BOT.
Actuators		Power train to run the articulated bot. Object detection algorithm to convey track awareness to bot. Safety anchor deployer

APPENDIX H

Tactical Support Drone

Performance Measure		Drone moves astride own troops. Slaved with RIFD IFF or QR code. Understand and takeover target indicated (illuminated) by own troops and provide covering fire. Stop fire if own troops with in danger area of drone fire. Regroup with own troops when ordered.
Environment	Fully & Partially Observable	Fully observable
	Episodic & Sequential	Sequential
	Static & Dynamic	Dynamic
	Discrete & Continuous	Continuous
	Deterministic & Stochastic	Stochastic, since it is live observation of the battle space.
	Single & Multiple Agent	IFF (Identify Friend of Foe) agent identifying own troops at all times. Navigation agent guiding the drone towards target area illuminated by own troops
	Know & Unknown	Yes, well known
Sensors		IFF agent identifying own troops at all times. Camera, radar, range finder, coaxial weapons. RFID distinguisher with own troops.
Actuators		Power train to run the bot. Weapon firing. Bot driving and navigation

APPENDIX I

Communication Relay Bot

Performance Measure		Monitor the VHF communication stability between designated tactical groups. Located near an anchor station, autonomously take off to manoeuvre and re-establish communication. Continue manoeuvring to keep the faraway tactical group connected to anchor station. Spare bots monitor power availability of deployed system and relieve it in time for autonomous return to base for recharge. Permit commander to deploy two systems simultaneously to increase the communication range
Environment	Fully & Partially Observable	Fully observable
	Episodic & Sequential	Sequential
	Static & Dynamic	Dynamic
	Discrete & Continuous	Continuous
	Deterministic & Stochastic	Stochastic, since deployed bot is monitoring the communication status between two stations and reserve bot is monitoring the power consumption of the deployed bot.
	Single & Multiple Agent	Multiple drones and bots monitoring communication status between tactical groups and power consumption of other systems. Return to base agent. Azimuth identification to created VHF communication relay between two radio stations.
	Know & Unknown	Yes, well known
Sensors		Communication strengths detector, Communication station azimuth detectors.
Actuators		Aerial or ground drones. Autonomous take off actuators, return to base actuators. Manoeuvring to establish communication actuators

APPENDIX J

Autonomous Weapon System

Task		Autonomous Weapon System Type 1
Performance Measure		Engage enemy armour in desert terrain. It is a fairly simple requirement and the AS is activated during hostilities. Deployed ahead of defences, it will detect the enemy at first instance, without prejudice. There is no concern of IFF, since hostilities are declared and the deployment is ahead of Forward Line of Own Troops (FLOT).
Environment	Fully & Partially Observable	Fully observable beyond IB
	Episodic & Sequential	Episodic
	Static & Dynamic	Static
	Discrete & Continuous	Discrete
	Deterministic & Stochastic	Deterministic, because enemy armour will reduce in numbers after engagements
	Single & Multiple Agent	Single
	Know & Unknown	Yes well known
Sensors	Camera/Range Finder/Drone	
Actuators		ATGM/30 mm AK 630/tracked carrier (remotely controlled)

Task		Autonomous Weapon System Type 2
Performance Measure		Engage enemy on LC and LAC. This a simple system, since the deployment area is devoid of non-military collateral. Pre-positioned deployment at 'most dangerous option' provided immediate destruction action. Proximity of own 'Vulnerable Points and Areas' near LC and LAC are ideal locations for such a deployment. The Galwan skirmish is a suitable site for such weapon system, which is weapon-free at all times in the designated fire lanes.
Environment	Fully & Partially Observable	Fully observable beyond LC and LAC
	Episodic & Sequential	Episodic
	Static & Dynamic	Static
	Discrete & Continuous	Discrete
	Deterministic & Stochastic	Deterministic, because enemy movements will be reduced after engagements

Task		Autonomous Weapon System Type 2
	Single & Multiple Agent	Single
	Know & Unknown	Yes, well known
Sensors	Camera/Range Finder/Drone	
Actuators		30 mm AK 630/This will be a static location and capable of sustained fire even under enemy fire

Task		Autonomous Weapon System Type 3
Performance Measure		Engage enemy surreptitiously. A sniper based, man-portable smaller autonomous system will provide additional capabilities during special operations, both during war and grey zone periods.
Environment	Fully & Partially Observable	Fully observable when deployed against a specific personnel or material target.
	Episodic & Sequential	Episodic
	Static & Dynamic	Static
	Discrete & Continuous	Discrete
	Deterministic & Stochastic	Stochastic, since it is trained for a specified mission
	Single & Multiple Agent	Single
	Know & Unknown	Yes, well known
Sensors		Specialised EO device
Actuators		Beretta .338 Lapua Magnum Scorpio TGT/Barrett .50 M95

APPENDIX K

Autonomous Prediction of Landslides and Rock Falls

Task		Predict landslides and rock falls which cause road blocks	
Performance Measure		What is expected to be done? What is the desired output or action? List out the desired tangible outcomes.	
		Generate an alert or warning if the model detects any signs of a potential landslide or rock fall.	
		Provide the user with an interface for interacting with the system, such as a mobile app or a web portal, to view alerts and take appropriate action.	
		Options	
		1	2
Environment	Fully & Partially Observable	Partial	Fully
	Episodic & Sequential	Yes, Episodic	Yes, Episodic
	Static & Dynamic	Yes Static, in terms of Agent deliberations	Yes Static, in terms of Agent deliberations
	Discrete & Continuous	Yes, it is Discrete	No, it is Continuous
	Deterministic & Stochastic	No it is stochastic	No it is stochastic
	Single & Multiple Agent	Yes, Single	No, Multiple
	Know & Unknown	Yes, known	Yes, known and supported with geological and engineering inputs
Sensors		Cameras	• Rainfall Gauge
			• Soil Moisture Sensor
			• Piezometers to identify groundwater pressure
			• Seismometers
			• GPS
			• Weather Station
			• Cameras
Actuators		Alert Generator	Alert Generator
Overall priority		II	I

Chapter Six

AI Revolution: Transforming Indian Military Operations

The technology development of advanced and expeditionary countries is driven by technology developers and highly versatile civil-military collaborations. High technology companies based in these countries share a common vision of national security and global aspirations. These conditions are absent in the Indian context and hence the relevance of this research. With the aim to look for bridges of common interest, the research proved its hypothesis and justifies documentation of a typical military AI landscape.

INDIAN ARMY REALITIES

The Indian Army is permanently deployed on the borders where the LC and LAC are called restive or contested and by many other variations of conflict. Small team operations, border action teams routinely violate border ceasefires and agreements. The LAC does not have similar characteristics, but preventive troop deployment has become a norm. This permanent deployment in combat zones is unique to the Indian Army. The physical strain of operating in difficult geographical condition increases the operational stress of soldiers. Troops in such areas and operational conditions (nomenclatured as field areas) are rotated to peace stations, where the geography is not challenging, and there are no combat duties. The troops rest, recuperate, and live with their families in peace stations. Historical experience and medical data define the rotation time of a field area between 18 and 24 months.

It is a volunteer force, where the officers and men are allotted units. All fighting units (infantry, artillery, armoured, etc.) are considered regimental, where unit and personal affiliation is permanent. In contrast to other military units, logistic units undergo rotations of troops and officers while remaining geographically fixed. This unique operational model reflects the specialised nature of the units, where unit processes are aligned to the operational requirement of a particular region. Similarly, all headquarters are static. They receive and despatch fighting units on a tenure basis mentioned earlier. Headquarters staff is a rotational force of soldiers and officers assigned from various regimental and logistic units.

All officers in the Indian Army are career officers who are promoted according to their service. Each promotion is mandatorily accompanied by a command assignment. Each newly-promoted officer is made a commanding officer responsible for operational, logistic, and administrative functioning of his unit. Each army officer without exception also performs additional duties and tenures. These tenures provide opportunities such as a staff officer or instructor or directing staff, military attaché or UN observer. In a volunteer army, such Extra Regimental Employment (ERE) provides additional knowledge, experience and improved promotion avenues. Peculiar certainties surround the Indian Army and special structures are essential to pole vault AI development. The two circumstances; peace-field profile and ERE are typical to the Indian Army and pose special challenges to AI developmental structures. A few other conditionalities and a way forward are discussed in this chapter. The succeeding sections are atypical but contextual suggestions for an Indian environment; an *Atamnirbharta* route to indigenous challenges.

DOCTRINAL ASPECTS

Indian Army doctrine provides strategic guidance to achieve national security objectives. As a conceptual process, it motivates military strategy, and operational art and as a force component it guides the means to fight. The combat applications of AI can undertake information warfare. In a tactical battlefield, it can assist in capturing high-value territory and selectively destroy enemy combat potential. The doctrine posits that future battlefields will be boundary less, simultaneous and unrestricted, with a healthy mix of lethality, contact, and high technology.

The doctrine is omnipotent; it does not need continuous change and cannot keep pace with rapid technology changes. At the same time, it cannot forgo the background of the conflict and remains constant even through evolving technology. War-fighting doctrines continue to evolve to involve the technology with concurrent changes to strategies and concepts that are essential to accommodate a new tool of warfare-AI.

Ethics and military AI will run parallel with compliance requirements to ensure they are fulfilled during the entire development cycle. Military risks arise from military ethics and revisionist principles. India has held a steadfast moral stance, and it will carry forward in AI development as well.

ORGANISATIONAL AND PERSONNEL ASPECTS

AI Cells. AI cells in the army serve as specialised offices to exploit AI technologies to enhance operational effectiveness, decision-making, and situational awareness. AI cells are better placed at regional levels to design AI systems as per operational construct. Its primary roles will be:

- Analyse operational requirements and identify areas where AI technologies can provide solutions.
- Evaluate existing AI technologies and determine their suitability for military applications. Incorporate AI laboratories to re-train the system specific to military needs.
- Collect feedback, requirements, and insights from the army and inform developers, ensuring that solutions are user-centric and aligned with operational needs.
- Ensure that AI tools and systems comply with operational standards, regulations, and ethical guidelines and oversee the integration while addressing legal and ethical, and security considerations.
- Monitor system and human failures, conduct post-deployment assessments, analyse feedback, and refine AI tools to improve performance and reliability over time.

Military AI Developers (Permanent or Contractors). Induction of specialist cadres on a permanent basis is an ideal situation where they understand the user perspectives and freely interact for product development. The frequent peace-field and rigid promotion and career progression criteria will have to be specially modified for them. Unless they regularly update their technical threshold, their skills might stagnate. A contractual person on the other hand requires infrastructural support when inducted for a project, and retain market skills due to frequent return after project completion.

TA Innovation Cell. As part of the Indian Army, the Territorial Army (TA) relieves army units from static duties and assists the civil administration in dealing with natural calamities and maintaining essential services when the life of the communities is affected. The TA is a 'part time concept' with mandatory two months training in a year. Pay and allowances and privileges remain the same as regular army officers when embodied for training and military service. The rest of the year, the officers carry on with their jobs or businesses. Innovatively, India has utilised this formal provision to recruit high technology knowhow into the army. This part-time job concept inducts expertise in various technical fields like AI, software development, cyber security, and financial services. Each of these inductees is a successful technology developers in their civilian field and recruiting them into the army brings in their specialities for innovative AI development. Such domain specialists have successfully developed a large number of technologies within the army. This arrangement also bridges the gap between developers and users. The part-time nature of this arrangement allows effective interaction between the military, industry, and academia.

AI Cadre. A discussion on establishing an AI cadre is a natural progression in the realm of military AI. The Indian Army operates across diverse operational geographies spanning the entire spectrum of conflict. With a rich history of combat experience, celebrated as 'Always in Operations', officers and soldiers possess a wealth of operational knowledge. This extensive experience transcends specialisation, with individuals and units routinely navigating various terrains and conflict scenarios. While certain specialised units possess niche expertise tailored to specific geographies, the Army as a whole maintains its combat edge through broad operational exposure. In the Indian context, this discussion extends to whether the cadre should be affiliated with a standard regiment, a specialist corps (such as the Corps of Signal or the Corps of Electrical and Mechanical Engineers), or whether it warrants the establishment of an entirely new corps. Historical decisions regarding similar specialised domains, such as cyber operations led by the Corps of Signals, highlight the significance of this debate. However, recent trends indicate a move towards inclusivity and broader scopes for such domains within the Army.

- The AI cadre necessitates a multidisciplinary approach, comprising expertise in data science, artificial intelligence, software development, and military operations. This blend of qualifications underscores the cross-functional nature of the cadre. Additionally, the reliance on compute capability and training datasets underscores the hardware-intensive nature of AI endeavours, necessitating access to robust cloud infrastructure. This dependency contrasts with the rotational obligations typical of tour of duty personnel, requiring a separate HR policy to ensure uninterrupted access to essential resources.
- Considerations regarding the hierarchy within the Army, including officers, Junior Commissioned Officers (JCOs), Non-Commissioned Officers (NCOs), and soldiers, are integral to the design of the AI cadre. Just as in technical tasks like computer hardware repair, where each rank has specified roles, the distribution of responsibilities within the AI cadre must align with existing military protocols. Drawing parallels with cyber organisations, where most tasks are performed by officers and JCOs, the structure of the AI cadre may initially mirror that of cyber manpower, evolving over time to accommodate specialised roles.

One aspect seems fairly certain – the academic pursuit of writing algorithms will remain the sole prerogative of officers only, unless an AI cadre is developed as a mix of military and civilian manpower with a new organisation structure. Military experts bring in tactical diligence and the civilian brings in the technical ability. The civilians, much like the Territorial Army Innovation Cell (TAIC) remain current with industry AI standards and commercial systems, bringing in the expertise for AI products. AI is a niche technology and will require niche manpower structures. Ultimately, the establishment of an AI cadre represents a significant organisational endeavour, blending military expertise with civilian technical proficiency to harness the full potential of AI technology. As AI continues to evolve, the structure of its supporting manpower must similarly adapt to meet the challenges of modern warfare effectively.

TRAINING ASPECTS

A widespread schedule to address all the training needs has to be established, initially wide and then gradually deepened. One officers' training model is discussed below and a parallel model can be extracted for soldiers.

- A selected school should undertake basic training for nominated officers to create a wide base. Common standards and definitions are achieved through a single-school model. The training could be in phases, with brighter candidates nominated for further phases.
  - Phase 1. Introduction of the AI theory with its operational construct.
  - Phase 2. Basic aspects of AI designing, algorithm selection, model making and data curation.
  - Phase 3. Algorithm engineering.
  - Self-learning shall be a mandatory part of each phase. Distance learning should recommend the Swayan portal credit courses for each phase of the training. Government initiatives to make education accessible can be fully exploited to train a wider base.
- Tactical schools of instructions should conduct training on tactical employment of AI in two phases. Initially, as theoretical knowledge and subsequently as skilful hands on training on AI tools released to the army. Indian Army soldiers and officers are imparted formal weapon and equipment training, a well-proven practice. On a similar premise, it is essential that AI user training forms a part of military skill development.
- Officers proceeding on Masters and PhD courses should especially focus on military AI systems and their subjects should be approved by an appropriate authority. The final subject for study should include all pillars of AI, algorithm engineering and data curation. Most importantly, their research and data should be stored in a higher knowledge repository for future use and reference.
- Paper presentation is another necessary process to identify core personnel for long-term AI deployment. After the call for papers (technical, theoretical, doctrinal, and conceptual), a peer review selects high-quality productions for presentation and discussion. All selected papers are compiled with commentaries and contact details of the presenter. These are also dumped in a higher knowledge repository for future use and reference.

- Trained leaders will be usefully exploited to enable them to take sound decisions on 'What is required and How to use it'. Focussed seminars could address; background discussions (to democratise AI), focussed discussions on use cases in India, discussion on exploratory usage and technical seminars to straighten out issues.
- A special set of war gaming could be organised for well-informed leaders. It can be introduced as part of seminars and war games, including a 'designing AI war game', dataset war gaming and form factor war gaming.
- Military AI White Papers can be released regularly to pre-emptively inform the civilian developers.

Lessons from Cyber Readiness. The Indian Army's HR training experiments with cyber threats provides a succinct case for AI training. The Army decided to work on Zero Trust networks and the end users were made responsible for the physical isolation of the browser machines. Cyber security instructions were conveyed to all the units and personnel in the form of an easy checklist, without putting too much of a technical load on end users. All were expected to blindly follow the checklist and the arrangement worked well. Cyber function is a software expertise at its basic level and the checklists abstracted the highly technical task to its lowest levels of compliance, whereas a cyber specialist's flair for software, algorithms and IT is but a natural progression. While the users sweated around checklists, specialist auditors (the people with the flair mentioned earlier), became cyber auditors, who checked all systems periodically and would recommend few add-ons routinely. Critical failure points were displayed prominently alongside all digital assets. The cyber security policy and instructions were promulgated and users made more skilful through various courses at military schools of instruction. Technical advisories began circulating to increased cyber resilience of digital devices, all this while the cyber audit continued as a formal security process. The scope of instructions increased to advise users on hardware vulnerabilities and restrictions on buying malicious devices. Cyber security instructions are now strongly implemented with robust mini detachments at all levels, but audited by well qualified personnel. The Indian Army also took a decision to manage the cyber specialist cadre separately. Over time, numerous audit teams were created to perform internal regulation and external audits within the organisational structure and hierarchy. Many of these specialists moved within cyber organisations under a fairly well-defined policy and thus retained their cyber knowhow. When such personnel moved out of the specialist organisations, they got grouped for cyber audit functions on a case-to-case basis. Sufficient literature now exists in the form of standard operating procedures, policies, and directives, while reprimanding cyber violators became a standard military-legal process.

AI training can follow a similar cycle, with a fundamental difference. Unlike cyber, AI percolation may not affect all military users and all digital devices. While the cyber speciality resides with IT, software and algorithm creators, AI functionality is much narrower as the AI systems are designed to provide a defined outcome (A to B matching) in a fixed environment (be it digital or physical). The AI workforce is cross-functional team consisting of software, data and ML engineers, grouped with subject matter experts (not necessarily IT specialists). The typical nature of AI development requires a variety of people to interact with each other during the entire cycle of development, testing and deployment. A specialist set of people (trained in military school of instructions) in cross-functional relationship combined with a variety of instructions, policies and directives will guide AI adoption.

Lessons from IW (Information Warfare) Training. AI development and the military is a system approach and cannot be allocated as a specialist attribute. The nature of AI development also advocates integration of AI awareness and understanding within training programs, commencing from foundational military training. Unless the military incorporates the employment of AI in conceptual discussions, it may just end up in a digital store, forgotten in time. An anecdotal analysis of the IW course at the Army War College (AWC) provides a compelling rationale for developing a composite AI course. The AWC was selected to conduct the course to bridge the gaps between psychological operations (military intelligence forte), perception management (organised by ADG Strategic Communication (SC) and media cells of formations), electronic warfare (Army Air Defence and Signal Intelligence specialty) and cyber warfare (Signals capability). It is now well understood that an IW operative must think across all realms and dimensions to design and launch an effective campaign. Similarly, AI development, deployment, and result delivery cycle will fall short if the user and developer are disconnected from each other.

MATERIAL AND FACILITY ASPECTS

AI Laboratory

AI cells and AI units in other armies function to design, induct and manage tools and systems. The Indian Army does not have a similar advantage due to the typical characteristics mentioned earlier in the chapter.

- Cyber specialists in the army move between same job role organisations and still abide by peace-field criteria, different from AI. A cyber specialist is a single person who can continue with his assignment if the infrastructure is provided. But an AI specialist is never a single person. AI is collaboration between a data curator, a machine-learning engineer, a software engineer and a subject matter expert. It is a collaborative enterprise, and thus requires a different approach.
- Unlike cyber, AI may not percolate to all units. Hence, hierarchical structures of AI along the military headquarters are not suitable. Critically, AI is a function of data to match A with B, which is strongly bound to a limit which may be geographical or digital. An autonomous AI load carrier for deserts will fail in urban combat zones and social media analytics will fail in formal text environments. These instances suggest that Indian AI units and offices should be matched with geographical parameters and theatre-specific developments.
- Percolation of AI is another decision dilemma for the Indian Army. How deep should it go and how wide should it spread? Before dwelling into a deep and wide discussion, there is a matter of traction. Like any technology, AI needs traction amongst military users and commanders. Military commanders will have to understand the capabilities and limits of such systems to decide the military role of AI. This traction also simulates interest and jump starts possibilities till there are sufficient practitioners, and then there is a necessity to formalise it. Appropriate policies, concepts and detailed battle drills have to be documented.

The primary role of the AI lab should be to ideate the AI design, with simple systems characteristics, create, train and quick delivery of ready-to-use system to military units. A combination manning structure is required, where the military personnel provide leadership and subject matter expertise in addition to technical qualifications, the contractual persons should fulfil the technical roles. It should possess project management capabilities with clear delivery timelines.

- It should focus on selecting a bouquet of ready algorithms from various dictionaries, test and combine them to make a suitable AI model. Like cyber tools, AI algorithm-making is a serious speciality and not easy to acquire; it is a long term academic pursuit. Selecting, testing and matching readymade algorithms from various dictionaries to create the necessary AI model is a prudent choice over algorithm engineering.
- It should also have the capability to analyse existing non-military models and re-train it to suit military needs. An AI lab should have the capability to take over MSME products for further development and deployment. Business expansion of all lab projects should be handed over to other organisations to relieve the lab only for research and product development.
- An AI lab should also be able to consume curated data from the data set repository, which can also be the part of the AI lab.
- Theoretical research should be limited to the operational focus of the sponsor headquarters.
- All used algorithms and the models should be moved into an AI artefact repository for future use.
- The laboratory should emphasise non-actuator, non-physical systems to ease developmental constraints of mechanical engineering. Instead, such projects can be handled by ADB for industry assimilation.

Testing and Evaluation. The framework helps to define the testing and evaluation criteria, including performance metrics and field-testing procedures. Military AI must always be effective robust and safe. Military AI applications are deployed in high-stake environments where precision, reliability, and adherence to safety standards are paramount. Thus, a comprehensive approach to testing and evaluation is essential to mitigate risks and maximize the utility of these systems.

- Clear testing and evaluation criteria with a framework provide a structured approach to assess the performance of military AI systems including specific metrics and benchmarks.
- Performance metrics encompass various aspects such as accuracy, precision, recall, response time, and resource utilisation, each carefully

chosen to reflect the system's ability to fulfil its intended functions accurately and efficiently.
- The system's resilience to adversarial attacks, noise, and unpredictable conditions, ensuring its reliability in real-world scenarios.
- Continuous improvement is a fundamental aspect of the framework, enabling military organisations to refine and enhance AI systems iteratively.

AI Artefact Repository. A centralised repository is required to store and manage projects, datasets, models, algorithms and share projects and assets within the military domain. It will serve as a comprehensive platform for storing, organising, and disseminating AI-related resources, knowledge, and best practices across the army. It will provide a centralised platform to facilitate collaboration and knowledge sharing among military personnel, researchers, and industry partners. It will also ensure that the security, integrity, and compliance of AI assets is recorded and informed in accordance with military standards and regulations.

AI as a Sector Store

Each army unit is provisioned with weapons, and equipment on a standard scale, since Indian units operate across multiple terrains and operational configurations. Overtime certain additional weapons and equipment got inducted to bolster the operational capability of the units. This new set of equipment was area specific and improved the local tactical ability of the units. Upon completion of a tour of duty, the specialised equipment called sector store is handed over to the relieving unit and thus it continues to stay in the operational area. The Indian Army utilises this concept of sector stores to provide specialist equipment, weapons and stores to improve mission success. By adapting this model to AI deployment, the Army can ensure rapid access to tailored AI capabilities for different operational sectors, facilitating quick adaptation to local challenges. This approach enables agility, flexibility, and enhanced mission effectiveness by leveraging AI capabilities.

The obvious pitfalls include inequitable distribution across different operational sectors. Compatibility issues, interoperability challenges, and training requirements may arise, requiring careful planning and investment in technical integration and personnel training. Interoperability, flexibility, and modularity of non-standardised interfaces and protocols in the sector stores may interfere with scaling the tools. AI tools are typically oriented to a specific output—operational region or a task and thus are more suitable to be developed as sector stores. A mountain sensor-based hypothetical decision support system will offer no assistance in a desert profile and thus is better left back in the mountains for continual use and improved learning. Overall, this concept is highly suited for AI tools and should find mention in the deployment and development policy.

Data

Data is critical, and its curation makes AI successful. Meticulous curation is directly linked to successful AI outcomes. Incorrect data or inaccurate curation risks making valuable resources and produces wasteful AI. One form of data could be numerical, sensor, geospatial data, and time-series data, which are generally produced by a combination of devices, sensors, instruments, systems, and human, so in some form almost curated but may require some calibration. Other forms are images, videos, formal and informal texts, and audios, which require major curation effort to make it AI usable. These data formats are rich in content and can vary widely in quality, context, and structure. These can be complex and diverse; while images and videos may contain varying resolutions or perspectives, the textual data varies in languages, writing styles and semantics. Noisy and irrelevant artefacts disturb the quality of audios, images and video data. Finally, all types have to be labelled and annotated to abstract data information for AI to understand. With known operational environments, the Indian Army is well placed to identify its datasets to serve the AI tools. Every type of dataset will provide input to the desired output and one simplistic use of datasets is detailed below.

- Image datasets can provide quick AI solutions to identify changes in the area of interest, differentiating non-military assets from military ones. It is most utilised for recognition detection, classification, and tracking of various things. Simple and efficient, it is the leanest option to enhance situation awareness and border management. Curating an image dataset is comparatively simple; removing all non-relevant images (images not required to be identified) as irrelevant artefacts, reducing noise and making images sharper for computer vision to identify and re-size images to a consistent size and normalise pixels or choose algorithms that can read through all such variations.

- Formal text data is available on a secured ecosystem and is most suitable for information retrieval functions. This can clarify background information or generate historical experiences as reference points. A prediction system could generate warning and opportunity alerts after analysing situation reports (SITREPS) and the background datat on secured ecosystem.
- Audio data is a typical format and highly useful for the AI system to analyse the persons of interest.
- Curating the video data is challenging due to size, noise, and unprecedented amount of irrelevant artefacts. It requires highly efficient processing methods to extract meaningful information. Combined strategies are required to make a useful dataset. Data compression to manage size, background removal feature extraction to isolate relevant information from the clutter is most essential for video datasets and by far most challenging. In spite of the challenges, video data provides a large canvas to extract information and thus is suitable for supporting all other types of data analyses. Due to the sheer technicality of its curation, it may be prudent to contract it out.
- Informal text is almost as noisy as the video data and similarly requires major curation efforts to fuel an AI system.
- Data curation is a continuous process and will get biased with time which requires continuous monitoring of the data pipelines.
- Unified data warehousing is important to ensure that common artefacts are available to all AI developers.

Internal Data Curation. AI Lab based data curation is a standard method, but there is also an 'Indian Way' for the Indian Army. In data collection, SSOT (Single Source of Truth) refers to a concept where all data within an organisation is stored, managed, and accessed from a single, centralised source. This approach ensures that all users within the organisation are accessing consistent and accurate data, eliminating the risk of discrepancies. Schematic data collection may look like the image below, with one SSOT, which offers data for further data curation.

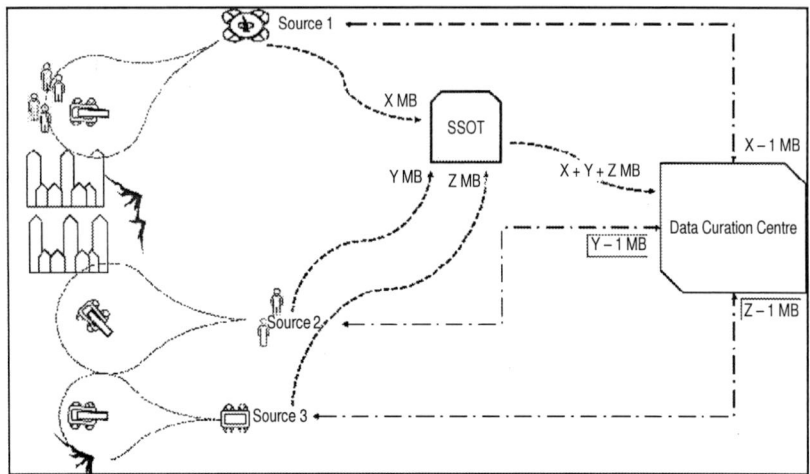

- Data curators receive X, Y and X MB data (curved line) from sources 1, 2 and 3 respectively, which is put through the entire sequence of cleaning, normalising and feature extraction.
- An innovative option (edged connectors in the layout above) distributes the work load where each of the sources conducts basic data curation of their respective data and then forwards the data to the data curators. SSOT continues to remain responsible for a consistent and accurate data, eliminating the risk of discrepancies, but now with a leaner data. This will reduce curation load since the delivered data is more consistent and accurate; result, a quick build and accurate AI solution.
- This proposal aligns with democratisation of AI and reaffirms an important fact that data creators have a high order of intrinsic knowledge, which lies wasted in the former standard format of internal data curation. The training lesson lies in the mass cyber training for the army, where all military schools ran ADP (Automated Data Processing), where a small and focussed 'just enough training' created a strong base. In a similar fashion, data curation workshops can be started to teach basic techniques. Image curation can commence with mean intensity, image resizing and cropping. Text curation could include word frequency, sentence length, and character frequency. Audios could include amplitude envelope, zero crossing rate (ZCR) and energy calculation. Video curation could include frame sampling, frame differencing and colour histograms (for still frames). This training methodology could

commence with the most efficient and the easiest of the datasets: image and formal text data curation.

Contracted Data Curation. For characteristically noisy data like videos, informal texts and audios, contracting could be a parallel decision. Such types of data curation require sorting of inconsistencies, errors, and missing values. Continuous monitoring and updates are also essential to maintain data quality over time, especially in dynamic environments where data distributions may change. Most importantly, military data curation is burdened with special security needs. The Indian Army could contract out the data curation to specialists much in line with the Ministry of External Affairs (MEA) and Tata Consultancy Services (TCS) relationship. TCS, the technology partner set up to manage the Passport Seva Kendras (PSK) (facilitation centres), data centres and end-to-end core passport application, across 93 PSK in 63 cities across India.[1] The arrangement works in two models, one where the data is curated offsite under scrutiny of MEA personnel and another model where TCS employees work in the MEA premises and security instructions for data integrity are endorsed in the contract. The Army regional data centres can act as onsite facilities for data curation and the army can benefit from the technical expertise of the contractors.

DRDO Data Curation. Data consciousness may also lead to curation contract to the DRDO. Since the DRDO has been the sole military product developer for a considerable time, it is well aware of the military data security concerns and is sufficiently able to provide security guarantees. Even though, as scientists, the DRDO possesses expertise and compute, however, the delivery mechanisms do not address the users' aspirations. During the December 2023 review by the Standing Committee on Defence,[2] the secretary DRDO acknowledged 23 projects were delayed out of 55 and the delay factors entail management issues like misreading industry supply capabilities, inaccurate manpower allocation and overoptimistic technical capabilities. The 2023 DRDO reorganisation committee is expected to look into the functioning and recommend changes, including administration, where the scientists are relieved of administrative duties to focus on research while offloading development and production to industry. Expertise on AI data curation is well established in the private industry (banks, air travel, etc.); this ready and proven capability may nudge the DRDO away from the prospect. The Army would do well to distribute the contracting between private players and the DRDO, creating the best balance between 'AI Now' and 'AI for the Future'.

LEADERSHIP AND EDUCATIONAL ASPECTS

As warfare becomes more complex and technology advances, military commanders must adapt to new strategies and tactics. Even with sufficient expertise, they may not be equipped to handle the newer aspects of warfare, and they will develop additional competencies to face these challenges successfully. They must have the cognitive flexibility to understand the intricacies of modern warfare and possess a variety of skills for multi-domain operations across land, air, sea, space, and cyber domains. They will need to integrate AI, unmanned systems, and autonomous weapons into their command structures, managing man-machine interfaces effectively. Commanders must constantly learn and acquire new skills to remain relevant in the changing military landscape. Future commanders must possess a diverse set of competencies to succeed in their roles.

Even though the research is focused on tactical employment of artificial intelligence, higher military leadership will still have to define the control measures for its employment. The leadership has a wide role in model development starting from design to deployment.

- Military leaders have to understand the development cycle and the typical A to B matching capability of the AI system. Highly focused AI output is a typical characteristic of these systems which is a huge advantage. Commanders will have orchestrated the AI tool to argument human capability from the contextual stage – defining the desired output in a specific geography. The leadership has to visualise the future battle field, identify opportunities to augment human soldiers and then order tool development. Senior military leaders have the most difficult task of undertaking tactical diligence while defining an AI tool. Unless it improves the mass or the effect, it is not essential for military business.
- Leadership also has the onerous responsibility to initiate ethical and legal discussions on AI systems. Since the same leader will undertake tactical diligence to quantify a design, it is upon him to define the legal and ethical boundaries and optimise the AI output. Leaders will have to possess mental mobility to extract best results from the tools within

the ethical and legal framework. Critical examinations of such control measures will assist in uninterrupted development and improved mission reliability of the AI tool.

Craig Reynolds, a famous programmer, creates computer models to understand natural systems. His most famous work simulating bird flights is called 'flocking algorithm.' He theorises; "A natural order exists where individuals follow simple rules in their movement, without external intervention. These rules enable movements and reactions much faster than if external interference were present." In the military AI context, this algorithm suggests that a force, when operating, will follow intrinsic flocking rules to achieve combined efficiency and effect without external intervention. A few interpretations related to this are as follows:

- Despite the transparency and communication present on the battlefield, external intervention may be wrongly perceived as a necessity to 'assist' the operating force. Military commanders may be tempted to take tactical decisions to intervene in the hope of quick and successful termination. However, this intervention acts as a 'flock spoiler' for the combat forces, questioning the ongoing nature of events and delaying the combatants' response instead of hastening it. Military commanders with access to such decision choices must be formally trained to take a 'Garur Drishti' or the long look, beyond the tactical battle, and at longer time frames.
- Generating isolation operations and coup de main options should become the ruling norms for future commanders. This approach of standoffish involvement is contradictory to standard military training and commanders must understand the natural nature of operations and suitably reorient their roles to adapt to it.

POLICY ASPECTS

AI Induction Policy. An induction policy outlining pan-army induction will be prohibitively expensive. Such a policy would also have to argue for full replacement of the existing military artefacts with the AI or retrofit them to attain compatibility. Similar arguments would ensue for partial replacements of military tools with AI. A better option would be to plan on augmenting existing capabilities rather than replacing. This option would augment a with well-defined depth and range of military operations. A few case analyses below justify the argument.

- Face recognition is an excellent access control function to restrict entry into military complexes. The outcome of such a complex function will benefit from a face recognition system but always with the human soldier sentry. The building itself will probably have further access control, where a standalone face recognition system will offer a successful solution. What about a military complex which receives large number of visitors? Does it justify face recognition as access control? This AI facial recognition solution is thus a restricted use technology and army policies could clearly indicate its deployment to specific complexes as it may not have pan army utility.
- Live translation NLP models similarly, have little value in the army, which boasts of a large volunteer force where officers pride in learning the language of the troops and adapting to their religion and cultural beliefs. It could but find some utility in schools of instruction where foreign military officers are trained or during multinational military exercises or military patrols along the LAC against China.
- Augmented and virtual reality application is yet another field which may not have pan-India utility. In addition to being exploratory technology, it is dependent on speciality infrastructure and prohibitively expensive. Training of an infantry section (the smallest fighting group of 10 soldiers), artillery gun crew (six soldiers), or a tank crew (four soldiers) may not be suitable justification for such a technology. In its simplest definition, it trains a human in a particular skill, like a neuro surgeon looking to slice the hypothalamus without degrading the patient's motor function or a ship clearance diver attempting a dangerous and difficult keel repairing. It is best used to train humans on rare events unavailable in the real world. No military skill is an individual intensive effort; rather, it is always a combined skill of a group of soldiers. The Army will have to define fair use of this technology.
- A medical infirmary (MI) room prescription for a specialist consultation is ideal application of an AI bot. Patient load in a traditional MI room slows the treatment cycle and medical aid. An expert system-based AI Bot could note the symptoms, look into the previous medical records

and recommend further tests and schedule an appointment with a specialist. Working in a highly controlled physical environment of a MI room with no background noise or useless artefacts, this type of AI system has universal applicability and could be endorsed as a policy.

Wide Development Base. National sponsorship of technology in general and AI in particular has found a large number of practitioners in India. These large numbers of developers create a wide base, which is an excellent provider of capable AI solutions. There is a catch though. The developers are focused on their product without analysing the military requirements. Numerous solutions; 'goggles which can look through fog, walls and dense foliage', or 'Mandarin translation into seven Indian languages', or 'vehicle number plate identification' and a few more are repeatedly displayed at military technology symposiums. It is unfortunate that such developers are oblivious to the actual army needs and try to sell their home-grown products instead, which obviously the army does not need. It would be prudent for the army to advertise its needs in terms of a broad range of outputs, like information retrieval or database realisation or decision support or load carriage or logistics management or autonomous systems. A product wish list may be the only way to focus development of military AI amongst the wide base rather than an informative document like Technology Perspective and Capability Roadmap (TPCR).

Traction Projects. Certain traction projects are essential to ensure that the developer and the user increase their confidence together. For the traction project, it is essential that 'zero risk' projects, specifically 'zero operational risk' are chosen. These projects should importantly align with user necessity rather than the developer's belief. Traction projects, if left untethered, will turn themselves into luxury decorations pieces, much coveted with least value. To avoid undue time burden, the AI laboratory should not deal with traction projects; instead, leave their development to the wider base of 'AI enthusiasts' in the army.

Commercial off the Shelf. The decision to develop AI for immediate use cannot act as an alternative to AI for the future. Both varieties are required and a good balance between the two is essential. As far as the 'AI now' is concerned, it is defined by immediate operational necessity. A policy decision to accelerate change by commissioning bespoke products should be considered only when commercial off-the-shelf projects cannot meet a particular demand.

Framework Consultation. Consultation for AI is necessary till the time in-house expertise is developed. The military may possess the user requirements; however, the technical diligence may not be available. Framework consultation by the Department of Science and Technology[3] and IITs will offer a good AI design. The Indian Army already has a number of MOUs with such specialist institutes and academia across the breadth of the country. This consultation service should become a mandatory condition during the tool designing stage.

Design Decision. Who should design the AI tool? Should it be the operational theatre or the arm or service headquarters; it all depends on various factors. If the PEAS analysis of the tool is well researched, theatre level tool designing is the ideal choice. Firsthand insights of the tactical environment, operating environment and terrain will design an ideal AI tool. It will ensure user-centric design and facilitate rapid prototyping and testing in real-world scenarios. In case of technical and resource limitation, service headquarters may design the product, promoting standardisation and interoperability. A collaborative approach will generally become the chosen option for combined benefits. Some other decision dilemmas are:

- Indian tactical battlefield will benefit from small, power efficient, Edge models, which can work without high dependence on a central server.
- While the cross-functional necessity of AI product development is proven, it is equally crucial that a combined effort be used for preparation of concepts and actual employment. Self-help, down to top and ad hoc AI combat conditions and tactics will not necessarily provide success in the military use of AI.
- Deep learning projects work without traditional learning, but are anchored by specialist algorithms. Machine learning projects meanwhile work on A to B matching, and thus are easily deployable.
- To safeguard against poor products, which may denude operational confidence, the products should be put on hold if minimal viable interest is not shown for it. A developer designed and hastily delivered product may lack usability in the tactical environment.

SEMINAR SUGGESTIONS

Numerous suggestions are available from seminars and panel presentations. As a single source, they may not evince confidence but are worth discussion and study.

- **MCEME Seminar (26 September 2023).** Image analysis systems trained on HD images produce an error when tested on low latency images. Despite all measures to re-train, the system failed to develop acceptable accuracy. A detailed discussion with human operators suggested a different pathway. The problem was converted to a statistical problem which stopped over engineering. The process was replaced by mathematical principles and a probabilistic model, resulting in acceptable accuracy. In addition, lightweight learning models trained on handcrafted features or simple feature representation provided viable efficient solutions.

- **Risk Mitigation in Military AI—International Panel Discussion at USI (20 November 2023).** Retired United States Air Force Lieutenant General John N.T. 'Jack' Shanahan provided a great 'guide for Indian AI Development' during the discussion. He observed that bureaucratic systems (military and civil) do not accept change with an argument 'why repair, if it is not broken'' as an alibi for self-preservation. Project Maven found it necessary to train the policy makers too! AI confidence was built in the USA with a low risk and low reward product and it was a tightly-controlled-top-down mandate. These initial projects also identified human disruptors, military and civilians, who were pooled in for special AI tasks later. These people disruptors could collapse the decisions and trepidations with focused ease. Some were technical geeks, but most, just enthusiasts who could speak for both sides (geeks and tacticians). The project hired civilian experts for commercial software decisions. After much deliberations to specifically save gestation delay, minimal viable product was accepted at 75 per cent. He emphasised robust military trials, without any compromise on hardware and weapons. Since a software could be changed easily it was allocated a lesser viability ratio.

- **GAPAI Conference (13 December 2023).** A large number of international practitioners and academics provided a positive tenor to the AI landscape. Acknowledging the permanence of generative AI, the speakers accepted that an expert human knowledge will become a premium as AI hallucinations become complex and unexplainable.

- **AICRA Conclave (18 January 2024).** With unstoppable permeability of AI into technology, the military cannot afford to slow down. Creative use of facial recognition during combat is an exciting possibility. Sophisticated employment of high-end technology by non-state, sub-state and private actors is troublesome as are the highly skilled Houthis in the Red Sea.

KEY TAKEAWAYS

This book started as an Indian Army research directive for Employment of AI at Tactical Levels. Despite a large volume of publications on the subject of AI and military use, there seemed to be a gap in describing Indian conditions (military and technical) which became the main theme for this book. Indian future battlefields are predictable and hence the demands for AI tools should also be well thought out, graduated and specific in nature, very different from the AI development curve elsewhere. A pragmatic analysis of the future battlefields and technology development speeds can assist the Indian Army to identify, define and demand its IA tools. The research so far validates the central theme and provides an action plan for typical Indian challenges.

Technology Trends

World technology trends indicate usable primacy of artificial narrow intelligence, far away from artificial general intelligence and artificial super intelligence, where the machine will exhibit self-awareness. ML is manpower intensive, but can performed most AI tasks, whereas multiple hidden layers of deep learning provide higher accuracy output, but is marred by black box paradox where the machine is unable to explain its decision logically!

Fog and Edge computing provide good options for security conscious organisations with inconsistent access to cloud or internet and this technology is ideally suited for the Indian Army in the present state. The AI development pipeline is collaborative rather than competitive and 'just enough' engineering is necessary to reduce inter-agent conflicts and make AI machines leaner and democratic. Just enough data is sufficient for AI machines but big data inputs will make the machine better.

All pillars of AI feed off each other and simultaneously transcend the growth too. Each of these pillars including machine learning, natural language processing, computer vision, and robotics contributes to the advancement of AI technologies, and their progress fuels other areas as well. ML ingests better data with computer vision (CV) and natural language processing (NLP), concurrently improving both CV and NLP. Generative AI like generative adversarial network (GAN) and generative pre-trained transformer (GPT) are disrupting AI in a manner not fully understood yet. There is a revival of old mathematical models like Expert Systems, which combined with fuzzy logic provides exciting opportunities. With a nomenclature of good old-fashioned AI (GOFAI) Expert Systems is the manifestation of actual man-machine interface which utilises the man to prepare a good machine which will assist the man and fuzzy logic provides a human like-context to this AI.

Military Trends

AI development in the military sector is intricately linked with national goals and aims, particularly in the context of perceived competitions and conflicts in the global geopolitical landscape. AI initiatives in the military sector support overarching objectives, be they deterrence, power projection, or regional aspirations.

Nations seek to leverage AI to gain competitive advantages, deter adversaries, and safeguard their interests. National context is derived from perceived conflict environment (against NATO on land borders by Russia and across vast oceans against the USA by China) or perceived adversary vulnerability (high communication dependence by NATO forces to be exploited by AI-driven EW, both by Russia and China). Landscaping, securing and attacking the cognitive domain to destroy or moderate adversary will is being examined by both Russia and China. Mitigating own vulnerabilities (lack of corporate development capability in Russia and under-experienced soldiers in China) force the nations to undertake special measures (public sector driven AI development in Russia and aggressive hiring of specialists in China).

- All countries are employing AI to study the adversary continuously, countries like Israel overtly, others covertly, but all of them surely. All countries are using AI to firstly collect data from multiple resources, fuse it and then weed out the noise.

- Human-Machine (man machine) teaming is progressing in all domains to increase combat mass without risking humans. Australia is typically attempting to cater for distanced operational areas and huge attack surfaces gifted by geography to the island country.
- International lobbying and parleys have stepped up alongside legal and ethical debates to stymie competition.

There is a natural survival instinct to prioritise development of AI tools in a manner most understood and most successful. No nation can afford to dwell into, out of context, prohibitively costly AI developments.

Indian Technology Trends

There is a positive AI development trend in the country and organisational attempts to develop AI tools and the effort resonates in all departments. Product identification and technology demonstration of AI products is impressive. Various committees and task forces have identified gaps in AI development and recommended corrective milestones.

The 'AI in Defence' Compendium is a good framework to understand the defence AI pathway in India. In-house development of products by the defence services is a typical insight and creates a unique pool of cross-functional teams with a technologist who is a tactician too. Various agencies are involved in AI product development reflecting the growing interest in the Indian defence and security sector. Product priorities appear to be set based on specific needs and capabilities.

ADB's CPDS 2023 addresses critical AI problem definitions, aligning with the army's focus on optimising information dominance and enhancing situational awareness across the full spectrum including open sources and tactical scenarios. Efforts also include refining tactical picture delivery for legacy observation devices, tactical logistic delivery via land and aerial drones. Almost similar focus emerged from DISC challenges of iDEX where 18 per cent are related to logistics, 30 per cent to decision support and 52 per cent provide force protection.

Tangible AI products are intrinsically linked to secured dataset and computational power which are finance-intensive and not commonly available. An executive decision has made such resources accessible through AI Research Analytics and Knowledge Dissemination Platform (AIRAWAT) and the Param Siddhi AI (PSAI) system under the management of the Centre for Development of Advanced Computing (C-DAC), MeitY. Combining Meghdoot cloud services (free and open-source Cloud suite based on Open Stack) with AIRAWAT and the PSAI system, creates a sophisticated ecosystem for high computing research across the country.

DAP is a complex decision-making process to enable expeditious procurements and indigenisation. It contains voluminous aspects of governance and compliance, which are difficult to comprehend, especially to the newly initiate—MSMEs and start-ups of the 'wider defence ecosystem' all fall under this inexperienced category. DAP stresses on probity more than operational factors like logistics, training and time-bound procurement and stake holders are attempting to create a time probity too.

Future Indian Military Conflicts

In line with the main theme of the research, prophesising conflict environment became necessary to identify the likelihood of future military confrontations and, subsequently, the AI tools required for successful battles. Content analysis, SME and surveys provided primary data to consolidate and rank the conflict environments. Priority AI developmental efforts should engage high impact and frequently occurring conflict possibilities.

- **Territorial and Maritime Conflicts.** AI tools to enhance situational awareness and decision-making and war fighting in territorial conflict scenarios.
- **Internal Security Situations.** Prioritise AI applications for surveillance, intelligence analysis, and threat detection in internal security operations.
- **Cyber Threats.** Develop AI-based cyber security solutions to detect, prevent, and respond to cyber threats targeting military infrastructure and communication networks.
- **Hybrid Threats.** Invest in AI technologies for analysing and countering hybrid threats, with a focus on non-kinetic and non-lethal approaches.

Designing AI Tools for the Army

A comprehensive framework called Indian Army Rapid AI Design Architecture (IRADA) has been created to assist tacticians to define AI tools and developers to understand the intrinsic environments in which such tools are expected to work. This framework provides a structured approach to understand the problem, identify key components, and develop AI solutions. Detailed discussions, validation and utilisation of this framework is essential for the Indian Army. The stages of this framework are simple to understand and flexible in utility.

- **Requirement Specification.** To specify the functional and non-functional requirements of AI tools, while determining the capabilities, performance criteria and interoperability requirements.
- **Problem Definition.** The framework guides military planners to 'over-define' the problem for clearer understanding of developers, who may not be aware of the military environment and its characteristics. AI solutions are highly focussed and thus cannot reflect generic needs of the military.
- **Design and Development.** A structured approach to developers in selecting an architecture design, algorithms, data acquisition, model training, validation, and deployment.
- **Technology Assessment.** A tool for developers to evaluate various AI technologies and methodologies, algorithms and techniques for their suitability to address the problem at hand.

ACTION PLAN

This research has demonstrated that the Indian Army possesses a unique understanding of its future conflicts, which are distinctly shaped by the country's geopolitical landscape, internal security challenges, and regional dynamics. Leveraging this assimilation of insights, the Indian Army is well-positioned to design and deploy AI tools tailored to its specific tactical requirements. By harnessing the power of AI technologies, the Indian Army can enhance its capabilities for intelligence gathering, situational awareness, and decision-making, thereby effectively addressing emerging security threats and safeguarding national interests. This underscores the importance of strategic foresight and indigenous innovation to shape the future of military operations and justify India's successful leveraging of cutting-edge technologies. In its true sense, this research was able to produce an AI Action Plan for the Indian Army. With sufficient deliberation, this plan can be modified to suit operational theatre requirements in line with the overall concept.

The Action Plan considers all the important aspects of doctrine, organisation and personnel, training, material and facilities, leadership and education, and finally, the policy. With SMART goal selection, the tasks have been allotted to the higher echelons of decision making. Though it is possible to assign overall priorities, however, 'Aspect Based' priorities make it more comprehensible. Detailed resource allocation, mile stoning and monitoring is deliberately avoided for this pioneering version of the Action Plan.

Aspect – Doctrine

Task Code	Task	Responsibility	Priority
Doctrine Goal 1 (DG1) – Document AI Land Warfare Doctrine for India			
DG1T1	Analyse existing AI doctrines	IHQ MoD (Army)	1
DG1T2	Analyse Indian conflicts	Theatre Command	2
DG1T3	Document and conduct internal communications	IHQ MoD (Army)	3
DG2 – Document AI Risk Compliance Doctrine			
DG2T1	Document concept paper for ideation	IHQ MoD (Army)	1
DG2T2	Seminar to understand the contours and complete the compliance document	IHQ MoD (Army)	2
DG2T3	Document and conduct internal communication	IHQ MoD (Army)	3

Aspect – Organisation and Personnel

Task Code	Task	Responsibility	Priority
Organisational Goal 1 (OG1) – Create AI Cells at Theatre Commands			
OG1T1	Concept paper and internal communications	Theatre Command	1
OG1T2	Identify infrastructure and appoint the core team for each theatre	IHQ MoD (Army)	2
OG1T3	Create AI Repository for Problem Statements and AI Designs	Theatre Command	3
OG2 – Recruit AI Specialists			
OG2T1	Concept paper on in-house developers (permanent recruitment and TA cadre)	IHQ MoD (Army)	1
OG2T2	Document communication protocol between user and in-house developers	IHQ MoD (Army)	2
OG2T3	Create in-house 'Make Capability'	IHQ MoD (Army)	4
OG2T4	Create AI Repository for AI Products	IHQ MoD (Army)	3
OG2T5	Create in-house 'Test Capability' for inducting COTS tools	IHQ MoD (Army)	5

Aspect -Training

Training Goal 1 (TG1) – Basic AI Course at MCTE and MCEME			
Task Code	Task	Responsibility	Priority
TG1T1	Commence Pilot course	ARTRAC	1
TG1T2	Commence regular training courses	ARTRAC	2
TG2 – Basic AI Course at Infantry, Armoured and Artillery Schools			
TG2T1	Commence one course during each training year	ARTARC	3 (with TG1)
TG2T2	Commence Tactical Employment of AI Tools (on hands training)	ARTRAC	4 (with TG1)
TG3 – Higher AI Education			
TG3T1	Document policy for Masters and PhD studies	ARTRAC	1
TG3T2	Create AI Repository for Higher AI Education	ARTRAC	3 (with TG4)
TG4 – Document Policy for AI Seminars, War Games and White Papers			
TG4T1	Concept paper on Seminars and War Games	ARTRAC	2 (with TG3)
TG4T2	Feedback on all aspects of military AI	ARTRAC	4 (with TG3)

Material and Facilities Aspect

Material Goal (MG1) – Create and Populate AI Laboratories			
Task Code	Task	Responsibility	Priority
MG1T1	Concept paper and internal communications	IHQ MoD (Army)	1
MG1T2	Identify infrastructure and appoint a core team for central and theatre lab	IHQ MoD (Army)	3
MG1T3	Create AI Repository for Algorithms and Models	Theatre Command	5
MG1T4	Concept paper and internal communications between all AI labs	IHQ MoD (Army)	2
MG1T5	Concept Paper on AI Sector Stores, including standardisation protocols	Theatre Command	4
MG2 – Finalise Data Dictionary for Images, Text, Video and Audio Datasets			
MG2T1	Concept paper with division of responsibility	Theatre Command	1
MG2T2	Data Curation training and facilities at AI laboratories	IHQ MoD (Army)	2
MG2T3	Data Curation training of the wide base of data collectors	IHQ MoD (Army)	3

Leadership and Education Aspects

Leadership Goals (LG1) – Create Awareness of AI Tools

Task Code	Task	Responsibility	Priority
LG1T1	Concept paper and internal communications	Theatre Command	1
LG1T2	Create advance and technical awareness of AI tools	IHQ MoD (Army)	2

Leadership Goals (LG2) – Create International Awareness of Military AI Tools

Task Code	Task	Responsibility	Priority
LG2T1	Concept paper on ethical and legal utilisation of Military AI tools	IHQ MoD (Army)	3
LG2T2	National seminar on ethical and legal utilisation of Military AI tools	IHQ MoD (Army)	3
LG2T3	Universal document on ethical and legal utilisation of Military AI tools	IHQ MoD (Army)	3

Policy Aspects

Policy Goals (PG1) – AI Products Induction Policy

Task Code	Task	Responsibility	Priority
PG1T1	Concept paper on AI Induction Policy	IHQ MoD (Army)	1
PG1T2	Regional and national seminars to improve communications	Theatre Command	2
PG1T3	Document Army AI products induction policy	IHQ MoD (Army)	4
PG1T4	Document policy on framework Consultation by technical specialists	IHQ MoD (Army)	3

Policy Goals (PG2) – Civil Military AI Technology Cooperation

Task Code	Task	Responsibility	Priority
PG2T1	SWOT analysis of existing structures	IHQ MoD (Army)	1
PG2T2	Military innovation and success analysis	Theatre Command	2
PG2T3	Military AI and COTS success analysis	Theatre Command	3
PG3T4	Document a partnership policy paper	IHQ MoD (Army)	4

Inter-Goal Priority

A comprehensive AI action plan for the Army comprises various aspects, each with distinct goals and associated tasks. Inter-goal priorities will ensure smooth execution and effectiveness of the plan, assisting the Army to allocate resources efficiently, address criticalities promptly, and maximise the overall impact of the action plan. Inter-goal priority will also define simultaneity and linearity of the plan, identifying critical paths and potential risks. Forteen goals identified by the study are listed below along with the supporting relationship and sequencing.

Goal Code	Goal
DG1	AI land warfare doctrine for India
DG2	AI risk compliance doctrine
OG1	Creation of AI cells
OG2	Recruit AI specialists
TG1	Basic AI course at MCTE and MCEME
TG2	Basic AI course at Infantry, Armoured and Artillery Schools
TG3	Higher AI Education
TG4	Policy document for AI seminars, war games and White Papers
MG1	Create and populate AI laboratories
MG2	Finalise data dictionary for image, text, video and audio datasets
LG1	Create awareness of AI tools
LG2	Create international awareness of Military AI tools
PG1	AI products induction policy
PG2	Civil-Military AI technology cooperation

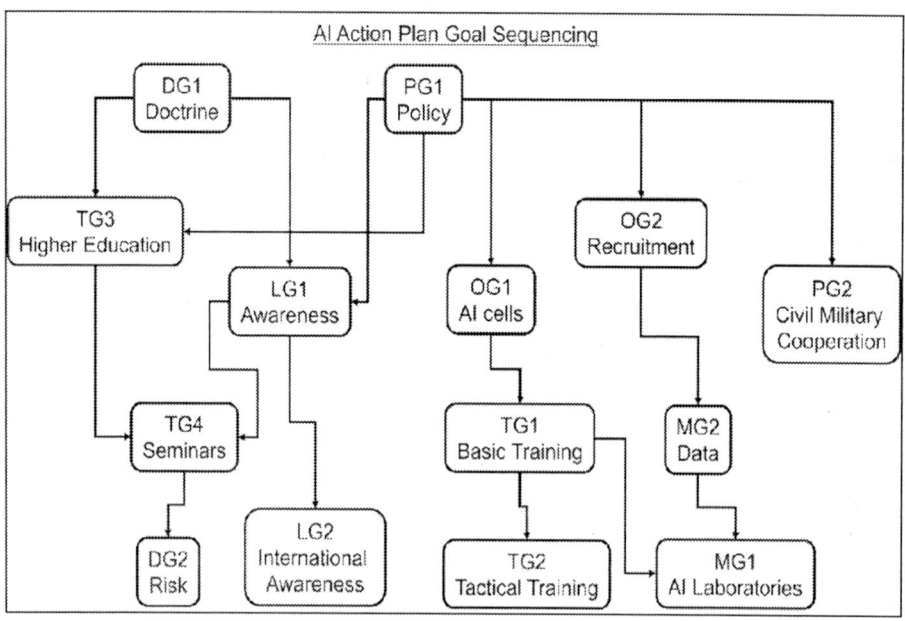

In the light of the Indian Army's tactical objectives and the dynamic nature of military technology, it is crucial to regard the proposed AI Action Plan not as a final solution, but as a strategic framework among various possible approaches. Continuous refinement and adaptation are essential to keep the AI strategy agile and responsive to evolving challenges and opportunities, aligning with emerging technologies, threats, and evolving mission priorities.

The iterative process of refining the Action Plan recognises the intricacy of integrating AI technologies into military operations, particularly considering India's unique manufacturing capabilities and technological landscape. This dynamic setting calls for a flexible approach that allows for adjustments at different levels of details, staying aligned with the pace of technological advancement.

To sum up, the AI Action Plan serves as a starting point, a structured framework that requires ongoing scrutiny and adaptation to remain relevant and effective. The journey towards AI integration is not a static destination but a continual process of evolution and refinement. The Indian Army must continue investing in fostering collaborations with academia, industry partners, and government agencies to drive innovation in military AI. This collaborative approach will empower armed forces to confront future challenges with confidence and resilience through AI-enabled capabilities, safeguarding national security and strategic interests.

Epilogue

HARMONY OF MILITARY AI INTEGRATION

In the research effort many crossroads of concepts and inquiries arose, seeking further examination of questions, each vying for attention and exploration. While some align closely with the central theme of the study, others diverge, branched off into tangential ideas. Each represented an interesting pathway towards deeper understanding; however, in the interest of maintaining focus and coherence within the study, many of these questions remained unexplored. These peripheral questions, while not directly related to the primary focus of research, possess intrinsic value and merit consideration for their potential to yield new insights and discoveries.

The importance of documenting and preserving these unanswered queries, acknowledging their significance as potential catalysts is important for future research endeavours. Readers are presented a compendium of these unattended questions, inviting them on exploratory journeys.

Unfathomable and Unchartered AI Battlespace

The contemporary theoretical framework of the military AI battle space encapsulates a dynamic environment marked by the seamless integration of AI technologies across diverse domains of warfare. At its essence, this construct is routed to harness AI capabilities such as data analysis and fusion, predictive analytics, NLP, image and video analysis, and cyber capabilities to bolster military operations.

Across the globe, nations are leveraging these intrinsic AI abilities to conduct in-depth searches of vast troves of data sourced from various channels, including open media platforms. AI systems excel in swiftly analysing and synthesising immense datasets to discern patterns, anomalies, and actionable intelligence, thereby facilitating predictive analysis and informed decision-making. Furthermore, NLP enhances human-machine interaction and machine learning, while image and video analysis tools automate the interpretation of visual data obtained from surveillance cameras, drones, and satellites, among others. These technologies empower nations to extract valuable intelligence for strategic planning and operational execution.

The same technologies assist in subterfuge, be it deep fake or cognitive warfare. The actors can create high quality deep fakes (the Pope in snow dress) or simple communication AI speech (Imran Khan addressing his followers before the 2023 elections from a prison). Contact battles are now fought in the minds too (Ukraine and Gaza), with leaders threatening nuclear war or offer surrender, depending on which side one is listening to.

AI is affecting physical domains by supporting autonomous systems, be it kamikaze drones (Black Sea Fleet, Israel, Iran) or MMT (Ukraine, Palestine). Actors are using low-cost autonomous systems to overwhelm and defeat high technology, high-cost systems, much like minnow chaff distracting radar in yesteryears.

AI is affecting the national electoral processes; Taiwan, India, Russia and the USA, all have reported some measure of concern. Generative AI creates realistic communications to address specific listeners and its mirage has astounding likeness to the real world. Cyber warfare and offences are a special concern; within itself, it was already a matter of importance and with AI, cyber tools get more sophisticated and attacks stealthier. Safeguarding critical infrastructure and networks from malicious actors is a serious concern and the quip, 'everyone has been hacked, some are not aware of it yet,' seems to be true.

The realm of military AI battle space is virtually boundless, limited only by the imagination—or lack thereof—of human actors. Within this expansive domain, the integration of AI technologies unleashes a bouquet of capabilities that transcend traditional constraints and redefine the very character of warfare.

Traditional Battlefield

With the integration of AI tools into military operations, the concept of traditional tactical battle space, characterised by geographical constraints and specified forces, will evolve into a more fluid and dynamic environment, transcending physical boundaries and conventional limitations. This hypothesis suggests that the utilisation of AI technologies will enable military forces to operate beyond traditional geographic confines, thereby reshaping the nature of warfare. A redefined concept of battle space will allow decentralised decision-making, adaptive tactics, enhanced situational awareness across broader and more complex operational domains. As AI technologies advance, military organisations will adapt their strategies, doctrines, and operational concepts to leverage the vast and unlimited capabilities afforded by AI systems. This may also entail rethinking traditional notions of territorial control, force structure, and command and control mechanisms

A Functional Research on AI at Tactical Level

This project undertook a geospatial analysis of the operational environment for a comprehensive understanding of the geographical factors that influence conflict dynamics. Identification and abstraction of strategic locations, vulnerabilities, and opportunities helps military planners to prepare for the future. Since the research was focused on offering a prescription, both to military users and developers, geospatial references became common orientations in the study.

During the research, other methodologies also came to the fore, which had their own advantages. This research adhered to temporal limits and disregarded other processes, one of which is mentioned here for future researchers to carry forth. Theoretically, a function is defined as the unity of the difference between a problem and various equivalent solutions.[4] Functional research in this context focuses on the existing solutions, which can be modified with inclusion of AI. Analysing the existing functions (processes and procedures) of the capabilities and limitations of military systems, tactics, and strategies is a fascinating way to study the future. Analysis of operational functionality of military equipment helps identify strengths, weaknesses, and areas for improvement, enabling the development of more effective and efficient military capabilities. This optimisation enhances battlefield performance and increases the likelihood of mission success in future conflicts. It also offers an opportunity to focus on emerging trends and opportunities, informed research and leads to development of newer equipment and strategies tailored to meet the challenges of future conflicts.

In the Indian scenario, the functionality of a military division is ideally suited to act as a research scope for AI at the tactical level. As fully functional military organisational level, a division can undertake its own missions under an overall strategy with its integral fire and manoeuvre elements. A fully AI immersive division could be understood to be the epitome of a military force, though this decision would depend on the national security decision maker—go deep or go wide. Augment soldiers at the pointy end across the entire spectrum or enable a manoeuvre force for expeditionary missions. The levels of AI ingress will be a final decision based on necessity, financials and degree of reliance the technology builds for itself. Assuming that building up an AI division is a long-term vision, the new research could focus on the following functional aspects:

- Intelligence preparation of battle field by analysing the vast trove of data streams could build up geospatial knowledge, enemy ability and indicate likely intentions. Co-sharing this analysis with subordinate intelligence centres, or more accurately suggesting tailored intelligence problems and gaps to military commanders will remove the 'fog of war' for all commanders in the chain. This arrangement will allow concurrent operations across the entire spectrum.
- Operational procedures vary across the spectrum of conflict and militaries with inadequate trigger analyses flounder before stabilising. Low-intensity operations merge with hybrid and unwittingly tie down military abilities, if not looked at holistically. AI offers this unique looking glass though its data analytics and prediction capabilities. A division equipped with such solutions will benefit from the early warning of the spectrum shift.
- Tactical level weapon systems and Man-Machine teaming (MMT) creates lethality, precision and accuracy for quick mission accomplishment.
- Smarter logistic solutions create flexible military organisations untethered to their bases, move faster and surprise the enemy across the entire spectrum. Predictive analysis and autonomous solutions will birth a unique 'on the go' logistic solution for such a divisional force.
- AI office automation systems will reduce the daily dullness and drudgery, which is critical but not given its due. Secured office environments, distanced from adversarial physical environment have to but struggle

with digital attacks only. It is an ideal low-hanging AI solution, which could be created on priority.
- Human resource development issues, recruitment, and education can be hooked to the vision of India Stack. It is a fully functional set of open APIs and digital public artefacts that supports identity, data, and payments on a population scale. It is no surprise that it is in use with many countries, offered free with independent access by India. If it is good for the population, it must be good for the Army as an organisation too.
- AI use in military communications has been a success. The Indian Army has already raised the Signals Technology Evaluation and Adaptation Group (STEAG),[5] to undertake research and evaluation of futuristic communication technologies for military purposes. Associated analysis to incorporate in operational responsibilities will modernise the division.
- Influence operations in a tactical battlefield is a major research area that requires deliberated analysis. Mind-proofing the soldiers and their families against malicious intend will rise up as a major operational constraint for future commanders.
- Volunteers join the Indian Army from society, which is continuously influenced by AI technology-enabled information and ideas. Could the same AI offer new recruitment solutions? Could the AI offer the military with AI decluttering of soldiers' minds?

These pointers are a mini milestone for the research, which can be scoped out on the researchers' interest and the limitations, but the functional approach to AI research will remain a fascinating field.

Peer Relationships

As AI technologies become integrated into military operations, traditional peer relationships based solely on conventional military capabilities, such as manpower and equipment, will evolve to encompass factors like technological prowess, data superiority, and adaptability to AI-driven tactics. This shift will blur the distinction between traditional military powers and emerging actors, as smaller technologically savvy entities gain the ability to disrupt or challenge larger, conventionally dominant forces. The decentralised nature of AI-enabled operations will facilitate collaboration and coordination among disparate actors, including state and non-state entities, leading to new forms of coalition-building, information sharing, and strategic alliances. As a result, traditional notions of peer relationships in the military will be supplanted by a more fluid and complex landscape characterised by agile networks of actors leveraging AI capabilities to achieve strategic objectives.

Future Soldiering

The introduction of AI into military operations will revolutionise traditional soldiering practices by streamlining routine tasks, improving accuracy, and augmenting soldiers' capabilities. AI-enabled automation will handle mundane activities such as equipment maintenance and logistics management, freeing soldiers to concentrate on higher-level responsibilities requiring human judgment and intuition. This shift will redefine the soldiering role, and emphasise the symbiotic relationship between human operators and AI systems, where soldiers leverage advanced technologies to enhance their effectiveness and decision-making abilities on the battlefield.

Mass or Effect

The introduction of military AI will present decision-makers with a dilemma between prioritising the increase in mass (deployment of a larger quantity of AI-enabled assets), and improvement of effect (focusing on enhancing the quality and precision of AI-driven capabilities). This decision dilemma will necessitate a nuanced understanding of strategic objectives, operational requirements, and resource constraints to effectively balance the trade-offs between quantity and quality in optimizing military effectiveness and achieving mission success. Considerations regarding the scalability, interoperability, and sustainability of AI systems will play a crucial role in determining whether to prioritise mass deployment for broad coverage or invest in improving the effectiveness of a select number of high-impact capabilities. Ultimately, decision-makers will need to carefully weigh the potential benefits and risks associated with each option to make informed choices that align with overarching strategic goals and operational priorities.

Man-Machine Teaming (MMT)/Man Unmanned Machine Teaming (MUMT)

This is a concept that describes the interaction between humans and machines at the tactical level to specific missions and tasks. This synchronised employment of soldiers, manned and unmanned vehicles achieve better success with improved situational awareness, greater lethality and improved survivability. This relationship seams up the best of machine ability (accuracy, precision, data handling, dangerous environment and vast data analysis) with that of human (cognitive decision making and ethics of human in the loop). MMT should importantly relieve the human operator from getting deluged with information, much of which may be superfluous.

MMT is sustained by improved sensor data collection, data analysis and uninterrupted, beyond line of sight, wideband, low latency communication between man and machine. Software has reached a higher level of sophistication and algorithms can perform a large range of human tasks and the autonomous control levels of the machine are of course the perquisites. MMT across all physical domains is a matter of decision and capability of the autonomous systems to fly, swim or drive. Depending on this ability, communication protocols between the machine and the man will be generated for specific mission parameters. Research shows that the MMT usability is driven by its suitability to increase effect (accuracy, precision, reduce collateral, reduced human casualty) and mass (across vast surfaces, conventional superiority and reduced human fighters).

In the Indian context, large coastline, EEZs and land borders are suitable spaces to increase mass. MMT can patrol vast borders, with slaved machines, 'controlled' by a human. It can carry payloads based on mission, reconnaissance, special or lethal. Specifically, MMT is launched on a mission and will not be able to mix up tasks as successfully as a human; hence tactical diligence to understand the mission will define the 'increased mass' of MMT. Australian Defence Forces (ADF) understand the necessity to control its periphery with integral resources, fully aware of the inability of allied partners to reach in time. Their reliance on MMT in the maritime and aerial domain is a pioneering effort in MMT utilisation. China has aggressively utilised the tactics of salami slicing along the LAC to create a 'new normal' and change the status quo. Continuous pinpricking and gradual step-up are a deliberate ploy to de-sensitise and create new 'boundaries'. MMT deployment allows appropriate reaction to each event without fatiguing decision makers or the others. A similarity also exists for India in the IOR and security of island territories. Piracy, illegal EEZ mining, and human trafficking is difficult to police with traditional force structures across such vast frontiers and the IOR is a suitable space for maritime and aerial MMT deployment.

MMT's role to increase effect is highly revered in the age of information sharing. Accuracy and precision of such teaming reduce the collateral and operational exposure. Quick in, accurate disposal and quick out from dangerous missions will de-escalate operational risks and complete the assignments successfully. Urban warfare, the bane of conventional operations and house-to-house intervention, the epitome of counter-terror operations will be excellently served by MMT deployment. Precise and discreet attacks on high-value targets where the air defence risk to manned aircraft is extremely high, or could cause unacceptable political fallout is yet another invite for MMT.

MMT can live stream images to distant humans and observe targets till final execution, potentially avoiding collateral damage. It could also launch a primitive strike on a possible threat to own forces in the operational sector. MMT allows machines to support deception and surprise, giving lead time to own forces to manoeuvre, as in the extended defences of the deserts or the vast barren mountains of the LAC.

There is an increased ubiquity to this technology which could place it in the hands of non-state actors and countries considered sub-peer category. India will have to participate in international debates and monitor such apprehensions, sharing concerns with similar- minded partners aggressively.

AI Artefacts

With the increasing democratisation of AI technologies, there arises a need to establish robust monitoring mechanisms for AI artefacts, akin to the monitoring of nuclear fissile materials, to mitigate potential risks associated with their misuse or proliferation. This hypothesis suggests that as AI technologies become more accessible and widely deployed across various domains the potential for misuse, unintended consequences, and malicious exploitation also increases. Therefore, implementing comprehensive monitoring frameworks for AI artefacts is essential to ensure accountability, transparency, and responsible use throughout their lifecycle.

Generative AI

Generative AI has evolved beyond its initial stages with the growth of variational autoencoders (VAEs), attention mechanisms, self-attention mechanisms and progressive growing techniques. VAEs offer an alternative approach to generative modelling, while attention mechanisms and self-attention mechanisms enable models to focus on relevant information and capture long-range dependencies, reducing latency. These developments collectively push the boundaries of generative AI, enabling the creation of diverse, coherent, and high-quality outputs across various applications and domains.

One of the gravest concerns surrounding generative AI in the realm of national security and defence is its potential to facilitate sophisticated deception and manipulation campaigns. Generative AI algorithms can produce convincingly realistic fake images, videos, audio recordings, and text, indistinguishable from genuine content. Adversaries could exploit this capability to disseminate false information, fabricate evidence, or manipulate public opinion on a massive scale, undermining trust in institutions and democratic processes. Additionally, generative AI cyber attacks pose challenges for detecting and mitigating risks to critical infrastructure, national security assets, and public safety.

Generative AI's capability to generate content with prompts also opens up innovative avenues for enhancing security and defence measures. Security analysts can leverage this feature by providing specific prompts to generate diverse scenarios and simulations tailored to their security needs. By inputting relevant prompts, analysts can generate realistic scenarios reflecting potential security threats, vulnerabilities, and adversarial tactics. These generated scenarios can then be used to simulate security gate operations, conduct training exercises, and evaluate the effectiveness of security protocols. Through iterative refinement and feedback, security professionals can fine tune the prompts to generate increasingly accurate and challenging scenarios, enabling security teams to proactively identify and address security gaps before they are exploited. Ultimately, this proactive approach empowers security organisations to stay ahead of evolving threats, bolster resilience, and safeguard critical assets and infrastructure against emerging security challenges.

Generative AI is a platform and a tool to learn from, and humans could learn from it to mitigate its ill effects. By studying generative AI, humans can gain insights into its inner working, identify vulnerabilities, and develop effective safeguards to prevent misuse or abuse. This approach could become a major focus for the Indian Army to 'Fight The AI as an AI' #FIGHTAIASAI (much like fighting a guerrilla like a guerrilla) and the importance of education, research, and collaboration in addressing the challenges posed by emerging AI technologies.

Securing AI through Legal Acts and Advisories

AI, being an international subject, is susceptible to replication; yet the absence of a universal monitoring body renders this replication uncontrollable, posing a concern for AI developers in India. In response to this concern, the Indian government issued an advisory during March 2024, mandating platforms to obtain explicit permission before implementing any 'unreliable Artificial Intelligence (AI) models/Large Language Models (LLM)/Generative AI, software, or algorithms' for users accessing the Indian internet. A subsequent clarification stated that the advisory only applies to 'significant large platforms' and excludes start-ups. The focus of the advisory was on untested AI platforms deployed on the Indian internet, with social media intermediaries directed to utilize consent pop-ups to explicitly inform users about the unreliability of AI-generated data. These regulations mark the beginning of AI governance in India, a necessary step considering AI's growing significance in business and its potential for misuse.

The governmental vision of #AIFORALL, coupled with unregulated dataset usage and controlled AI development, could lead to 'AI inflation' if not accompanied by appropriate regulations. This unchecked proliferation of AI technologies risks compromising quality, reliability, and ethical considerations, potentially undermining trust in AI systems and increasing associated risks. The concept of property rights in India differs from the global perspective, with the association of '*daan*' (donation) with community service being unique to the Indian psyche. The Indian concept of '*Bhash Daan*' under the National Language Translation Mission aims to crowd source translation models through open data and open-source software, with projects like Project Vaani funded by Google in IISc, Bangalore, and ARTPARK underway. Without regulations, these donations will fraudulently convert to commercial ventures and commercial datasets.

The DPDP Act 2023 governs the processing of digital personal data in

India, irrespective of its original format, and can tackle some of the privacy issues related to AI platforms. The Information Technology (Intermediary Guidelines and Digital Media Ethics Code) Act 2021 oversees various entities, including social media intermediaries, OTT platforms, and digital news media. Section 66E of the Information Technology Act, 2000, addresses deep fake crimes related to privacy violations. Section 66D targets malicious use of communication devices or computer resources, with penalties including imprisonment and/or fines. The IT Rules mandate social media platforms to swiftly remove such content, risking loss of 'safe harbour' protection otherwise.

AI relies on regulated datasets for its functionality, and unless these datasets are secure, the country's security may be compromised. To address this concern, the government plans to release a national data governance framework policy to promote ownership safety and trust in non-personal data, with access restricted to designated and authorised offices. India also aims to safeguard against bias, discrimination, or compromise in the electoral process by labelling all artificially generated media and text with unique identifiers or metadata.

Military AI development in India must be mindful of the risk that even minor breaches could compromise national security while consolidating datasets. The eagerness to develop AI products through unverified cottage industry experts must be tempered with allegiance to national digital infrastructure and standardised security parameters thus becoming crucial facets of military AI development.

Ethics—User or Developer Driven

The debate over whether ethics in technology, particularly in the realm of AI, should be driven by users or developers is a complex and multifaceted one, touching upon various philosophical and practical considerations. Users are the ultimate stakeholders in AI technologies, and therefore should have a significant say in defining ethical standards. It is users who interact with AI systems on a daily basis, whether digital or the physical realm and they will possess valuable insights into real-world impacts.

Developer-driven ethical standards on the technical capabilities of AI systems may make the system inherently weak especially in the military arena. Scientific knowledge of ethics may not be suitable for the battlefield. Since the decision to create an AI tool has been taken with due deliberations, building additional constraints into the system will render it less useful in battle. In short the technologists and developers are expected to remove the technical problem from the table and allow unrestricted use of the final tool.

A balanced approach that incorporates insights from both users and developers is essential to fostering ethical AI innovation and ensuring that AI technologies align with societal values and priorities. The Indian Army will have to create its own expert group on ethical standards for international discussions. This group will remain responsible to understand the ethical standards in the correct perspective, suggest modifications to own systems to retain their edge on the battle space, rather than act as a tether.

This integration of ethics into military AI systems will mitigate risks and ensure compliance with moral principles but also strengthen the overall quality and effectiveness of AI products. By prioritising ethical considerations, military AI developers will demonstrate their commitment in upholding moral values and societal norms, thereby justifying the usefulness of AI systems. This trust is essential for the widespread adoption and acceptance of military AI technologies, ultimately strengthening the battle space. This proactive risk management will also reduce the likelihood of adverse outcomes and operational failures.

THE BEGINNING

In the context of India, the military AI voyage unfolds against the backdrop of a growing technological landscape. A collaborative approach, reinforced by stringent reliance on governmental regulations, emerges as a guiding principle towards realising the full benefits of AI while mitigating associated risks. The military AI also desires the imperative for cooperation over competition. In diverse threats and challenges, collaboration among stakeholders becomes paramount for fostering innovation, enhancing operational efficiency, and strengthening defence capabilities.

Central to India's approach towards military, AI requires formulation and implementation of robust regulatory frameworks that prioritise transparency, accountability, and ethical conduct. Governmental regulations serve as the cornerstone, providing clear guidelines and oversight mechanisms to oversee the development, deployment, and utilisation of AI technologies. With rigorous vetting procedures, adherence to industry standards and collaboration with trusted entities that demonstrate a commitment to uphold safety parameters, India will mitigate the risks associated with commercial ventures and safeguard its strategic interests. Embracing industry standards will enhance the quality and reliability of AI systems but also facilitate seamless integration and collaboration across diverse operational environments.

India's journey towards harnessing the potential of military AI is guided by a commitment to cooperation, innovation, and regulatory compliance. Through collaborative partnerships, stringent reliance on governmental regulations and adherence to industry standards, India seeks to maximise the benefits of AI while mitigating risks to national security. By embracing a collaborative ethos and leveraging the collective expertise of the global AI community, India endeavours to strengthen its defence capabilities and safeguard its strategic interests in an increasingly complex and dynamic security environment.

karmaṇy-evādhikārastemāphaleṣhukadāchana
mā karma-phala-heturbhūrmātesaṅgo'stvakarmaṇi

You have a right to perform your prescribed duties,
but you are not entitled to the fruits of your actions.
Never consider yourself to be the cause of the results of your activities,
nor be attached to inaction.

Bhagavad Gita: Chapter 2, Verse 47

NOTES

1. TCS Transforms Passport Services for Indians across the Globe, https://www.tcs.com/what-we-do/industries/public-services/case-study/e-governance-passport-transform-indian-passport-office, accessed on 8 March 2024.
2. Standing Committee Report Summary, Defence Research and Development Organisation, https://prsindia.org/files/policy/policy_committee_reports/Report_Summary-DRDO.pdf, accessed on 15 January 2024.
3. National Programme on Artificial Intelligence (NPAI) Skilling Framework, 23 June 2023, https://www.ugc.gov.in/pdfnews/5732498_Report-on-NPAI-Skilling-Framework.pdf, accessed on 15 February 2024.
4. Knudsen, Morten (2010). Surprised by Method—Functional Method and Systems Theory [56 paragraphs]. Forum Qualitative Sozialforschung/Forum: Qualitative Social Research, 11(3), Art. 12, http://nbn-resolving.de/urn:nbn:de:0114-fqs1003122.
5. Indian Army's Tech Revolution: The Rise of Signals Technology Evaluation and Adaptation Group (STEAG), https://www.financialexpress.com/business/defence-indian-armys-tech-revolution-the-rise-of-signals-technology-evaluation-and-adaptation-group-steag-3430022/, accessed on 24 March 2024.

Bibliography

Books

Ai, Zhong. Role of Second Strike in No First Use Doctrine: A Study of China and India. New Delhi, Knowledge World Publishers Private Limited, 2018.

Bhalla, AS. Asia's troubled Spots: The leadership Questions in Conflict Resolution. USA, Rowman and Littlefield International, 2019.

Black, Jeremy. Military Strategy: A Global History, London, Yale University Press, 2020.

Bostrum, Nick. Super Intelligence: Paths, Dangers and Strategies. UK: Oxford University Press, 2017.

Chakravorty, PK. Future of Land warfare: Beyond the Horizon. New Delhi, Pentagon Press LLP, 2020.

Davis, Elizabeth Van Wei. Shadow Warfare. UK, Roeman and Littlefield, 2021.

Jaishankar, S. The Indian Way: Strategy for Uncertain World. Noida, Harper Collins, 2020.

Jasper, Scott. Russian Cyber Operations. Washington DC Georgetown University Press, 2020.

Jorisch, Avi. Thou Shalt Innovate. Mumbai, India, Jaico Publishing House, 2019.

Kaplan, Fred. Dark Territory: The Secret History of Cyber War. New York, Simon and Schuster Paperbacks, 2016.

Levy, Adrian and Cathy Scott-Clark. Spy Stories: Inside Secret World of RAW and ISI, New Delhi, Juggernaut, 2021.

Mallick, PK. Artificial Intelligence in Armed Forces: An Analysis. New Delhi CLAWS Journal, 2018.

Mallik, PK. Armenia- Azeri conflict over Nagorno Karabakh: Geopolitical Implications. New Delhi, Vivekananda International Foundation, 2021.

Menon, Shivshankar. Choices: Inside Making of India's Foreign Policy. UK, Penguin Books, 2016.

Mueller, John Paul and Massoron, Luca. AI for Dummies. India: Nice Printing Press, 2020.

Pandalai, Shruti. Combatting Terrorism: Evolving Asian Perspective. India, IDSA, 2019.
Pant, Harsh V. Indian Foreign Policy. India, Orient Blackswan Private Limited, 2019.
Riordan, Shaun. The Geopolitics of Cyberspace: A Diplomatic Perspective. Leiden, The Netherlands, Brill Research Perspective, 2019.
Russel, Stuart J. and Norvig, Peter. AI A Modern Approach. India: Pearson India Education Ltd, 2010.
Samuel, Cherian and Sharma, Munish. India's Strategic Options in Changing Cyberspace. India, IDSA, Pentagon Press, 2019.
Schroeter, John, ed. Aftershock, USA, Abundant World Institute, 2020.
Subramaniam, Arjun. India's Wars: A Military History, 1947-1971. New Delhi, Harper Collins, 2016.
Svendsen, Adam DM. Intelligence Engineering. USA, Rowman and Littlefield International, 2017.
Ucko, David, and Thomas Marks. Crafting Strategy for Irregular Warfare. Washington DC, National Defence University Press, 2020.
Woolley, Samuel. The Reality Game, India Hachette India, 2020.
Malhotra, Rajiv and Viswanathan Vijaya. Snakes in the Ganga. India, Garuda Prakashan Pvt Ltd, 2022.

Online News

Adam, Bertie. Royal Navy's first unmanned submarine will be built in Plymouth. Accessed on 2 March 2023. https://www.plymouthherald.co.uk/news/plymouth-news/royal-navys-first-unmanned-submarine-7883582#
Bekdi!, Burak Ege. Turkish firm develops AI-powered software for drone swarms. Accessed on 27 February 2023. https://www.defensenews.com/unmanned/2020/11/24/turkish-firm-develops-ai-powered-software-for-drone-swarms/
British Navy announces arrival of autonomous mine hunter in Gulf. Accessed on 2 March 2023. https://www.naval-technology.com/news/british-autonomous-mine-hunter-gulf/
Chen, Stephen. Chinese AI plays war games like a human, with military strategists unable to identify it as a machine, developers say in Beijing. Accessed on 13 March 2023. https://www.scmp.com/news/china/science/article/3211142/chinese-ai-plays-war-games-human-military-strategists-unable-identify-it-machine-say-developers
Chen. Stephen. In China, AI warship designer did nearly a year's work in a day. Accessed on 14 March 2023. https://www.scmp.com/news/china/science/article/3213056/china-ai-warship-designer-did-nearly-years-work-day
China deploying 'en masse' underwater drones in Indian Ocean. Accessed on 14

March 2023. https://www.livemint.com/news/india/china-deploying-en-masse-underwater-drones-in-indian-ocean-report-11609373452869.html

Ebrahimi, Soraya. What are the Malloy drones the UK is sending to Ukraine? Accessed on 2 March 2023. https://www.thenationalnews.com/world/uk-news/2022/05/03/what-are-the-malloy-drones-the-uk-is-sending-to-ukraine/

Frantzman, Seth J. and Kelsey D. Atherton. Israel's Rafael integrates artificial intelligence into Spice bombs. Accessed on 23 February 2023. https://www.defensenews.com/artificial-intelligence/2019/06/17/israels-rafael-integrates-artificial-intelligence-into-spice-bombs/

French army first experiments GACI mule UGVs in external operation. Accessed on 27 February 2023. https://www.armyrecognition.com/defense_news_april_2021_global_security_army_industry/french_army_first_experiments_gaci_mule_ugvs_in_external_operation.html

Greene, Andrew. Air Force flags need for low-cost drones to prepare Australia for future conflicts. Accessed on 7 March 2023. https://www.abc.net.au/news/2023-02-27/air-force-need-low-cost-drones-australia-future-conflict/102028934

Grevatt, Jon. Lockheed Martin partners on AI training system for Australia. Accessed on 7 March 2023. https://www.janes.com/defence-news/news-detail/lockheed-martin-partners-on-ai-training-system-for-australia

Heikkilä, Melissa. AI: Decoded: Putin's high-tech war – Making sense of AI systems – Deepmind controls nuclear fusion reactor. Accessed on 13 February 2023. https://www.politico.eu/newsletter/ai-decoded/putins-high-tech-war-making-sense-of-ai-systems-deepmind-controls-nuclear-fusion-reactor-2/,

Microsoft shares these examples to show how Iran, North Korea, China and Russia are using AI for Cyber War. Accessed on 30 March 2024. http://timesofindia.indiatimes.com/articleshow/107705046.cms?utm_source=contentofinterest&utm_medium=text&utm_campaign=cppst.

https://economictimes.indiatimes.com/news/defence/chinas-pla-aims-to-leverage-advanced-technology-for-use-of-unmanned-weapons-artificial-intelligence-says-report/articleshow/97431564.cms?from=mdr. Accessed on 8 March 2023.

https://hitex.co.in/news/iit-hyderabad-and-drdo-collaborate-for-advanced-technologies.html. Accessed on 4 September 2023.

https://timesofindia.indiatimes.com/blogs/ChanakyaCode/a-national-security-strategy-for-india-documenting-strategic-vision-prudently/. Accessed on 15 January 2024.

https://timesofindia.indiatimes.com/business/india-business/cyber-crime-losses-rise-to-rs-63-crore-in-fy20-21-govt/articleshow/90532711.cms?utm_source=contentofinte&from=mdr. Accessed on 26 December 2023.

https://timesofindia.indiatimes.com/city/bengaluru/tcs-to-face-125-million-loss-in-ip-infringement-case/articleshow/105400072.cms. Accessed on 28 November 2023.

https://www.ainonline.com/aviation-news/defense/2018-02-21/new-chinese-armed-uav-project-unveiled. Accessed on 14 March 2023.

https://www.armyrecognition.com/april_2014_global_defence_security_news_uk/russian_army_to_use_unmanned_ground_robot_taifun_to_protect_yars_and_topol-_missile_sites_2304143.html. Accessed on 21 February 2023.

https://www.intelligence-airbusds.com/newsroom/news/airbus-zephyr-solar-high-altitude-pseudo-satellite/. Accessed on 2 March 2023.

https://www.janes.com/defence-news/news-detail/update-taiwans-ncsist-unveils-new-male-class-uav-development. Accessed on 28 February 2023.

https://www.moneycontrol.com/news/technology/indias-cto-nilekani-unveils-ai-strategy-focus-on-use-cases-not-my-model-bigger-than-yours-11853961.html. Accessed on 5 December 2023.

Ju, Juan. Norinco's Sharp Claw I UGV in service with Chinese army. Accessed on 14 March 2023. https://www.janes.com/defence-news/news-detail/norincos-sharp-claw-i-ugv-in-service-with-chinese-army

Lau, Jack. US should use AI to beat Chinese censors in case of Taiwan attack. Accessed on 27 February 2023. https://www.scmp.com/news/china/military/article/3197496/us-should-use-ai-beat-chinese-censors-case-taiwan-attack-think-tank-chaired-former-google-ceo-says

Lin, Chia-nan. Formosat-7 to bolster national security. Accessed on 28 February 2023. https://www.taipeitimes.com/News/taiwan/archives/2019/02/22/2003710204

Min, Roselyne. Israel deploys AI-powered robot guns that can track targets in the West Bank. Accessed on 9 December 2022. https://www.euronews.com/next/2022/10/17/israel-deploys-ai-powered-robot-guns-that-can-track-targets-in-the-west-bank#:~:text=Amid%20increasing%20tension%20between%20Israel,camp%20in%20the%20West%20Bank

Pradhan, S. D. Transformation of the Chinese influence operations into a Cognitive war: Warning for India. Accessed on 13 March 2023. https://timesofindia.indiatimes.com/blogs/ChanakyaCode/transformation-of-the-chinese-influence-operations-into-a-cognitive-war-warning-for-india/

Presidential Address to the Federal Assembly at Saint Petersburg, Russia, on 1 March 2018. http://en.kremlin.ru/events/president/news/56957

Rheinmetall Wins Bid for Spiral 3 of UK's Robotic Platoon Vehicles Programme. Accessed on 2 March 2023. https://www.asdnews.com/news/defense/2022/05/02/rheinmetall-wins-bid-spiral-3-uks-robotic-platoon-vehicles-programme

Press Release

AI task force hands over Final Report to RM. Accessed on 12 July 2023. https://pib.gov.in/newsite/PrintRelease.aspx?relid=180322

Bibliography

Cabinet Approves Ambitious IndiaAI Mission. Accessed 19 April 2024. https://pib.gov.in/PressReleaseIframePage.aspx?PRID=2012355

Commerce and Industry Minister Sets up Task Force on AI for Economic Transformation. Accessed on 11 July 2023. https://pib.gov.in/newsite/PrintRelease.aspx?relid=170231

Dare to Dream' scheme for promoting start-ups. Accessed on 17 July 2023. https://pib.gov.in/PressReleasePage.aspx?PRID=1882706

Defence India Start-up Challenge. Press Information Bureau. Accessed on 13 July 2023. https://pib.gov.in/Pressreleaseshare.aspx?PRID=1547093

Defence Innovation Hubs. Press Information Bureau. Accessed on 13 July 2023. https://pib.gov.in/newsite/PrintRelease.aspx?relid=188372

Defence Research Institutes, Ministry of Defence. Accessed on 17 July 2023. https://pib.gov.in/PressReleaseIframePage.aspx?PRID=1848675

English rendering of Prime Minister's address after the successful landing of Chandrayan-3 on Moon via video conferencing. Accessed on 25 September 2023. https://pib.gov.in/PressReleasePage.aspx?PRID=1951491#:~:text=I%20am%20confident%20that%20all,journey%20beyond%20the%20Moon's%20orbit.

Greater Civil-Military Jointness must to Further Strengthen National Security. Accessed on 15 January 2024. https://pib.gov.in/PressReleaseIframePage.aspx?PRID=1833513

iDEX Initiative. Accessed on 13 July 2023. https://pib.gov.in/PressReleaseIframePage.aspx?PRID=1849786

Innovation in Defence Production Projects. Accessed on 17 July 2023. https://pib.gov.in/PressReleseDetailm.aspx?PRID=1906339

MeitYStartup Hub and Meta shortlists 120 Start-ups and Innovators for the XR Start-up Program. Accessed on 7 September 2023. https://pib.gov.in/PressReleasePage.aspx?PRID=1894084

Raksha Mantri Inaugurates Workshop on AI in National Security and Defence. Accessed on 11 July 2023. https://pib.gov.in/newsite/PrintRelease.aspx?relid=179445

Task Force for Implementation of AI. Accessed on 12 July 2023. https://pib.gov.in/PressReleaseIframePage.aspx?PRID=1810442

Vice-President's address at the 2023 edition of the Indo-Pacific Regional Dialogue. Accessed on 1 January 2024. https://pib.gov.in/PressReleasePage.aspx?PRID=1977077

Year End Review 2022 – Ministry of Defence. Accessed on 12 July 2023. https://pib.gov.in/PressReleasePage.aspx?PRID=1884353

Government Websites

AI/ML Technology. Accessed on 17 July 2023. https://drdo.gov.in/aiml-technology

AI-Based Intelligent COVID-19 detector Technology for Medical Assistance (ATMAN). Accessed on 17 July 2023. https://drdo.gov.in/ai-based-intelligent-covid-19-detector-technology-medical-assistance-atman

Army Robotic & Autonomous Systems Strategy v2.0. Accessed on 3 March 2023. https://researchcentre.army.gov.au/sites/default/files/Robotic%20and%20Autonomous%20Systems%20Strategy%20V2.0.pdf

Artificial Intelligence in Support of Defence. Accessed on 25 February 2023. https://www.defense.gouv.fr/sites/default/files/aid/Report%20of%20the%20AI%20Task%20Force%20September%202019.pdf

Artificial Intelligence Task Force Constituted by Ministry of Commerce and Industry. Accessed on 11 July 2023. https://www.aitf.org.in/

Artificial Intelligence, UN Chief Executives Board for Coordination. Accessed on 6 February 2023. https://unsceb.org/topics/artificial-intelligence

Chapter II, Updated Version of DAP 2020. Accessed on 8 November 2023. https://mod.gov.in/dod/sites/default/files/wn25423.pdf

Defence Acquisition Procedure 2020. Accessed on 13 July 2023. https://www.mod.gov.in/sites/default/files/DAP2030new.pdf

Defence Artificial Intelligence research network contracts signed. Accessed on 8 March 2023. https://www.defence.gov.au/news-events/releases/2023-03-06/defence-artificial-intelligence-research-network-contracts-signed

Defence Artificial Intelligence Strategy of UK, June 2022. Accessed on 10 January 2023. https://assets.publishing.service.gov.uk/government/uploads/system/uploads/attachment_data/file/1082416/Defence_Artificial_Intelligence_Strategy.pdf

Defence Artificial Intelligence Strategy. Accessed on 13 February 2023. https://www.gov.uk/government/publications/defence-artificial-intelligence-strategy/defence-artificial-intelligence-strategy#:~:text=The%20DAIC%20achieved%20initial%20operating,AI%20capabilities%20across%20the%20Department

Defence Research and Development Organisation—DRDO. Accessed on 17 July 2023. https://www.drdo.gov.in/labs-establishment/technologies/centre-artificial-intelligence-robotics-cair

Department of Defence outside the Continental United States (OCONUS). Accessed on 19 March 2023. https://dodcio.defense.gov/Portals/0/Documents/Library/DoD-OCONUSCloudStrategy.pdf

DoD Chief Digital and Artificial Intelligence Office Hosts Global Information Dominance Experiments. Accessed on 19 March 2023. https://www.defense.gov/News/Releases/Release/Article/3282376/dod-chief-digital-and-artificial-intelligence-office-hosts-global-information-d/

Hogenhout, Lambert. A Framework for Ethical AI. Accessed on 6 February 2023. https://unite.un.org/sites/unite.un.org/files/unite paper-_ethical_ai_at_the_un.pdf

http://geneva.china-mission.gov.cn/eng/dbdt/202112/t20211213_10467517.htm. Accessed on 10 January 2023.

https://assets.publishing.service.gov.uk/government/uploads/system/uploads/attachment_data/file/1082416/Defence _Artificial_Intelligence_Strategy.pdf. Accessed on 6 February 2023.

https://crsreports.congress.gov/product/pdf/IF/IF11150. Accessed on 10 January 2023.

https://dst.gov.in/indias-ai-supercomputer-param-siddhi-63rd-among-top-500-most-powerful-non-distributed-computer. Accessed on 13 September 2023.

https://idex.gov.in/challenge-categories. Accessed on 23 March 2023.

https://idex.gov.in/sites/default/files/2022-04/i4D%20guidelines%2020-4-22.pdf. Accessed on 5 September 2023.

https://indiaai.gov.in/government/government-of-karnataka. Accessed on 9 September 2023.

https://indiaai.gov.in/government/government-of-tamil-nadu. Accessed on 9 September 2023.

https://indiaai.gov.in/government/government-of-uttar-pradesh. Accessed on 9 September 2023.

https://indiaai.gov.in/news/israel-claims-to-have-fought-the-world-s-first-ai-war. Accessed on 23 February 2023.

https://indiaai.gov.in/research-reports/airawat-establishing-an-ai-specific-cloud-computing-infrastructure-in-india/. Accessed on 9 September 2023.

https://ipindia.gov.in/writereaddata/Portal/ev/sections/ps35.html. Accessed on 25 August 2023.

https://mod.gov.in/sites/default/files/DRDOall1412220.pdf. Accessed on 25 August 2023.

https://msh.meity.gov.in/assets/Scheme-Report.pdf. Accessed on 6 September 2023.

https://ndap.niti.gov.in/. Accessed on 13 November 2023.

https://pro.similarweb.com/#/digitalsuite/websiteanalysis/audience-geography/*/999/3m?key=indiaai.gov.in&webSource=Total. Accessed on 13 November 2023.

https://sdgs.un.org/sites/default/files/2021-04/Resource%20 Guide%20on%20AI%20Strategies_April%202021_rev_0.pdf. Accessed on 6 February 2023.

https://srijandefence.gov.in/ProductList. Accessed on 14 November 2023.

https://tdf.drdo.gov.in/daretodream. Accessed on 5 September 2023.

https://www.airforce.gov.au/our-work/projects-and-programs/ghost-bat. Accessed on 7 March 2023.

https://www.ddpmod.gov.in/sites/default/files/ai.pdf. Accessed 23 March 2023.
https://www.defense.gov/News/News-Stories/Article/Article/2667212/hicks-announces-new-artificial-intelligence-initiative/. Accessed on 19 March 2023.
https://www.defense.gov/News/Transcripts/Transcript/Article/2672391/joint-artificial-intelligence-center-press-briefing. Accessed on 19 March 2023.
https://www.drdo.gov.in/adv-tech-center. Accessed on 25 August 2023.
https://www.drdo.gov.in/sites/default/files/inline-files/Guidelines-For-DRDO-Grants-in-Aid-Scheme.pdf. Accessed on 25 August 2023.
https://www.drdo.gov.in/transfer-technologies. Accessed on 10 September 2023.
https://www.investindia.gov.in/innovation-challenge-for-development-of-machine-aided-translation-system. Accessed on 9 September 2023.
https://www.investindia.gov.in/pm-stiac#:~:text=The%20Artificial%20Intelligence%20(AI)%20Mission,which%20will%20include%20international%20collaborations. Accessed on 19 April 2024.
https://www.meity.gov.in/writereaddata/files/Committes_A Report_ on_ Platforms.pdf. Accessed on 6 September 2023.
https://www.meity.gov.in/writereaddata/files/Committes_B-Report-on-Key-Sector.pdf. Accessed on 6 September 2023.
https://www.meity.gov.in/writereaddata/files/Committes_C-Report-on_RnD.pdf. Accessed on 6 September 2023.
https://www.meity.gov.in/writereaddata/files/Committes_D-Cyber-n-Legal-and-Ethical.pdf. Accessed on 6 September 2023.
https://www.meity.gov.in/writereaddata/files/constitution_ of_ four_ committees_on_artificial_intelligence_0.pdf. Accessed on 6 September 2023.
https://www.meity.gov.in/writereaddata/files/Digital%20Personal%20Data%20Protection%20Act%202023.pdf. Accessed on 28 November 2023.
https://www.meitystartuphub.in/about/. Accessed on 26 March 2023.
https://www.meitystartuphub.in/tide-1-0/. Accessed on 26 March 2023.
https://www.meitystartuphub.in/tide-2-0/. Accessed on 26 March 2023.
https://www.mha.gov.in/sites/default/files/BMIntro-1011.pdf. Accessed on 22 November 2023.
https://www.psa.gov.in/pm-stiac. Accessed on 9 September 2023.
https://www.similarweb.com/website/indiaai.gov.in/#overview. Accessed on 13 November 2023.
https://www.similarweb.com/website/ndap.niti.gov.in/#display-ads. Accessed on 13 November 2023.
https://www.similarweb.com/website/srijandefence.gov.in/#geography. Accessed on 14 November 2023.
India's AI supercomputer Param Siddhi 63rd among top 500 most powerful non-distributed computer systems in the world. Accessed on 9 September 2023.

https://dst.gov.in/indias-ai-supercomputer-param-siddhi-63rd-among-top-500-most-powerful-non-distributed-computer

Kainikara, Sanu. Artificial Intelligence and the Future of Air Power. Accessed on 2 March 2023. https://airpower.airforce.gov.au/sites/default/files/2021-03/WP45-Artificial-Intelligence-and-the-Future-of-Air-Power.pdf

Landslide Monitoring and Early Warning System. Accessed on 25 April 2024. https://www.indiascienceandtechnology.gov.in/node/167820

Lee, Jeirou. Artificial Intelligence Technology and China's Defence System. Accessed on 15 December 2022. https://media.defense.gov/2022/Mar/28/2002964034/-1/-1/1/FEATURE_LI.PDF

National Artificial Intelligence Strategy. Accessed on 27 February 2023. https://cbddo.gov.tr/SharedFolderServer/Genel/File/TRNationalAIStrategy2021-2025.pdf

National RARAM Super Computing Facility (NPSF) Annual Report 2021. Accessed on 24 September 2023. https://www.cdac.in/index.aspx?id=pdf_annual_report_npsf_2021

National Strategy for AI. Accessed on 9 September 2023. https://www.niti.gov.in/sites/default/files/2023-03/National-Strategy-for-Artificial-Intelligence.pdf

Open Challenge. Accessed on 16 July 2023. https://idex.gov.in/disc-category/5

Optimising Defence Acquisition Procedure. Accessed on 9 November 2023. https://www.iadb.in/2023/07/30/optimising-defence-acquisition-procedure/

RAS-AI strategy 2040. Accessed on 3 March 2023. https://www.navy.gov.au/sites/default/files/documents/RAN_WIN_RASAI_Strategy_2040f2_hi.pdf

Request for Information by Directorate General of Information System Army Integrated Decision Support System. Accessed on 7 March 2024. https://mod.gov.in/sites/default/files/Final-RFI-of-Proj-AIDSS-on-27-Oct-22.pdf

SAPIENT autonomous sensor system, 22 June 2022. Accessed on 2 March 2023. https://www.gov.uk/guidance/sapient-autonomous-sensor-system

SPARK—Support for Prototype and Research Kickstart. Accessed on 16 July 2023. https://idex.gov.in/sites/default/files/2020-09/5d5fc4f2c701def4b72aad9c_SPARK_-Support_for_Prototype_and_Research_Kickstart_in_Defence_framework_under_iDEX.pdf

Technology Development Fund. Accessed on 17 July 2023. https://tdf.drdo.gov.in/scheme

The Armed Forces (Special Powers) Act, 1958. Accessed on 8 March 2024. https://www.mha.gov.in/sites/default/files/armed_forces_special_powers_act1958.pdf

The British Army has used Artificial Intelligence (AI) for the first time during Exercise Spring Storm, as part of Operation Cabrit in Estonia. Accessed on 1 March 2023. https://www.gov.uk/government/news/artificial-intelligence-used-on-army-operation-for-the-first-time

The Disturbed Areas (Special Courts) Act, 1976. Accessed on 8 March 2024. https://lddashboard.legislative.gov.in/sites/default/files/A1976-77.pdf

TPCR 2018. Accessed on 8 November 2023. https://www.mod.gov.in/sites/default/files/tpcr.pdf

Web Articles

AI reports for duty in the Australian military. AAccessed on 2 March 2023. https://www.deloitte.com/global/en/services/consulting/perspectives/AI-reports-for-duty-in-the-australian-military.html

Aksungur Medium-Altitude Long Endurance (MALE) UAV. Accessed on 27 February 2023. https://www.naval-technology.com/projects/aksungur-medium-altitude-long-endurance-male-uav-Turkiye/

An overview of Britain's drones and drone development projects. Accessed on 2 March 2023. https://dronewars.net/british-drones-an-overview/

Anka S. Unmanned Aerial Vehicle. Accessed on 27 February 2023. https://www.airforce-technology.com/projects/anka-s-unmanned-aerial-vehicle/

Arul, Akashdeep. How China is using AI for Warfare. Accessed on 15 December 2022. https://analyticsindiamag.com/how-china-is-using-ai-for-warfare/

Ball, Mike. New Developments for UK MOD Nano-UAS Program. Accessed on 2 March 2023. https://www.unmannedsystemstechnology.com/2022/05/new-developments-for-uk-mod-nano-uas-program/

Bendett, Samuel, Stephen Blank, Joe Cheravitch and Michael B. Petersen. Russian Unmanned Vehicle Developments: Syria and Beyond. Research Report, Centre for Strategic and International Studies (CSIS), 2020. Accessed on 16 February 2023. https://www.jstor.org/stable/pdf/resrep24241.9.pdf?refreqid=excelsior%3Acb84ae3db9884c6124737659bae8e07b&ab_segments=&origin=&initiator=&acceptTC=1

Bendett, Samuel. Red Robots Rising: Behind the Rapid Development of Russian Unmanned Military Systems. Accessed on 18 February 2023. https://thestrategybridge.org/the-bridge/2017/12/12/red-robots-rising-behind-the-rapid-development-of-russian-unmanned-military-systems,

Bengio, Y. Low precision arithmetic for deep learning. Accessed on 27 September 2022. https://www.researchgate.net/publication/269932963_Low_precision_arithmetic_for_deep_learning

Bhardwaj, Pawan. India's Geopolitical Evolution: A Rise to Transformational Growth. 26 September 2023. https://www.usiofindia.org/strategic-perspective/India-Geopolitical-Evolution-A-Rise-to-Transformational-Growth.html

Biswas, Debolina. Israel's Iron Dome Puts AI At The Forefront Of Modern Warfare. Accessed on 25 February 2023. https://analyticsindiamag.com/israels-iron-dome-puts-ai-at-the-forefront-of-modern-warfare/

Pamuk, Humeyra. U.S. says potential F-16 sale to Turkiye would serve U.S. interests. Accessed on 16 August 2022. https://www.reuters.com/world/us-says-potential-f-16-sale-Turkiye-would-serve-us-interests-nato-letter-2022-04-06/

Panday, Jyoti and Mila T. Samdub. Promises and Pitfalls of India's AI Industrial Policy. Accessed on 20 April 2024. https://ainowinstitute.org/wp-content/uploads/2024/03/AI-Nationalisms-Chapter-4.pdf

Petrella Stephanie, Chris Miller, and Benjamin Cooper. Russia's Artificial Intelligence Strategy: The Role of State-Owned Firms, FPRI, Orbis, volume 65, issue 1, November 2020, p 76. https://www.fpri.org/article/2021/01/russias-artificial-intelligence-strategy-the-role-of-state-owned-firms/

Poncet, Guerric. The French Ministry of the Army will entrust all its satellite images to Preligens' AI. Accessed on 26 February 2023. https://preligens.com/resources/press/french-ministry-army-will-entrusts-all-its-satellite-images-preligens-ai

Rajaraman, V. Multi-core Microprocessors. Accessed on 27 September 2022. https://www.ias.ac.in/article/fulltext/reso/022/12/1175-1192#:~:text=Multi%2Dcore%20microprocessor%20is%20an,than%20a%20single%20core%20processor

Report of Task Force on Artificial Intelligence. Accessed on 11 July 2023. https://dipp.nic.in/whats-new/report-task-force-artificial-intelligence

Respond Basket. Accessed on 9 September 2023. https://www.nitt.edu/home/Respond-Basket.pdf

Robotic boats making waves. Accessed on 14 March 2023. https://www.nature.com/articles/d42473-022-00140-y

Root, Philip. Squad X Core Technologies (SXCT). Accessed on 20 March 2023. https://www.darpa.mil/program/squad-x-core-technologies

Royal Navy Madfox unmanned vessel launches missile on NATO drill. Accessed on 2 March 2023. https://defbrief.com/2021/10/14/royal-navy-madfox-unmanned-vessel-launches-missile-on-nato-drill/

Rubin, Michael. Iran Claims Development of Cruise Missiles Guided by Artificial Intelligence. Accessed on 30 March 2024. https://fmso.tradoc.army.mil/2023/iran-claims-development-of-cruise-missiles-guided-by-artificial-intelligence/#_edn.1

Sahu, Akash. AUKUS, Quad key to Australia's AI-powered defence. Accessed on 2 March 2023. https://asiatimes.com/2022/12/aukus-quad-key-to-australias-ai-powered-defence/

Sciforec Report. Accessed on 27 September 2022. https://medium.com/sciforce/ai-hardware-and-the-battle-for-more-computational-power-3272045160 a6#:~:text=As%20AI%20systems%20become%20more,and%20reduce%20the%20power%20consumption

Stepovoy, Alexey Ramm Bogdan. Sees the target: without the participation of the operator'. Accessed 16 February 2023. https://iz.ru/1000101/aleksei-ramm-bogdan-stepovoi/vidit-tcel-bylina-smozhet-atakovat-protivnika-bez-uchastiia-operatora

Stock, Petra. Lethal drones: The future of the Air Force could be un-crewed. Accessed on 7 March 2023. https://cosmosmagazine.com/technology/lethal-drones-the-future-of-the-air-force-could-be-un-crewed/

Sullivan, Lee. Artificial Intelligence in Russia. Accessed on 19 February 2023. https://geohistory.today/ artificial-intelligence-in-russia/

The curtain of Era of Machine Warfare is opening. Accessed on 10 March 2023. http://www.xinhuanet.com //mil/2017-05/04/c_129588629.htm

The IDF Sees Artificial Intelligence as the Key to Modern-Day Survival. Accessed on 23 February 2023. https://www.idf.il/en/mini-sites/technology-and-innovation/the-idf-sees-artificial-intelligence-as-the-key-to-modern-day-survival/#:~:text=Why%20the%20IDF%20 needs%20artificial,and%20automatically%20flag%20suspicious%20activity

Thomas Timothy. Russia's Electronic Warfare Force-Blending Concepts with Capabilities. Accessed on 18 February 2023. https://www.mitre.org/sites/default/files/2021-11/prs-19-2714-russias-electronic-warfare-force-blending-concepts-with-capabilities.pdf

Thomas, Timothy. Russian Robotics: A Look at Definitions, Principles, Uses, and Other Trends. Accessed on 13 February 2023. https://irp.fas.org/eprint/russian-robotics.pdf

TOGAF® Fundamental Content. Accessed on 7 March 2024. https://pubs.opengroup.org/togaf-standard/introduction/chap03.html

Tran, Tony Ho. An AI Successfully Flew an F-16 Fighter Jet for 17 Hours. Accessed on 27 February 2023. https://www.yahoo.com/now/ai-successfully-flew-f-16-182649243.html?guccounter=1&guce_referrer=aHR0cHM6Ly93d3cuZ29vZ2xlLmNvbS8 &guce_ referrer_sig= AQAAAMVK-4wAQdy-xexCp0cMUYdqRIV4 BE2wi_p PO8r8 NiU19j5w2BVMXtK9nnzaytF4zzs7Fb5s8xZmcX5rtqpwdRsIHEsw Sb8 gm4aU-wWqLZWIUe0yOExvgl-IUlgyebrO069 LJjLws GQq6ua Ok GuwKeJJ9VyQ3pSjfrcCyKdSuexH

Trevithick, Joseph. RAF Tests Swarm Loaded With BriteCloud Electronic Warfare Decoys To Overwhelm Air Defences. Accessed on 2 March 2023. https://www.thedrive.com/the-war-zone/36950/raf-tests-swarm-loaded-with-britecloud-electronic-warfare-decoys-to-overwhelm-air-defences

Tubert, Sandra and Laura Ziegler. France: Artificial Intelligence Comparative Guide. Accessed on 23 February 2023. https://www.mondaq.com/france/technology/1059760/artificial-intelligence-comparative-guide

Turek, Dr. Matt. Explainable Artiûcial Intelligence (XAI). Accessed on 5 September 2022. https://www.darpa.mil/program/explainable-artificial-intelligence

Turkiye to be among pioneers of AI-controlled warplane: Erdoðan. Accessed on 27

February 2023. https://www.dailysabah.com/business/defense/Turkiye-to-be-among-pioneers-of-ai-controlled-warplane-erdogan

Turkiye's 'Fully Loaded' Baykar Akinci Combat Drone Completes Test Flight Armed With All Weapon Stations. Accessed on 27 February 2023. https://eurasiantimes.com/Turkiyes-fully-loaded-baykar-akinci-combat-drone-completes-test-flight-armed-with-all-weapon-stations/

Tzeng, Yisuo. Prospect for Artificial Intelligence in Taiwan's Defence. Accessed on 27 February 2023. https://www.jewishpolicycenter.org/2019/01/11/prospect-for-artificial-intelligence-in-taiwans-defence/

UN Convention on Certain Conventional Weapons (CCW). Accessed on 6 February 2023. https://www.auswaertiges-amt.de/en/aussenpolitik/themen/-/218382#:~:text=Article,which%20may%20cause%20unjustifiable%20suffering,

Uppal, Rajesh. Russia Deployed Family of Killer Robots. Accessed 10 February 2023 https://idstch.com/military/army/russia-developing-family-of-killer-robots-conduct-war-games/,

Veisdal, Jørgen. The Birthplace of AI. Accessed on 13 October 2022. https://www.cantorsparadise.com/the-birthplace-of-ai-9ab7d4e5fb00

Ward-Foxton, Sally. Top 10 Processors for AI Acceleration at the Endpoint. Accessed on 20 April 2020. https://www.eetimes.eu/top-10-processors-for-ai-acceleration-at-the-endpoint/2/

Weizenbaum, Professor Joseph. Accessed on https://web.njit.edu/~ronkowit/eliza.html

What is Training Data and Why Is It Important for AI and Computer Vision? Find Out Here. Accessed on 16 December 2022. https://www.programsbuzz.com/article/what-training-data-and-why-it-important-ai-and-computer-vision-find-out-here

White House says Turkiye's involvement in F-35 program impossible. Accessed on 16 August 2022. https://www.reuters.com/article/Turkiye-security-usa/white-house-says-Turkiyes-involvement-in-f-35-program-impossible-idINFWN24I0I3

Whittaker, Zack. 'Documents reveal how Russia taps phone companies for surveillance'. Accessed on 21 February 2023. https://tcrn.ch/32MaCYb

Work starts on shaping first national security strategy. Accessed on 15 January 2024. https://indianexpress.com/article/india/work-starts-on-shaping-first-national-security-strategy-long-wait-ends-9012566/

Yildirim, Goksel. STM Defence Technologies Engineering produces autonomous 'Kamikaze' Kargu drones for military use. Accessed on 22 February 2023. https://www.aa.com.tr/en/economy/anadolu-agency-tours-state-of-the-art-turkish-uav-maker/1877808

Zhang, Michael. The Explorer is a Tactical 360 Ball Camera That Can Be Thrown

Into Danger. Accessed on 24 September 2023. https://petapixel.com/2015/07/02/the-explorer-is-a-tactical-360-ball-camera-that-can-be-thrown-into-danger/

Blogs

Bernstein, Steven D. Taiwan's Defence Strategy and Artificial Intelligence. Accessed on 27 February 2023. https://bpb-us-e1.wpmucdn.com/blogs.gwu.edu/dist/b/3853/files/2022/06/Copy-of-Deans-Scholars-Thesis-Edited-for-Journal_BernsteinSteven.docx-2.pdf

Carpenter, Cameron. University of Waterloo: Artificial Intelligence Without Internet Now Possible, The University Network, 12 March 2019, https://www.tun.com/blog/university-of-waterloo-artificial-intelligence-without-internet-now-possible/, Accessed on 4 February 2023.

Gill, Jagreet Kaur. What's the Difference Between Cloud, Edge, and Fog Computing? Accessed on 1 November 2022. https://www.akira.ai/blog/difference-between-cloud-edge-and-fog-computing/

Grant, Matthew. AI and Expert Systems: Powering the Future of Business. Accessed on 9 January 2023. https://www.leanix.net/en/blog/artificial-intelligence-expert-systems#:~:text=AI%20and%20Humans%3A%20Better%20Together&text=But%20AI%20and%20expert%20systems,action s%20is%20revolu tionizing%20that%20discipline,

https://www.concentric.io/blog/the-global-landscape-of-ai-development-china-russia-and-irans-strategies-and-impacts#:~:text=the%20Ukraine%20war.-,Iran,the%20AI%20market%20by%. Accessed on 22 November 2023

https://www.concentric.io/blog/the-global-landscape-of-ai-development-china-russia-and-irans-strategies-and-impacts#:~:text=the%20Ukraine%20war.-,Iran,the%20AI%20market%20by%202032. Accessed on 22 November 2023.

https://www.edureka.co/blog/fuzzy-logic-ai/#:~:text=Fuzzy%20logic%20is%20used%20in%20Natural%20language%20processing%20and%20various,use%20it%20with%20Neural%20Networks. Accessed on 9 January 2023.

https://www.superannotate.com/blog/guide-to-training-data. Accessed on 16 December 2022.

Huellmann, Thilo. Precision vs. Recall in Machine Learning. Accessed on 04 February 2023. https://levity.ai/blog/precision-vs-recall

Importance of Training Data for Machine Learning. Accessed on 16 December 2022. https://www.dataentryoutsourced.com/blog/importance-of-training-data-for-machine-learning/

Owen-Hill, Alex. What's the Difference Between Robotics and Artificial Intelligence? updated on 27 July 2021. Accessed on 28 November 2022. https://blog.robotiq.com/whats-the-difference-between-robotics-and-artificial-intelligence

Blasko, Dennis J. The Chinese Military Speaks to Itself, Revealing Doubts. Accessed on 10 March 2023. https://warontherocks.com/2019/02/the-chinese-military-speaks-to-itself-revealing-doubts/,

Brown, Roger. Understanding the Importance of Training Data. Accessed on 16 December 2022. https://medium.com/the-ai-technology/understanding-the-importance-of-training-data-in-machine-learning-da4235332904#:~:text=It%20helps%20them%20to%20recognize,the%20failure%20of%20AI%20project

Bybelezer, C. 'How Russia is using Syria as a military 'guinea pig', The Jerusalem Post, 28 February 2018. Accessed on 18 February 2023. https://www.jpost.com/Middle-East/How-Russia-is-using-Syria-as-a-military-guinea-pig-543839,

Chitti, Rajesh, (ed.). What is GAN in AI?', K-tech Centre of Excellence. Accessed on 20 December 2022. https://coe-dsai.nasscom.in/what-is-gan-in-artificial-intelligenceai/

Cole, J. Michael. How Taiwan Can Defend Its Coastline Against China. Accessed on 28 February 2023. https://macdonaldlaurier.ca/taiwan-can-defend-coastline-china-j-michael-cole/

Concept for Robotic and Autonomous Systems. Accessed on 3 March 2023. https://tasdcrc.com.au/wp-content/uploads/2020/12/ADF-Concept-Robotics.pdf

Conger, Kate. The Fight for a Massive Pentagon Cloud Contract Is Heating Up. Accessed on 19 March 2023. https://gizmodo.com/the-fight-for-a-massive-pentagon-cloud-contract-is-heat-1825517332

Congressional Research Service. Artificial Intelligence and National Security. Accessed on 11 June 2022. https://sgp.fas.org/crs/natsec/R45178.pdf

Conte, Claudia and et al. Using Drone Swarms as a Countermeasure of Radar Detection. Accessed on 28 February 2023. https://doi.org/10.2514/1.I011131

CPDS 2023. Accessed on 4 November 2023. https://indianarmy.nic.in/writereaddata/adb-documents/Compendium%20of%20 Problem%20Definition%20Statement%202023.pdf

Difference between Cloud, Fog and Edge Computing in IoT. Accessed on 10 November 2022. https://www.digiteum.com/cloud-fog-edge-computing-iot/

Dul'nev, P. A. The Employment of Robotic Complexes During the Assault of a Town (Fortified Area), Vestnik Akademii Voennykh Nauk (Journal of the Academy of Military Science), no. 3, 2017, p. 27, referred to by Timothy Thomas, Russian Robotics: A Look at Definitions, Principles, Uses, and other Trends. Accessed on 13 February 2023. https://irp.fas.org/eprint/russian-robotics.pdf,

Easton, Ian. Taiwan Defence Strategy in an Age of Precision Strike. Accessed on 28 February 2023. https://project2049.net/wp-content/uploads/2018/06/Easton_Able_Archers_Taiwan_Defense_Strategy.pdf

Eliaçýk, Eray. Guns and Codes: The Era of AI-Wars Begins. Accessed on 9 December 2022. https://dataconomy.com/2022/08/how-is-artificial-intelligence-used-in-the-military/#:~:text=Artificial %20intelligence%20is%20 used%20in,driven %20systems%20in%20the%20military,

Eran Ortal 'Going on the Attack: The Theoretical Foundation of the Israel Defence Forces' Momentum Plan'. Accessed on 23 February 2023. https://www.idf.il/en/mini-sites/dado-center/vol-28-30-military-superiority-and-the-momentum-multi-year-plan/going-on-the-attack-the-theoretical-foundation-of-the-israel-defense-forces-momentum-plan-1/#:~:text=The%20challenge%20of%20the %20Momentum,Israel%2C%20that%20of%20rocket%20fire.

Fathi, Mahan. Collecting a Nation's Points of Interest: Computer Vision to the Rescue. Accessed on 29 March 2024. https://medium.com/@MahanFathi/collecting-a-nations-points-of-interest-computer-vision-to-the-rescue-41026053bdf6

Fink, Anya. Book Review, Russian Studies Series 5/21, 'Russian Thinking on the Role of AI in Future Warfare. Review of V.M. Burenok, "IskusstvennyyIntellekt v VoennomProtivostoyaniiBudushchevo" ("Artificial intelligence in the military confrontation of the future"). Accessed on 13 February https://www.ndc.nato.int/research/research.php?icode=712,

Foote, Keith D. A Brief History of Deep Learning. Accessed on 13 October 2022. https://www.dataversity.net/brief-history-deep-learning/

Frantzman, Seth J. Iran reveals bizarre new AI-powered miniature tank robots—analysis. Accessed on 22 November 2023. https://www.jpost.com/middle-east/article-730433

Freedberg, Sydney Jr. A Slew To A Kill: Project Convergence. Accessed on 7 March 2023. https://breakingdefense.com/2020/09/a-slew-to-a-kill-project-convergence/

Ghoshal, Anirban. India's advisory on LLM usage causes consternation. Accessed on 20 April 2024. https://www.cio.com/article/1311757/indias-advisory-on-llm-usage-causes-consternation.html

Gill, Jaspreet. Project Convergence. Accessed on 7 March 2023. https://breakingdefense.com/2022/11/how-project-convergence-is-informing-british-australian-military-modernization/

Gill, Jaspreet. Say goodbye to JAIC and DDS, as offices cease to exist as independent bodies. Accessed on 19 May 2023. https://breakingdefense.com/2022/05/say-goodbye-to-jaic-and-dds-as-offices-cease-to-exist-as-independent-bodies-june-1/

Goled, Shraddha. What Are the Scope and Challenges of Using AI in Military Operation. Accessed on 23 March 2022. https://analyticsindiamag.com/what-are-the-scope-and-challenges-of-using-ai-in-military-operations/,

Harris, Dustin. What Is Artificial Intelligence (AI)? How Does AI Work? Accessed on 10 Oct 2022. https://builtin.com/artificial-intelligence

Hintze, Arend. Understanding the four types of AI. Accessed on 27 September 2022. https://theconversation.com/understanding-the-four-types-of-ai-from-reactive-robots-to-self-aware-beings-67616

Hoffman, Frank. The Contemporary Spectrum of Conflict: Protracted, Gray Zone, Ambiguous, and Hybrid Modes of War. Accessed on 14 April 2024. https://www.heritage.org/sites/default/files/2019-10/2016_IndexOfUSMilitaryStrength_The%20Contemporary%20Spectrum%20of%20Conflict_Pro tracted % 20Gray%20Zone%20Am

http://www.army-guide.com/eng/product.php?prodID=5525&printmode=1. Accessed on 21 February 2023.

http://www.beta.iitkgp.ac.in/files/RD0520235.pdf. Accessed on 4 September 2023.

https://acqnotes.com/acqnote/acquisitions/dotmlpf-analysis#google_vignette. Accessed on 3 February 2024.

https://ai.iith.ac.in/research/ai-research-centre.html. Accessed on 4 September 2023.

https://aiindex.stanford.edu/wp-content/uploads/2022/03/2022-AI-Index-Report_Master.pdf. Accessed on 23 November 2023.

https://andrei-bt.livejournal.com/949786.html. Accessed on 21 February 2023.

https://article36.org/wp-content/uploads/2016/04/UK-and-LAWS.pdf. Accessed on 6 February 2023.

https://carnegietsinghua.org/2018/10/24/tides-of-change-china-s-nuclear-ballistic-missile-submarines-and-strategic-stability-pub-77490. Accessed on 14 March 2023.

https://chat.openai.com/chat. Accessed on 3 February 2023.

https://cpdm.iisc.ac.in/ril/research/. Accessed on 4 September 2023.

https://docs.google.com/forms/d/e/1FAIpQLScg2ooU1ErIV6ZYg Nq5JQHeJ3gm Osh1lQSm613w9Dr2TkmO2A/viewform?usp=sf_link, created on 14 October 2023.

https://economictimes.indiatimes.com/tech/ites/meet-professor-hn-mahabala-the-man-who-mentored-indias-it-icons/articleshow/53346662.cms?utm_source=contentofinterest&utm_medium=text&utm_campaign=cppst. Accessed on 11 July 2023.

https://en.wikipedia.org/wiki/Buoyancy_engine. Accessed on 14 March 2023.

https://en.wikipedia.org/wiki/Hongdu_GJ-11. Accessed on 14 March 2023.

https://fedtechmagazine.com/article/2022/10/intelligence-community-developing-new-uses-ai-perfcon#:~:text=At%20the%20National%20Security%20Agency,National%20Security%20Alliance's%20spring%20symposium. Accessed on 20 March 2023.

https://futureskillsprime.in/. Accessed on 9 September 2023.

https://hai.stanford.edu/sites/default/files/2020-10/AI%20Index%202017%20Annual%20Report.pdf. Accessed on 23 November 2023.

https://hbr.org/2023/09/is-india-the-worlds-next-great-economic-power. Accessed on 31 December 2023.

https://indiaeducationdiary.in/iit-jodhpur-awarded-516-degrees-in-various-academic-programs-in-its-8th-convocation/. Accessed on 4 September 2023.

https://marineregions.org/gazetteer.php?p=details&id=8480. Accessed on 21 January 2024.

https://milremrobotics.com/defence/. Accessed on 2 March 2023.

https://ria.ru/20171015/1506649786.html. Accessed on 17 February 2023.

https://ria.ru/20171017/1507003664.html. Translated by Chat GTP. Accessed on 21 February 2023,

https://tass.com/defense/1277657. Accessed on 18 February 2023.

https://www.armyrecognition.com/china_chinese_unmanned_aerial_ground_systems_uk/sharp_claw_2_ugv_6x6_unmanned_ground_vehicle_technical_data_sheet_specifications_pictures_video_11412165.html. Accessed on 14 March 2023.

https://www.army-technology.com/projects/bug-nano-drone/. Accessed on 2 March 2023.

https://www.army-technology.com/projects/tien-kung-iii-sky-bow-iii-surface-to-air-missile-system/. Accessed on 28 February 2023.

https://www.britannica.com/biography/Alan-Turing/Computer-designer. Accessed on 13 October 2022.

https://www.britannica.com/print/article/405392. Accessed on 20 March 2023.

https://www.business-standard.com/article/pti-stories/india-has-reached-out-to-countries-across-the-world-on-caa-nrc-mea-120010201069_1.html. Accessed on 16 January 2024.

https://www.businesstoday.in/latest/trends/story/israel-warns-india-on-ip-related-issues-while-pledging-to-share-tech-on-make-in-india-339830-2022-06-30. Accessed on 28 November 2023.

https://www.c4isrnet.com/unmanned/2019/07/31/is-this-russias-gateway-drone-to-better-armed-robots/. Accessed on 17 February 2023.

https://www.cdac.in/index.aspx?id=cloud_ci_cloud_computing. Accessed on 24 September 2023.

https://www.cdac.in/index.aspx?id=hpc_nsf_npsfidx. Accessed on 11 September 2023.

https://www.cdac.in/index.aspx?id=hpc_nsf_siddhi-CP. Accessed on 22 September 2023.

https://www.cdac.in/index.aspx?id=hpc_nsf_siddhi-spec. Accessed on 11 September 2023.

https://www.economist.com/international/2023/06/12/indias-diaspora-is-bigger-and-more-influential-than-any-in-history. Accessed on 31 December 2023.

https://www.eeas.europa.eu/search_en?fulltext=artificial+intelligence. Accessed on 16 February 2023.

https://www.globalsecurity.org/military/world/china/anjian.htm. Accessed on 14 February 2023.

https://www.globalsecurity.org/military/world/china/d3000.htm. Accessed on 13 March 2023.

https://www.globalsecurity.org/military/world/russia/soratnik.htm. Accessed on 21 February 2023.

https://www.globalsecurity.org/military/world/russia/uran-9.htm. Accessed on 21 February 2023.

https://www.guru99.com/what-is-fuzzy-logic.html#5. Accessed on 4 February 2023.

https://www.horiba-mira.com/unmanned-ground-vehicles/media-centre/case_study/viking-multirole-ugv-platform/. Accessed on 2 March 2023.

https://www.ibm.com. Accessed on 4 November 2022.

https://www.ibm.com/in-en/topics/computer-vision#:~:text=Computer%20vision%20is%20a%20field,recommendations%20based%20on%20that%20information. Accessed on 4 November 2022.

https://www.idsa.in/system/files/book/book_NetSecurityProvider.pdf. Accessed on 15 January 2024.

https://www.indiatoday.in/business/story/cabinets-nod-to-india-ai-mission-with-outlay-of-rs-10372-crore-2512045-2024-03-07. Accessed on 20 April 2024.

https://www.insightsonindia.com/2023/02/24/giving-data its-due-national-data-and-analytics-platform-ndap/#:~:text=NDAP%20aims %20to%20democratize %20 access,innovators%2C%20and%20civil%20society%20groups. Accessed on 9 September 2023.

https://www.jasondavies.com/wordcloud/. Accessed on 2 February 2023.

https://www.jstor.org/stable/pdf/resrep19585.26.pdf?refreqid= excelsior% 3A28f037d2b7aaa39b25be6e923051aced&ab_segments=&origin=&initiator=&acceptTC=1. Accessed on 9 March 2023.

https://www.jstor.org/stable/pdf/resrep19585.26.pdf?refreqid=excelsior %3A28f037d2b7aaa39b25be6e923051aced&ab_segments=&origin=&initiator=&acceptTC=1. Accessed on 9 March 2023.

https://www.lntsmartworld.com/defence#:~:text=We%20are%20executing%20a%20state,Centres%20which%20are%20under%20implementation. Accessed on 7 March 2024.

https://www.mathworks.com/discovery/deep-learning.html#:~:text=Deep%20learning%20is% 20a% 20machine,a%20pedestrian%20from%20a%20lamppost. Accessed on 18 October 2022.

https://www.newindianexpress.com/galleries/nation/2021/Feb/04/twitter-war-who-are-the-international-celebrities-supporting-the-farmers-protest-in-delhi-meet-ri-103073.html. Accessed on 16 January 2024.

https://www.prophesee.ai/about-prophesee/. Accessed on 4 November 2022.

https://www.rand.org/content/dam/rand/pubs/research reports/RR3100/RR3139-1/RAND_RR3139-1.pdf. Accessed on 23 January 2023.

https://www.scimagojr.com/countryrank.php?category=1702&order=it&ord=desc. Accessed on 29 March 2024.

https://www.smart-shooter.com/. Accessed on 23 February 2023.

https://www.techtarget.com/searchenterpriseai/definition/neural-network. Accessed on 18 October 2022.

https://www.techtarget.com/whatis/definition/optical-computer-photonic-computer. Accessed on 4 November 2022.

https://www.youtube.com/watch?v=34k7UI-DR_8. Accessed on 8 March 2023.

https://www.youtube.com/watch?v=DgM7hbCNMmU. Accessed on 20 March 2023.

https://www2.deloitte.com/ca/en/pages/deloitte-analytics/articles/age-with-ai-advantage-defence-security.html. Accessed on 15 December 2022.

Husain, Amir. Turkiye Builds a Hyperwar Capable Military. Accessed on 27 February 2023. https://www.forbes.com/sites/amirhusain/2022/06/30/Turkiye-builds-a-hyperwar-capable-military/?sh=1c0f0e855e11

IBM Cloud Learning Hub. Natural Language Processing (NLP). Accessed on 22 November 2022. https://www.ibm.com/cloud/learn/natural-language-processing#:~:text=Natural%20language%20 processing%20(NLP)%20refers,same%20way%20human%20beings%20can

Jha, Manish Kumar and Amit Das. AI Technology in Military Will Transform Future Warfare. Accessed on 10 November 2021. https://www.businessworld.in/article/AI-Technology-In-Military-Will-Transform-Future-Warfare/13-08-2021-400525/

Jing, Yuan-Chou. How Does China Aim to Use AI in Warfare? Accessed on 15 December 2022. https://thediplomat.com/2021/12/how-does-china-aim-to-use-ai-in-warfare/

Kania, Elsa B. Chinese Sub Commanders May Get AI Help for Decision-Making. Accessed on 8 March 2023. https://www.defenseone.com/ideas/2018/02/chinese-sub-commanders-may-get-ai-help-decision-making/145906/

Kasapoðlu, Can and Barý' Kýrdemir. The Rising Drone Power: Turkiye on the Eve of Its Military Breakthrough. Accessed on 27 February 2023. http://www.jstor.com/stable/resrep21043

Konaev, Margarita and Samuel Bendett. Russian AI-Enabled Combat: Coming to a City Near You. Accessed on 13 February 2023 https://warontherocks.com/2019/07/russian-ai-enabled-combat-coming-to-a-city-near-you/,

Konaev, Margarita. U.S. Military Investments in Autonomy and AI Costs, Benefits, and Strategic Effects. Accessed on 9 December 2022. https://cset.georgetown.edu/publication/u-s-military-investments-in-autonomy-and-ai-executive-summary/

Kumar, P R. For India's National Security, Time for Civil and Military Synergy. Accessed on 15 January 2024. https://www.indiandefencereview.com/spotlights/for-indias-national-security-time-for-civil-and-military-synergy/

Launch of the Man Machine Teaming advanced study programme. Accessed on 27 February 2023. https://www.dassault-aviation.com/en/group/press/press-kits/launch-man-machine-teaming-advanced-study-programme/

Layton, Peter. The ADF could be doing much more with artificial intelligence. Accessed on 2 March 2023. https://www.aspistrategist.org.au/the-adf-could-be-doing-much-more-with-artificial-intelligence/

Lisman, Evan Omeed. Iran's Bet on Autonomous Weapons. Accessed on 29 March 2021. https://warontherocks.com/2021/08/irans-bet-on-autonomous-weapons

Lowman, Ron. Why In-Memory Computing Will Disrupt Your AI SoC Development." Accessed on 3 November 2022. https://www.synopsys.com/designware-ip/technical-bulletin/in-memory-computing-ai.html

Lt. Saurabh Kalia: The Patrol Leader. Accessed on 7 march 2024. https://www.indiandefencereview.com/spotlights/the-patrol-leader/

Lutkevich, Ben. What is natural language processing? Accessed on 22 November 2022. https://www.techtarget.com/searchenterpriseai/definition/natural-language-processing-NLP

Luzin, Pavel. Artificial intelligence in the Russian Army. Accessed on 13 February 2023. https://ridl.io/artificial-intelligence-in-the-russian-army/,

MADFOX—USV Designed for Surveillance and Force Protection Missions. Accessed on 2 March 2023. https://www.bairdmaritime.com/work-boat-world/maritime-security-world/unmanned-systems/vessel-review-madfox-usv-designed-for-surveillance-and-force-protection-missions/

Mahajan, Kapil. 'Role of Artificial Intelligence and Machine Learning in Robotics'. Accessed on 22 November 2022. https://emeritus.org/in/learn/role-of-artificial-intelligence-and-machine-learning-in-robotics/

Mahony, Niall O' et al. Deep Learning vs. Traditional Computer Vision. Accessed on 4 November 2022. https://arxiv.org/ftp/arxiv/papers/1910/1910.13796.pdf

Majumdar, Rumki. India economic outlook. Accessed on 31 December 2023. https://www2.deloitte.com/xe/en/insights/economy/asia-pacific/india-economic-outlook.html

Manuel, Rojoef. Turkiye Develops Tech Reducing Small Drones Reliance on GPS. Accessed on 27 February 2023. https://www.thedefensepost.com/2022/10/13/Turkiye-drones-gps-navigation/

Martin, Alan. Robotics and artificial intelligence: The role of AI in robots. Accessed on 26 November 2022. https://aibusiness.com/verticals/robotics-and-artificial-intelligence-the-role-of-ai-in-robots

McDermott, Roger. Russian UAV Technology and Loitering Munitions. Accessed on 21 February 2023. https://www.realcleardefense.com/articles/2021/05/06/russianuavtechnology_ and loitering munitions_775980.html

Mihajlovic, Ilija. Everything You Ever Wanted to Know about Computer Vision. Accessed on 4 November 2022. https://towardsdatascience.com/everything-you-ever-wanted-to-know-about-computer-vision-heres-a-look-why-it-s-so-awesome-e8a58dfb641e

Military Applications of Artificial Intelligence. Accessed on 24 January 2023. https://www.rand.org/content/dam/rand/pubs/ research_reports/RR3100/RR3139-1/RAND_RR3139-1.pdf

Military Applications of Artificial Intelligence: Ethical Concerns in an Uncertain World. RAND. Accessed on 24 January 2023. https://www.rand.org/pubs/research_reports/RR3139-1.html

Mizokami, Kyle. Everything We Know about Israel's Robotic Machine Gun. Accessed on 25 February 2023. https://www.popularmechanics.com/military/weapons/a37708762/robotic-machine-gun-kills-iranian-nuclear-scientist/

Nadibaidze, Anna. Russian Perceptions of military AI, automation, and autonomy. Accessed on 10 January 2023. https://www.fpri.org/wp-content/uploads/2022/01/012622-russia-ai-.pdf

Nair, Sharathkumar. A closer look at Russia's AI-powered artillery. Accessed on 21 February 2023. https:// analyticsindiamag.com/a-closer-look-at-russias-ai-powered-artillery/

Natural Language Processing (NLP). Accessed on 22 November 2022. https://www.sas.com/en_us/insights/analytics/what-is-natural-language-processing-nlp.html

Nelson, Amy J and Gerald L Epstein. The PLA's Strategic Support Force and AI Innovation. Accessed on 8 March 2023. https://www.brookings.edu/techstream/the-plas-strategic-support-force-and-ai-innovation-china-military-tech/

New Locally-Built Armed USV for Turkish Navy. Accessed on 27 February 2023. https://www.bairdmaritime.com/work-boat-world/maritime-security-world/unmanned-systems/vessel-review-sancar-new-locally-built-armed-usv-for-turkish-navy/

Nikolsky, Aleksey. Lofty Goals. Accessed on 19 February 2023. https://forceindia.net/cover-story/lofty-goals/

Ntoutsi, Eirini and others. Bias in Data-Driven Artificial Intelligence Systems—An Introductory Survey. Accessed on 18 October 2022. https://www.researchgate.net/publication/338998132_Bias_in_data-driven_artificial_intelligence_systems-An_introductory_survey

Peck, Michael. Russia Wants to Use AI-Sea Mines to Sink America's Navy. Accessed on 28 February 2023. https://nationalinterest.org/blog/buzz/russia-wants-use-ai-sea-mines-sink-americas-navy-120951

Säuberlich, Frank Dr. and Danko Nikoliæ, Prof. 'AI Without Machine Learning', https://www.teradata.com/Blogs/AI-without-machine-learning, Accessed on 4 February 2023.

Selig, Jay. What You Need to Know About Natural Language Processing (NLP). Accessed on 18 October 2022. https://www.expert.ai/blog/natural-language-processing/

Singh, Tejaswi and Amit Gulhane. 8 Key Military Applications for Artificial Intelligence. Accessed on 23 March 2022. https://blog.marketresearch.com/8-key-military-applications-for-artificial-intelligence-in-2018.

Vallath, Sekhar. The What, Where, and Why of Contextual AI. Accessed on 1 November 2022. https://symbl.ai/blog/the-what-where-and-why-of-contextual-ai/

What is Federated Learning? Accessed on 23 February 2023. https://research.ibm.com/blog/what-is-federated-learning

@NewsIADN, Indian Aerospace Defence News (IADN), 14 March 2023. Accessed on 14 March 2023. https://twitter.com/NewsIADN/status/1635369382120673280?t=Q8fDFFQVY1TBmUjoc2obQA&s=03

https://www.radware.com/blog/security/2023/05/india-one-of-the-most-targeted-countries-for-hacktivist-groups/. Accessed on 26 December 2023.

@IranObserver0, 3 August 2023. Accessed on 31 March 2024., https://twitter.com/IranObserver0/status/1687093036764008449

Index

Accelerating Growth of New India Innovation (AGNIi), 146
Acceptance of Necessity (AON), 118
Access/Area Denial (A2/AD), 57
Accuracy, 14
Active Protection System (APS), 119
Actuator, 215
Adpix Tech Private Limited, 126
Advanced Research Foundation (ARF), 43
Advancing Analytics (ADVANA), 86
Aether, 79
Afghanistan, 80, 164, 183
Africa, 176
AI and Data Acceleration (ADA), 86
AI Research Analytics and Knowledge Dissemination Platform, 140, 264
AI Research, Analytics and Knowledge Assimilation (AIRAWAT), 132, 149, 152, 265
AICRA Conclave, 262
   AI Drone Private Limited, 127
AI-Powered Smart Shooter – SMASH, 96
Airborne Robotic Systems, 61
Akinci (Raider), 66
Aksungur, 66
Algorithms, 6
AlphaGo, 58
AlphaWar, 58
Altius, 54
Always in Operations, 244
Android Team Awareness Kit (ATAK), 80
Anka-S, 65
Annual Acquisition Plan (AAP), 118
Anti-Submarine Warfare (ASW), 66
Application Specific Integrated Circuits (ASICs), 15

Architecture for Massive Information Processing and Exploitation from Multi-Sources and Artificial Intelligence (ARTEMIS.IA), 83
Area of Responsibility (AOR), 195, 197
ArgoAmphibious Vehicle, 54
Armoured Fist of S-300 and S-400, 53
Army Design Bureau (ADB), 121-22
   Compendium of Problem Definition Statement 2023, 121, 264
Army War College (AWC), 248
Artificial General Intelligence (AGI), 3, 31
Artificial Intelligence (AI), 1, 30, 36, 42, 44, 69, 87, 122, 125, 142, 151, 281
   Artefact Repository, 251, 279
   as a Sector Store, 251
   at Rest, 95, 99
   Black Box, 13
   Bot, 258
   Cadre, 244
   Cells, 243
   Contextual, 13
   Flyaway Teams, 86
   for the Future, 212, 220
   Generative, 280-81
   Governance, 173
   in Defence, 137, 264
   in Motion, 96
   Induction Policy, 257
   Laboratory, 249-50
   Military Developers, 243
   Military Risk Mitigation, 261
   Now, 211, 220
   Output, 218
   Play Book, 216
   Problems, Defining, 212
   R&D, 90

Required, 210
Artificial Narrow Intelligence (ANI), 3, 31
Artificial Neural Network (ANN), 134
Artificial Super Intelligence (ASI), 3, 31
Asia, 176
Assam Rifles (AR), 198
Asymmetric Technologies, 125
ATMAN, 140
Atmanirbhar Bharat (self-reliant India), 119
AUC-ROC, 14
Audio Data, 253
AUKUS, 71
Australia, 42, 71, 99
Australia Air Force, 72
Australian Defence Force (ADF), 71-72, 74, 278
Australian Navy, 71
Automata, 4
Automated Command and Control Systems, 49
Automated Data Processing (ADP), 254
Automatic Speech Recognition (ASR), 142, 149
Autonomous
 Command Decision Making, 59
 Movement Navigation, 55
 Prediction of Landslides and Rock Falls, 240
 Tracking POI, 234
 Vehicles, 40
Autonomous Systems (AS), 212
Autonomous Weapons Systems (AWS), 46, 226, 238
Avtomatizirovannaya Sistema Upravleniya (ASU), 50

Baidu, Alibaba, Tencent and Xiaomi (BATX), 83
Baramulla, 160
Battle
 of Hearts and Minds, 173
 Changing Nature of, 172
Battlefield Healthcare, 41
Battlespace Analysis, 37
Bayraktar TB2, 65
Bengaluru Police, 135
Bharat Electronics Limited (BEL), 116, 137
Black Sea Fleet, 273
Blitz on Taiwan, 57
Block Chain, 128
BRICS, 183
Bug nano UAV, 80

Build-up Areas, 204
Burenok, V.M., 45
Bylina, 50
BZK-005 UAV, 61

Cafe Bazaar, 69
Canada, 142
Caracal UGV, 71
Carbon Fibre Winding (CFW), 119
Carlos Del Toro, 64
CCTV, 116
CDAO (Office of the Chief Digital and Artificial Intelligence Officer), 87
Central Asian Republics, 164
Central Processing Units (CPU), 15
Centre for Artificial Intelligence and Robotics (CAIR), 125, 140
Centre for Development of Advanced Computing (C-DAC), 132, 140, 265
Chandrasekaran, N., 116
ChatGPT (Chat Generative Pre-Trained Transformer), 23-24
Chief of Defence Staff (CDS), 179
Chile, 121
China, 29, 42, 56, 58, 60-61, 63-64, 83, 86, 90, 99, 143, 160, 164, 173, 181, 227, 258, 263
China-Myanmar Economic Corridor, 176
China-Nepal Economic Corridor, 176
Chinese Hanzi, 56
Cloud, 218
Cognitive Health, 173
Cognitive Technologies, 125
Cognitive Warfare, 57
Colombia, 121
Combat Missions, 92
Combat Search and Rescue (CSR), 73
Command and Control (C2), 49
Command, Control, Communications, Computer, Intelligence, and Information (C4I2), 46
Command, Control, Computers, Communications, Intelligence, Surveillance and Reconnaissance (C4ISR), 51, 137-38
Commercial off the Shelf, 259
Communication Intelligence (COMINT), 116
Communication Relay Bot, 237

Competent Financial Authority (CFA), 118
Composite Border Surveillance System, 214
Comprehensive National Power, 180
Computer Vision (CV), 6, 8, 12, 16-18, 21, 31, 69, 77, 263
Continental United States (CONUS), 87
Contract Negotiations by a Committee (CNC), 118
Contracted Data Curation, 255
Convolutional Neural Network (CNN), 8, 17, 22, 49
Copyright Act, 1957, 145
Core Capabilities, 174
Counter-Insurgency (CI), 172-73, 198
CPDS, 122, 138
CPEC, 176
Creation of Prospective Military Robotics through 2025, 43
Cross-Functional Teams (CFT), 216
Curating Video Data, 253
Cyber Readiness, Lessons from, 247
Cyber Security, 23, 25, 41, 116, 130, 247
Cyber Threats, 164, 166, 184-85, 188, 193, 265
Cyber Warfare, 60
Cyberspace, 45, 98
Cycle-Consistent Adversarial Networks (CLGAN), 23
CycleGAN, 22
Cyran AI Solutions, 126
Czech Republic, 142

Dare to Dream, 126, 135
Dark Sword UAV, 61
Dartmouth Summer Research Project, 4
Dassault Aviation, 84
Data, 5, 252
    Curation, 253
    Labelling, 65
Data Protection and Privacy Act (DPDP Act), 2023, 144-46, 151-52, 281
Data Protection and Privacy Draft (DPDP) Bill, 145
Dataflow Processing Unit (DPU), 15
Dataset, 219
Decision-Making Support Systems, 41
Decluttered Border Surveillance System, 215
Deconvolutional Neural Network, 22

Dedicated Deep Neural Network, 15
Deep Learning (DL), 6-9, 12, 18, 24, 31, 65, 70, 217, 260
Deep Neural Network, 7
Defence Acquisition Council (DAC), 118, 121
Defence Acquisition Policy 2020 (DAP 20), 117-19, 136-37
Defence Acquisition Procedure (DAP), 151, 265
Defence Advanced Research Projects Agency (DARPA), 88
Defence AI Council (DAIC), 117, 135
Defence AI Project Agency (DAIPA), 117, 135
Defence Capital Acquisition Plan (DCAP), 118
Defence Digital Service (DSS), 86
Defence India Start-up Challenge (DISC), 119
Defence Industry Academia Centre of Excellence (DIA-CoE), 123
Defence Innovation Hubs (DIH), 119
Defence Innovation Organisation (DIO), 119
Defence Procurement Procedure (DPP), 117
Defence Public Sector Undertakings (DPSUs), 121
Defence Research and Development Organisation (DRDO), 116, 121-27, 135, 137, 140, 150
Department of Defence (DoD), 122, 150
Department of Defence Production (DDP), 117, 121-22, 135
Department of Ex-Servicemen Welfare (DESW), 122
Department of Military Affairs (DMA), 122, 150, 179
Department of Science and Technology, 260
Desert Hawk III, 79
Deserts, 202
Design and Development, 210
Design Decision, 260
Designing AI Tools for the Army, 265
    Design and Development, 266
    Problem Definition, 266
    Requirement Specification, 266
    Technology Assessment, 266
Designing AI War Game, 247
Developing Military Warfighting Concepts, 56
Dhankhar, Jagdeep, Vice-President of India, 177
Digital Data Banks, 115
Digital Economy Programme, 47

Digital India Bhashini, 149
Digital India Corporation (DIC), 147
Digital Personal Data Protection Act, 2023, 136
Digital Public Infrastructure (DPI), 180
Distributed Denial-of-Service (DDoS), 48
Doctrine, Organisation, Training, Material, Leadership and Education, Personnel, Facilities, and Policy (DOTMLPFP), 228
DRDO Data Curation, 255
DRDO Young Scientists Laboratory (DYSL), 125
Dull, Dirty and Dangerous (3D), 84
DYST-CT (Cognitive Technology), 125

EA-03 Xianglong, 61
East China Sea, 63
Economic Threats, 164, 168-69
Education and Science (MES), 44
EEZs, 174, 177, 278
Electronic Warfare (EW), 50, 60
Emergency Credit Line Guarantee (ECLG) Scheme, 180
Enhance, Augment, Replace (EAR) Framework, 72
Environment, 215, 218
Environment, Values, and Resources (EVR), 159
Estonian THeMIS UGV, 68
Eta Compute ECM3532, 16
Ethics: User or Developer Driven, 282
Event-Condition-Action (ECA), 10
Exercise Spring Storm, 78
Expert Systems (ES), 25-26, 28
Explainable AI (XAI), 13
Explorer Ball Cameras, 222
Extra Regimental Employment (ERE), 242
Extramural Research & Intellectual Property Rights (ER&IPR), 125

Face Recognition, 258
Face Recognition System (FRS), 135
Fakhrizadeh, Mohsen, 97
Feed Forward Neural Networks (FFN), 8
Field Evaluation Trials (FET), 118
Fight the AI as an AI, 281
Final Experimental Demonstration Object Research (FEDOR), 55
FindFace, 48

Finland, 42
Fire Weaver, 97
FIRESTORM, 74
Five Incapables, 56
Flexible Tactical Unmanned Air System (FTUAS), 80
Flocking Algorithm, 257
Flow-GAN, 23
Force Protection, 72
Foreign Policy, 169
Forests, 205
Formal Text Data, 253
Framework Consultation, 260
France, 42, 83, 99, 181
France-based PropheSee, 16
French motto, Liberty, Equality, Fraternity (Liberté, Égalité, Fraternité), 83
Frshr Tech Private Limited, 126
Fully Programmable Gate Array (FPGA), 15
Future Indian Military Conflicts, 265
    Cyber Threats, 265
    Hybrid Threats, 265
    Internal Security Situations, 265
    Territorial and Maritime Conflicts, 265
Future Soldiering, 277
FutureSkills Prime, 132
Fuzzy Logic (FL), 25

G20, 183
GAFA (Google, Apple, Facebook and Amazon), 83
Galtel, 47
Galwan, 202
GAPAI Conference, 261
Garur Drishti, 257
Gaza, 97
    Hamas in, 95
Gazprom Neft, 44-45
Generative Adversarial Networks (GAN), 22, 31, 49, 263
Generative Pre-trained Transformer (GPT), 23-24, 32, 263
Geographical Dynamics, 171
Geographical Information System (GIS), 127
Geopolitical Balance, 171
Georgia, 49
Germany, 42, 144

Ghost Bat, 72
GitHub Projects, 144
Glaciers, 200, 206
Global AI Index, 143
Global Information Dominance Experiments (GIDE), 87
Global Positioning System (GPS), 50, 227
Global South, 113, 183
Goal-Based Agent, 11
Goldwater–Nichols Act (1986), 87
Good Old-Fashioned AI (GOFAI), 27, 32
Goods and Services Tax (GST) Bill, 160
Gopalakrishnan, 114
Government of India, 115, 160
Graphics Processing Units (GPUs), 15, 141-42
Grey Zone Campaigns, 172
Ground Robotic Systems, 62
Guardian of the Walls, 95

Haider Smaller Vehicle, 70
Handwara, 160
Heavy RTK, 55
High-altitude, 206
High-Performance Computing (HPC), 140
High-Performance Computing-Artificial Intelligence (HPC-AI), 132, 140
Hints, Arendt, 'Theory of Mind Machine', 3
Honduras, 121
Hsiung Feng IIE (HF2E), 92
Human Intelligence (HUMINT), 116
Human Level Machine Intelligence (HLMI), 4
Human Machine (Man-Machine), 99
Human Resource Development, 276
Human Resources Management, 208
Humanitarian Assistance and Disaster Relief (HADR), 184
Human-Machine (man machine), 264
Human-to-Human Interactions, 194
Hybrid, 164-65, 168-69, 187, 193, 165
    Operations, 199
    Threats, 167, 177, 188, 265

Identification of Friend/Foe, 56
Image
    Classification, 17
    Datasets, 252
    Restoration, 18
    Retrieval, 18
Image & Video Processing for Net Centric Operations (IVP NCO), 127
Independent Business Division (IBD), 147
India, 75, 113-14, 143-44, 178, 180-81, 185, 264, 284
INDIAai (The national AI portal of India), 142, 147
    Application Development Initiative, 148, 150
    Compute Capacity, 147
    Datasets Platform, 148
    Future Skills, 148, 150
INDIAai Innovation Centre (IAIC), 147, 150
INDIAai Start-up Financing, 148
Indian Army, 121, 135, 138, 143, 198-99, 209, 211, 217, 219, 221, 223, 227, 24142, 244, 246-47, 249, 251-53, 255, 260, 262, 266, 271, 276, 281, 283
Indian Army Rapid AI Design Architecture (IRADA), 209, 219-21, 265
Indian
    Desert Battles, 203
    Economy, 164-65, 180-81
    Foreign Policy, 182-83
Indian Institute of Remote Sensing (IIRS), 134
Indian Navy, 64, 120, 137
Indian Ocean Region (IOR), 63, 164-65, 171, 176-77, 193, 278-79
Indian Space Research Organisation (ISRO), 134
Indian Technology Trends, 264
Indian Way, 253
Indonesia, 63
Indo-Pacific Region, 71, 75, 164-67, 176-77, 193
Indo-Pacific Regional Dialogue (2023), 177
InfiniBand HDR, 141
Influence Operations, 276
Informal Text, 253
Information & Communication Technology (ICT), 117, 129
Information Retrieval (IR) System, 220, 232
Information Technology (IT), 248, 282
Information Technology Act, 2000, 144, 282
Information Warfare (IW) Training, 91, 248
Infrastructure as a service (IAAS), 142
Inkers Technologies, 127

Index ❑ 313

In-Memory Computing (IMC), 15
Innovate4Defence (i4D), 120
Innovation and Indigenisation Organisation (IIO), 118-19
Innovations for Defence Excellence (iDEX), 119-20, 135-37, 139, 152, 264
Integrated Air and Missile Defence (IAMD), 74
Integrated Capability Development Plan (ICDP), 118
Intelligence Preparation, 275
Intelligence Processing Unit (IPU), 15
Intelligence, Surveillance, and Reconnaissance (ISR), 38, 58, 63, 65, 76, 80, 172
Intelligent Agent, 10
Intelligent Character Recognition (ICR), 16, 127
Intelligent Traffic Management System (ITMS), 135
Internal Data Curation, 253
Internal Security Situations, 188
Internal Security, 164, 166-69, 177
Internet of Things (IoT), 45, 128
Iran, 42, 68, 70, 99, 273
Iraq, 70, 183
Ireland, 42
Iron Dome, 97
Islamic Revolutionary Guard Corps (IRGC), 70
Islamic State of Iraq and the Levant (ISIS), 88
Islamic State, 46
Israel, 93-94, 97, 273
Israel Defence Forces (IDF), 95
Israel-Hamas War, 178

Japan, 42, 75, 181
Japan International Cooperation Agency (JICA), 124
Joint Artificial Intelligence Centre (JAIC), 86
JWP02 UAV, 61

Kamakoti, Prof V., 114
Kamikaze Drones, 66, 273
Kargil War, 202
KERKES, 66
KILPECK, 77
Knowledge-Based Computing Systems, 114
Known Employment, 187

Language Translator, 220, 231

Lantset, 54
Large Language Models (LLM), 150, 281
Launch & Recovery Systems (LARS), 72
Learner Algorithm, 7
Learning Agents, 11
Lebanon, 97
Left Wing Extremism (LWE), 164, 168, 177, 193
Lenz, Callen, 80
Lethal Autonomous Weapon Systems (LAWS), 28-30, 40, 57, 76, 86, 117
LiDAR, 15
Light RTK, 56
Line of Actual Control (LAC), 172, 176, 200-01, 227, 241, 258, 278-79
Line of Control (LC), 172, 199, 201, 223, 227, 241
Live translation, 258
Logistics, 200, 207
Loyal Wingman (Mosquito), 78, 88

Machine Learning (ML), 12-13, 15, 21, 31, 37, 41, 45, 77, 90, 248, 262
Machine Translation, 19
Machines are intelligent if they
    Act Humanly, 1
    Act Rationally, 2
    are Thinking Humanly, 2
Mahabala, Prof H.N., 114
Mahdi, Abu, 70
Mail.ru, 47
Make in India, 137
Mali, 79
Man Unmanned Machine Teaming (MUMT), 277
Manhattan Project, 212
Man-Machine Teaming (MMT), 84, 273, 275, 277-79
Manned-Unmanned Teaming (MUT), 64, 138
Manohar Parrikar Institute for Defence Studies and Analyses (MP-IDSA), 161, 163, 166, 184
Maritime, 168-69
    Challenge, 165, 171, 176
    Diplomacy, 171
    Islands, 204
    Robotic Systems, 62

Maritime Demonstrator for Operational eXperimentation (MADFOX), 81
Marker, 55
Mass or Effect, 277
McCarthy, John, 4
MCEME Seminar, 261
Medical Infirmary (MI), 258
Meghdoot Cloud, 142
MeitY Start-up Hub (MSH), 128, 131-32
Mellanox ConnectX-6 VPI, 141
Micro, Small & Medium Enterprises (MSMEs), 116, 119-20, 136, 141, 151, 180-81, 250, 265
Military
    Leaders, 256
    Robustness, 218
    Trends, 263
Military-AI-Ecosystem, 36
Military-to-Military Training, 182
MilSOFT, 67
Minimum Viable Product (MVP), 132
Ministry of Commerce and Industry, 121, 136
Ministry of Defence (MoD), 44, 120-21, 123, 135, 137, 150
Ministry of Education, 128
Ministry of Electronics & Information Technology (MeitY), 128-32, 140, 142, 146-47, 265
Ministry of External Affairs, 123, 199
Ministry of National Defence (MND), 90
Missile and Air Defence, 92
Missing the Bigger Picture, 215
Model Based Reflex Agent, 10
Modi, Narendra, Prime Minister of India, 130, 183
Mountains, 200, 206
Msta-SM 2S19M2, 53
Mule Robots, 85
Multi-Core Microprocessor, 15
Multi-Robot Cooperation, 84
Multi-Role Ocean Surveillance Ship (MROSS), 82
Murthy, N.R. Narayana, 114
Myanmar, 176

Nagorno-Karabakh, 178
Nagrota Military Camp, 160
Nano Uncrewed Air Systems (nUAS), 79
National Artificial Intelligence Mission (N-AIM), 115
National Artificial Intelligence Resource Platform (NAIRP), 130, 132, 142, 152
National Association of Software and Service Companies (NASSCOM), 132
National Combat Aircraft (MMU), 67
National Data & Analytics Platform (NDAP), 133, 142, 152
National Defence Management Centre, 49
National Digital Library of India (NDLI), 130
National Grid (NATGRID), 116
National Institution for Transforming India (NITI) Aayog, 128, 132-33
National Occupational Standards (NOS), 132
National PARAM Supercomputing Facility (NPSF), 140-41
National Security Agency (NSA), 88
National Security Commission on Artificial Intelligence (NSCAI), 86
National Security Strategy (NSS), 180
National Skills Qualification Framework (NSQF), 132
National Strategy for Artificial Intelligence (NSAI), 132
Natsional'nyi Tsentr Upravleniya Oboronoi (NTUO), 49
Natural Language Processing (NLP), 6, 18-20, 25, 27, 31, 38, 138, 142, 263, 272-73
Natural Language Translation, 133
Naval Innovation and Indigenisation Organisation (NIIO), 120
Nazir, 70
Nerekhta, 52
Nerekhta, Kamikaze, 52
Neural Networks, 9
Ng, Andrew, 216
NGIS, 131
Nilekani, Nandan, 151
Non-kinetic Conflict, 186
Non-military Conflict, 186
Non-state actors, 98-99, 171
North Atlantic Treaty Organization (NATO), 30, 50-51, 67, 71, 76, 78, 99, 263
North Eastern Space Applications Centre (NESAC), 134

Object Detection, 17
Object Tracking, 18
Oceania, 176
Open Government Data Portal, 65
OpenAI, 23-24
Operational
    Aspects, 197
    Logistics, 93
    Procedures, 275
Oppenheimer, 212
Opportunities for Make in India in Defence, 121
Optical Character Recognition (OCR), 16
Optical Computing Chip, 16
Ortal, Brig Gen Eran, 93
Out of Area Contingencies (OOAC), 161, 164-66, 168-69, 184, 197
Over Engineering, 217
Overcoming Operational Inexperience through Hybrid Intelligence, 58

Pakistan, 164, 173, 176, 227
Pakistan-China Strategic Nexus, 176
Palestine, 273
Pampore, 160
Param Siddhi AI (PSAI), 140, 149, 152, 264-65
Part of Speech Tagging, 19
Partially Known Employment, 188
Patents Act, 1970, 145
Pathankot Air Force Station, 160
Peace Disease, 56
Pearl Harbour, 183
Peer Relationships, 276
People's Liberation Army (PLA), 56-58, 62, 91-93
    Air Force, 60
    Navy, 62-63, 93
    Rocket Force, 61
    Strategic Force, 61
Performance Measure, 214-15
Performance Measure, Environment, Actuators, Sensor (PEAS), 10, 31, 214, 219, 221, 230, 260
Persons of Interest (POI), 199, 212, 222-23
Platform as a Service (PAAS), 142
Platforma M, 52
Points of Interest, 69

Poseidon AUV, 54
Precision, 14
Prediction of Landslides and Rock Falls, 226
Predictive Maintenance and Logistics, 60
Preligens, 84
Preliminary Services Qualitative Requirements (PSQRs), 118
Prime Minister's Science, Technology & Innovation Advisory Council (PM-STIAC), 133, 146
PRIME SPRINT, 120, 137
Principal Scientific Advisor (PSA), 128
PRISM, 88
Production-linked Incentive (PLI) Schemes, 180
Project Carmel, 97
Project Convergence, 73
Putin, Vladimir, 43

QinetiQ Banshee Jet 80+ drones, 78
QR Codes, 227
QUAD, 183
Quantum Technologies, 125

R&D, 119, 121, 123, 127, 130, 132
RADAR, 15
Radio (wireless) Communication, 201
Raketno-YadernoyeNapadenie (RYAN), 49
RAND, 86
Rashtriya Rifles (RR), 198
Realistic/Virtual Simulation, 40
Recall (Sensitivity), 14
Reconnaissance, 91, 197, 209
Reconnaissance Strike Contour (RSC), 50
Recurrent Neural Networks (RNN), 8
Regional Technology Node (RTN), 121
Reinforcement, 7
Request for Information (RFI), 118
Request for Proposal (RFP), 118
Requirement Specification, 210
Reynolds, Craig, 257
Rheinmetall Mission Master SP, 81
Right to Information Act, 2005, 145
RNMB Hebe, 82
ROBOIK, 68
Robotekhnicheskie Kompleksy (RTK), 51
Robotics, 20
    Manipulator, 127

Strike Complexes, 51
Urban Warfare, 55
Robotic Platoon Vehicles (RPV), 80
Robotics and Autonomous Systems (RAS), 71
Robotization of the Armed Forces of the Russian Federation, 43
Rostec, 44
Royal Navy, 78-79, 81-82
Royal Navy Motor Boat (RNMB), 81
Russia, 30, 42-45, 47-50, 69, 86, 99, 263
Russian Academy of Sciences (RAS), 44
Russian Military Robots, 50
Russia-Ukraine War, 31, 178

S-70 Okhotnik-B, 54
Safe & Trusted AI, 148
SALT, 90
Sancar USV, 66
Sberbank, 44
Scenario Generation, 40
Scene Reconstruction, 18
Schwartz, Riva, 12
Science, Technology, Engineering and Mathematics (STEM) Graduates, 180
Sea Wing (Haiyi) Glider, 63
Selayar Islands, 63
Sensing for Asset Protection with Integrated Electronic Networked Technology (SAPIENT), 77
Sensor, 215
Services Qualitative Requirements (SQRs), 118, 136-37, 151
Shanghai University, 62
Sharing of High Computing Facilities, 140
Sharp Claw, 62
Sharp Sword, 62
Shopian, 160
Shtrum, 53
Signal Intelligence (SIGINT), 116
Signals Technology Evaluation and Adaptation Group (STEAG), 276
Simplex Agents, 10
Singapore, 42
Singh, Rajnath, the Defence Minister of India, 178
Single Source of Truth (SSOT), 253-54
Situational Awareness, 39

Sky Bow III, 92
SkyGuardian, 78
SMART, 267
Smart Helmet, 127
Smart Materials, 125
Smart Quad Rack, 97
Smarter Logistic Solutions, 275
Social Media Sentiment Analysis, 19
Software as a Service (SAAS), 142
Soratnik, 52
South Korea, 42
South-East Asia, 164
Soviet Union, 49
Space Operations, 58
SPOTTER, 77
Squad Mission Support System (SMSS), 68
Squad X, 88
SQUINTER, 77
Stages of Conflict, 171
Standing Committee on Defence, 256
Star Shadow, 61
START, 90
Start-up Accelerators of MeitY for Product Innovation, Development and Growth (SAMRIDH), 131
Statements of Case (SoC), 118
STELaRLab, 74
Stop Word Removal, 19
StratCom Task Force, 49
Strategic Choice. Australia, 75
Strategic Partners, 166-69, 181
Strategic Reach and Global Power Status, 171
Strix, 73
StyleGAN (Generative Adversarial Network), 22
Subject Matter Experts (SME), 159, 162, 170
Suicide Robot, 70
Supervised, 7
Supply Autonomous Robotic Assistant Hardware (SARAH), 74
Support for Prototype and Research Kick-start (SPARK), 120
Surveillance, 84, 202
Surveillance and Intelligence, 206
Swarm Robots, 55
Swarm Technology, 55
Swarm UAV, 67
Swarming, 40, 63, 99

Sweden, 42
Switzerland, 42
Synergy, 167, 178, 189
    Challenge, 164, 166
    Needs, 169
Synthetic Aperture Radar (SAR), 61
Syria, 70, 99
Syriatel, 46
System for Operative Investigative Activities (SORM), 48
System on Chip (SoC), 15

TA Innovation Cell, 244
Tactical
    Battles, 172
    Conditions, 197
    Landscape, 197
    Level Weapon Systems, 275
    Logistic Bot, 223, 235
    Support Bot, 224
    Support Drone, 236
Tactical Battlefields Characteristics, 196
    Integration of Manoeuvre and Firepower, 196
    Lethality, 196
    Objective Orientation, 196
    Technology Characteristics, 196
    Tempo, 196
    Terrain and Weather, 196
    Uncertainty, 196
Taifun-M, 53
Taiwan, 90, 91, 92, 99
Taiwan's National Space Organization (NSO), 92
Tamil Nadu Government, 135
Targeting, 39
Task Force for 'Strategic Implementation of AI for National Security and Defence', 116
Tata Constancy Services (TCS), 145
TDA4VM, 15
Technical Affiliation Scheme, 141
Technical Evaluation Committee (TEC), 118
Technical Offset Evaluation Committee (TOEC), 118
Technical Oversight Committee (TOC), 118
Technology Assessment, 210, 217
Technology Development Acceleration Cell (TDAC), 120
Technology Development Fund (TDF), 135-36, 152
Technology Incubation and Development of Entrepreneurs (TIDE), 129, 131
Technology Perspective and Capability Roadmap (TPCR), 118, 259
Technology Perspective and Capability, 137
Teller, Edward, 212
Tensor Processing Units (TPUs), 141
Tensor Processing Unit of Google, 15
Tera Operations per Second (TOPS), 15
Territorial and Maritime Conflicts, 188
Territorial Army Innovation Cell (TAIC), 245
Territorial Threat, 42, 164, 166-69, 175
Testing and Evaluation, 250
Text Analytics, 19
Text-to-Speech (TTS), 142, 149
Thales, 84
The Netherlands, 42
Tnufa (Momentum), 94
Togan UAV, 66
Tracked Hybrid Modular Infantry System, 81
Traction Projects, 259
Training and Simulation, 40
Transfer of Technology (ToT), 127
Transport RTK, 56
Transportation and Logistics, 41
Triton Unmanned Aircraft System, 73
TÜBITAK (The Scientific and Technological Research Institution of Turkiye), 65
Turkish Aerospace Industries (TAI), 65
Turkiye, 42, 65, 67, 68, 69, 121
Twin-gun Autonomous Turrets, 96

UGVs, 51, 55, 62
Ukraine, 79-80, 273
UN Convention on Certain Conventional Weapons (CCW), 28, 30
UN General Assembly, 28
UN Sustainable Development Goals, 146
Underwater Mini Torpedoes, 54
Unforgiving Nature of Military Conflicts, 172
Unicum, 54
Unified Data, 253
United Kingdom (UK), 4, 30, 42, 73, 79-80, 82, 99, 142, 181
    Defence AI Strategy, 76

Drones, 78
United Service Institution (USI) of India, 160, 162-63, 165, 168, 187, 261
University of South Australia, 74
University of Waterloo, 28
Unknown Employment, 188
Unmanned Aerial Systems, 98
Unmanned Aerial Vehicles (UAVs), 31, 51, 54-55, 58, 60, 94, 96
Unmanned Surface Vehicle (USV), 66
Unmanned Underwater Vehicle (UUV), 47, 51
UNSC, 181
Unsupervised, 7
UP Police, 134
Uran 9, 46, 52
Urban Combat Buddy, 221, 233
Uri Military Camp, 160, 222
USA, 29-30, 42-43, 49-51, 67, 73, 83, 85-86, 88, 90, 99-100, 142, 144, 181, 261, 263, 273
  Army, 40, 89
  Navy, 63-64
Utility-Based Agent, 11
Uttar Pradesh Government, 134

Vaccine Maitri (Vaccine Friendship), 119
Variable-Buoyancy Propulsion, 63
Variational Autoencoders (VAEs), 280
Vasudhaiva Kutumbakam, 183
Very Low Precision Computation, 16
Video Motion Analysis, 18
Vikhr, 52
VIKING, 81
Virtual and Augmented Reality (AR/VR), 138
Voenno Kosmicheskie Sily (VKS), 47
Volk 2, 53

War-gaming Analysis, 40
Watchkeeper, 78
WaveGAN, 23
West Asia, 164
West Bank, 96
Wide Development Base, 259
World Heritage Palmyra, 46
World War II, 212

XLUUV (Extra Large Uncrewed Underwater Vehicle), 82
XR (extended reality), 131-32

Yandex, 44-45, 47
Young Scientists Laboratories (DYSLs), 125

Zakura, 160
Zephyr, 79
Zero Crossing Rate (ZCR), 254
Zero Operational Risk, 259
Zhuhai Air Show, 61